原版影印说明

1.《聚合物百科词典》(5册) 是 Springer Reference *Encyclopedic Dictionary of Polymers*(2nd Edition) 的影印版。为使用方便，由原版 2 卷改为 5 册：

第 1 册 收录 A-C 开头的词组；

第 2 册 收录 D-I 开头的词组；

第 3 册 收录 J-Q 开头的词组；

第 4 册 收录 R-Z 开头的词组；

第 5 册 为原书的附录部分及参考文献。

2. 缩写及符号、数学符号、字母对照表、元素符号等查阅说明各册均完整给出。

由 Jan W. Gooch 主编的《聚合物百科词典》是关于高分子科学与工程领域的参考书，2007 年出版第一版，2011 年再版。本书收录了 7 500 多个高分子材料方面的术语，涉及高分子材料的各个方面，如粘合剂、涂料、油墨、弹性体、塑料、纤维等，还包括生物化学和微生物学方面的术语，以及与新材料、新工艺相关的术语；并且不仅包括其物理、电子和磁学性能方面的术语，还增加了数据处理的统计和数值分析以及实验设计方面的术语。每个词条方便查找，并给出了简洁的定义，以及相互参照的相关术语。为了说明得更清晰，全书给出 1 160 个图、73 个表。有的词条还给出方程式、化学结构等。

材料科学与工程图书工作室

联系电话 0451-86412421
　　　　 0451-86414559

邮　　箱 yh_bj@aliyun.com
　　　　 xuyaying81823@gmail.com
　　　　 zhxh6414559@aliyun.com

 Springer 词典精选原版系列

聚合物百科词典

Jan W. Gooch

Encyclopedic Dictionary of Polymers

2nd Edition

VOLUME 4
R-Z

哈尔滨工业大学出版社
HARBIN INSTITUTE OF TECHNOLOGY PRESS

黑版贸审字08-2014-010号

Reprint from English language edition:
Encyclopedic Dictionary of Polymers
by Jan W.Gooch
Copyright © 2011 Springer New York
Springer New York is a part of Springer Science+Business Media
All Rights Reserved

This reprint has been authorized by Springer Science & Business Media for distribution in China Mainland only and not for export therefrom.

图书在版编目（CIP）数据

聚合物百科词典.4,R～Z:英文/（美）古驰（Gooch,J.W.）主编.—哈尔滨:哈尔滨工业大学出版社,2014.3

（Springer词典精选原版系列）

ISBN 978-7-5603-4445-4

Ⅰ.①聚… Ⅱ.①古… Ⅲ.①聚合物–词典–英文 Ⅳ.①O63-61

中国版本图书馆CIP数据核字（2013）第304143号

材料科学与工程
图书工作室

责任编辑　杨　桦　许雅莹　张秀华
出版发行　哈尔滨工业大学出版社
社　　址　哈尔滨市南岗区复华四道街10号　邮编150006
传　　真　0451-86414749
网　　址　http://hitpress.hit.edu.cn
印　　刷　哈尔滨市石桥印务有限公司
开　　本　787mm×1092mm　1/16　印张　15.75
版　　次　2014年3月第1版　2014年3月第1次印刷
书　　号　ISBN 978-7-5603-4445-4
定　　价　118.00元

（如因印刷质量问题影响阅读，我社负责调换）

Acknowledgements

The editor wishes to express his gratitude to all individuals who made available their time and resources for the preparation of this book: James W. Larsen (Georgia Institute of Technology), for his innovations, scientific knowledge and computer programming expertise that were invaluable for the preparation of the Interactive Polymer Technology Programs that accompany this book; Judith Wiesman (graphics artist), for the many graphical presentations that assist the reader for interpreting the many complex entries in this publication; Kenneth Howell (Springer, New York), for his continued support for polymer science and engineering publications; and Daniel Quinones and Lydia Mueller (Springer, Heidelberg) for supporting the printed book and making available the electronic version and accompanying electronic interactive programs that are important to the scientific and engineering readers.

Preface

The second edition of Encyclopedic Dictionary of Polymers provides 40% more entries and information for the reader. A Polymers Properties section has been added to provide quick reference for thermal properties, crystallinity, density, solubility parameters, infrared and nuclear magnetic spectra. Interactive Polymer Technology is available in the electronic version, and provides templates for the user to insert values and instantly calculate unknowns for equations and hundreds of other polymer science and engineering relationships. The editor offers scientists, engineers, academia and others interested in adhesives, coatings, elastomers, inks, plastics and textiles a valuable communication tool within this book. In addition, the more recent innovations and biocompatible polymers and adhesives products have necessitated inclusion into any lexicon that addresses polymeric materials. Communication among scientific and engineering personnel has always been of critical importance, and as in any technical field, the terms and descriptions of materials and processes lag the availability of a manual or handbook that would benefit individuals working and studying in scientific and engineering disciplines. There is often a challenge when conveying an idea from one individual to another due to its complexity, and sometimes even the pronunciation of a word is different not only in different countries, but in industries. Colloquialisms and trivial terms that find their way into technical language for materials and products tend to create a communications fog, thus unacceptable in today's global markets and technical communities.

The editor wishes to make a distinction between this book and traditional dictionaries, which provide a word and definition. The present book provides for each term a complete expression, chemical structures and mathematic expression where applicable, phonetic pronunciation, etymology, translations into German, French and Spanish, and related figures if appropriate. This is a complete book of terminology never before attempted or published.

The information for each chemical entry is given as it is relevant to polymeric materials. Individual chemical species (e.g., ethanol) were taken from he *CRC Handbook of Chemistry and Physics*, 2004 Version, the Merck Index and other reference materials. The reader may refer to these references for additional physical properties and written chemical formulae. Extensive use was made of ChemDraw®, CambridgeSoft Corporation, for naming and drawing chemical structures (conversion of structure to name and vice versa) which are included with each chemical entry where possible. Special attention was given to the IUPAC name that is often given with the common name for the convenience of the reader.

The editor assembled notes over a combined career in the chemical industries and academic institutions regarding technical communication among numerous colleagues and helpful acquaintances concerning expressions and associated anomalies. Presently, multiple methods of nomenclature are employed to describe identical chemical compounds by common and IUPAC names (eg. acetone and 2-propanone) because the old systems (19th century European and trivial) methods of nomenclature exists with the modern International Union of Pure and Applied Chemistry, and the conflicts between them are not likely to relent in the near future including the weights and measures systems because some nations are reluctant to convert from English to metric and, and more recently, the International Systems of Units (SI). Conversion tables for converting other systems to the SI units are included in this book for this purpose. In addition, there are always the differences in verbal pronunciation, but the reasons not acceptable to prevent cogent communication between people sharing common interests.

In consideration of the many challenges confronting the reader who must economize time investment, the structure of this book is optimized with regard the convenience of the reader as follows:

- Comprehensive table of contents
- Abbreviations and symbols
- Mathematics signs
- English, Greek, Latin and Russian alphabets
- Pronunciation/phonetic symbols
- Main body of terms with entry term in English, French German and Italian
- Conversion factors

- Microbiology nomenclature and terminology
- References

The editor acknowledges the utilization of many international sources of information including journals, books, dictionaries, communications, and conversations with people experienced in materials, polymer science and engineering. A comprehensive reference section contains all of the sources of information used in this publication. Pronunciation, etymological, cross-reference and related information is presented in the style of the 11th Edition of the Meriam-Webster Dictionary, where known, for each term. The spelling for each term is presented in German, French, and Spanish where translation is possible. Each term in this book includes the following useful information:

- Spelling (in **bold** face) of each term and alternative spellings where more than one derivation is commonly used
- Phonetic spelling \-\ using internationally published phonetic symbols, and this is the first book that includes phonetic pronunciation information missing in technical dictionaries that allows the reader to pronounce the term
- Parts of speech in English following each phonetic spelling, eg. *n.*, *adj.*
- Cross-references in CAPITALS letters
- Also called *example* in italics
- Etymological information [-] for old and new terms that provides the reader the national origins of terms including root words, prefixes and suffixes; historical information is critical to the appreciation of a term and its true meaning
- French, German, Italian and Spanish spellings of the term { - }
- A comprehensive explanation of the term
- Mathematical expressions where applicable
- Figures and tables where applicable
- A comprehensive reference section is included for further research

References are included for individual entries where a publication(s) is directly attributable to a definition or description. Not all of the references listed in the Reference section are directly attributable to entries, but they were reviewed for information and listed for the reader's information. Published dictionaries and glossaries of materials were very helpful for collecting information in the many diverse and smaller technologies of the huge field of polymers. The editor is grateful that so much work has been done by other people interested in polymers.

The editor has attempted to utilize all relevant methods to convey the meaning of terms to the reader, because a term often requires more information than a standard entry in a textbook dictionary, so this book is dedicated to a complete expression. Terminology and correct pronunciation of technical terms is continuously evolving in scientific and industrial fields and too often undocumented or published, and therefore, not shared with others sometimes leading to misunderstandings. Engineering and scientific terms describe a material, procedure, test, theory or process, and communication between technical people must involve similar jargon or much will be lost in the translation as often has been the editor's experience. The editor has made an attempt to provide the reader who has an interested in the industries that have evolved from adhesives, coatings, inks, elastomers, plastics and textiles with the proper terminology to communicate with other parties whether or not directly involved in the industries. This publication is a single volume in the form of a desk-handbook that is hoped will be an invaluable tool for communicating in the spoken and written media.

Physics, electronic and magnetic terms because they are related to materials and processes (e.g., *ampere*).

Biomolecular materials and processes have in the recent decade overlapped with polymer science and engineering. Advancements in polymeric materials research for biomolecular and medical applications are rapidly becoming commercialized, examples include biocompatible adhesives for sutureless tissue bonding, liquid dressings for wounds and many other materials used for *in vitro* and *in vivo* medical applications. To keep pace with these advancements, the editor has included useful terms in the main body that are commonly used in the material sciences for these new industries.

A microbiology section has been included to assist the reader in becoming familiar with the proper nomenclature of bacteria, fungi, mildew, and yeasts – organisms that affect materials and processes because they are ubiquitous in our environment. Corrosion of materials by microorganisms is commonplace, and identification of a specific organism is critical to prevent its occurrence. Engineers and materials scientists will appreciate the extensive sections on different types of microorganisms together with a section dedicated to microbiology terminology that is useful for communicating in the jargon of biologists instead of referring to all organisms as "bugs."

New materials and processes, and therefore new terms, are constantly evolving with research, development and global commercialization. The editor will periodically update this publication for the convenience of the reader.

Statistics, numerical analysis other data processing and experimental design terms are addressed as individual terms and as a separate section in the appendix, but only as probability and statistics relate to polymer technology and not the broad field of this mathematical science. The interactive equations are listed in the Statistics section of the Interactive Polymer Technology program.

Interactive Polymer Technology Programs

Along with this book we are happy to provide a collection of unique and useful tools and interactive programs along with this Springer Reference. You will find short descriptions of the different functions below. Please download the software at the following website: http://extras.springer.com/2011/978-1-4419-6247-8

Please note that the file is more than 200 MB. Download the ZIP file and unzip it. It is strongly recommended to read the **ReadMe.txt** before installing. The software is started by opening the file InPolyTech.pdf and following the instructions. Detailed instructions can be found under 'Help Instructions'.

The software consists of 15 programs and tools that are briefly described in the appendix.

Abbreviations and Symbols

Abbreviations	Symbols
An	absorption (formerly extinction) (= $\log t_i^{-1}$)
A	Area
A	surface
A	Helmholtz energy ($A = U - TS$)
A	preexponential constant [in $k = A \exp(-E^{\ddagger}/RT)$]
A_2	second virial coefficient
a	exponent in the property/molecular weight relationship ($E^{\ddagger} = KM^a$); always with an index, e.g., a_η, a_s, etc.
a	linear absorption coefficient, $a = l^{-1}$
absolute	abs
acre	spell out
acre-foot	acre-ft
air horsepower	air hp
alternating-current (as adjective)	a-c
A^m	molar Helmholtz energy
American Society for Testing and Materials	ASTM
amount of a substance (mole)	n
ampere	A or amp
ampere-hour	amp-hr
amplitude, an elliptic function	am.
angle	β
angle, especially angle of rotation in optical activity	α
Angstrom unit	Å
antilogarithm	antilog
a_0	constant in the Moffit–Yang equation
Area	A
Atactic	at
atomic weight	at. wt
Association	Assn.
atmosphere	atm

Abbreviations	Symbols
average	avg
Avogadro number	N_L
avoirdupois	avdp
azimuth	az or α
barometer	bar.
barrel	bbl
Baumé	Bé
b_0	constant in the Mofit–Yang equation
board fee (feet board measure)	fbm
boiler pressure	spell out
boiling point	bp
Boltzmann constant	k
brake horsepower	bhp
brake horsepower-hour	bhp-hr
Brinell hardness number	Bhn
British Standards Institute	BSI
British thermal unit[1]	Btu or B
bushel	bu
C	heat capacity
c	specific heat capacity (formerly; specific heat); c_p = specific isobaric heat capacity, c_v = specific isochore heat capacity
c	"weight" concentration (= weight of solute divided by volume of solvent); IUPAC suggests the symbol ρ for this quantity, which could lead to confusion with the same IUPAC symbol for density
c	speed of light in a vacuum
c	speed of sound
calorie	cal
candle	c
candle-hour	c-hr
candlepower	cp
ceiling temperature of polymerization, °C	T_c

Abbreviations	Symbols
cent	c or ¢
center to center	c to c
centigram	cg
centiliter	cl
centimeter or centimeter	cm
centimeter-gram-second (system)	cgs
centipoise	cP
centistokes	cSt
characteristic temperature	Θ
chemical	chem.
chemical potential	μ
chemical shift	δ
chemically pure	cp
circa, about, approximate	ca.
circular	cir
circular mils	cir mils
cis-tactic	ct
C^m	molar heat capacity
coefficient	coef
cologarithm	colog
compare	cf.
concentrate	conc
conductivity	cond, λ
constant	const
continental housepower	cont hp
cord	cd
cosecant	csc
cosine	cos
cosine of the amplitude, an elliptic function	cn
cost, insurance, and freight	cif
cotangent	cot
coulomb	spell out
counter electromotive force	cemf
C_{tr}	transfer constant ($C_{tr} = k_{tr}/k_p$)
cubic	cu
cubic centimeter (liquid, meaning milliliter. ml)	cu, cm, cm³
cubic centimeter	cm³ cubic expansion coefficient ∝
cubic foot	cu ft
cubic feet per minute	cfm
cubic feet per second	cfs

Abbreviations	Symbols
cubic inch	cu in.
cubic meter	cu m or m³
cubic micron	cu μ or cu mu or μ³
cubic millimeter	cu mm or mm³
cubic yard	cu yd
current density	spell out
cycles per second	spell out or c
cylinder	cyl
D	diffusion coefficient
D_{rot}	rotational diffusion coefficient
day	spell out
decibel	db
decigram	d.g.
decomposition, °C	T_{dc}
degree	deg or °
degree Celsius	°C
degree centigrade	C
degree Fahrenheit	F or °
degree Kelvin	K or none
degree of crystallinity	∝
degree of polymerization	X
degree Réaumur	R
delta amplitude, an elliptic function	dn
depolymerization temperature	T_{dp}
density	ρ
diameter	diam
Dictionary of Architecture and Construction	DAC
diffusion coefficient	D
dipole moment	p
direct-current (as adjective)	d-c
dollar	$
dozen	doz
dram	dr
dynamic viscosity	η
E	energy (E_k = kinetic energy, E_p = potential energy, $E^{\ddagger}$ = energy of activation)
E	electronegativity
E	modulus of elasticity, Young's modulus ($E = \sigma_{ii}/\varepsilon_{ii}$)
E	general property

Abbreviations	Symbols
E	electrical field strength
e	elementary charge
e	parameter in the Q-e copolymerize-tion theory
e	cohesive energy density (always with an index)
edition	Ed.
Editor, edited	ed.
efficiency	eff
electric	elec
electric polarizability of a molecule	$\propto$
electrical current strength	I
electrical potential	V
electrical resistance	R or X
electromotive force	emf
electronegativity	E
elevation	el
energy	E
enthalpy	H
entropy	S
equation	eq
equivalent weight	equiv wt
et alii (and others)	et al.
et cetera	etc.
excluded volume	u
excluded volume cluster integral	β
exempli gratia (for example)	e.g.
expansion coefficient	$\propto$
external	ext
F	force
f	fraction (excluding molar fraction, mass fraction, volume fraction)
f	molecular coefficient of friction (e.g., f_s, f_D, f_{rot})
f	functionality
farad	spell out or f
Federal	Fed.
feet board measure (board feet)	fbm
feet per minute	fpm
feet per second	fps
flash point	flp

Abbreviations	Symbols
fluid	fl
foot	ft
foot-candle	ft-c
foot-Lambert	ft-L
foot-pound	ft-lb
foot-pound-second (system)	fps
foot-second (see cubic feet per second)	
fraction	$\int$
franc	fr
free aboard ship	spell out
free alongside ship	spell out
free on board	fob
freezing point	fp
frequency	spell out
fusion point	fnp
G	Gibbs energy (formerly free energy or free enthalpy) ($G = H - TS$)
G	shear modulus ($G = \sigma_{ij}$/angle of shear)
G	statistical weight fraction ($G_i = g_i/\Sigma_i\, g_i$)
g	gravitational acceleration
g	statistical weight
g	*gauche* conformation
g	parameter for the dimensions of branched macromolecules
G^m	molar Gibbs energy
gallon	gal
gallons per minute	gpm
gallons per second	gps
gauche conformation	g
Gibbs energy	G
grain	spell out
gram	g
gram-calorie	g-cal
greatest common divisor	gcd
H	enthalpy
H^m	molar enthalpy
h	height
h	Plank constant
haversine	hav

Abbreviations	Symbols
heat	Q
heat capacity	C
hectare	ha
henry	H
high pressure (adjective)	h-p
hogshead	hhd
horsepower	hp
horsepower-hour	hp-hr
hour	h or hr
hundred	C
hundredweight (112 lb)	cwt
hydrogen ion concentration, negative logarithm of	pH
hyperbolic cosine	cosh
hyperbolic sine	sinh
hyperbolic tangent	tanh
I	electrical current strength
I	radiation intensity of a system
i	radiation intensity of a molecule
ibidem (in the same place)	ibid.
id est (that is)	i.e.
inch	in.
inch-pound	in-lb
inches per second	ips
indicated horsepower	ihp
indicated horsepower-hour	ihp-hr
infrared	IR
inside diameter	ID
intermediate-pressure (adjective)	i-p
internal	int
International Union of Pure and Applied Chemistry	IUPAC
isotactic	it
J	flow (of mass, volume, energy, etc.), always with a corresponding index
joule	J
K	general constant
K	equilibrium constant
K	compression modulus ($p = -K \Delta V/V_o$)
k	Boltzmann constant

Abbreviations	Symbols
k	rate constant for chemical reactions (always with an index)
Kelvin	K (Not °K)
kilocalorie	kcal
kilocycles per second	kc
kilogram	kg
kilogram-calorie	kg-al
kilogram-meter	kg-m
kilograms per cubic meter	kg per cu m or kg/m^3
kilograms per second	kgps
kiloliter	Kl
kilometer or kilometer	km
kilometers per second	kmps
kilovolt	kv
kilovolt-ampere	kva
kilowatt	kw
kilowatthour	kwhr
Knoop hardness number	KHN
L	chain end-to-end distance
L	phenomenological coefficient
l	length
lambert	L
latitude	lat or ϕ
least common multiple	lcm
length	l
lin*ear* expansion coefficient	Y
linear foot	lin ft
liquid	liq
lira	spell out
liter	l
logarithm (common)	log
logarithm (natural)	log. or ln
kibgutyde	kibg. or λ
loss angle	δ
low-pressure (as adjective)	l-p
lumen	1*
lumen-hour	1-hr*
luments per watt	lpw
M	"molecular weight" (IUPAC molar mass)
m	mass
mass	spell out or m
mass fraction	w

Abbreviations	Symbols
mathematics (ical)	math
maximum	max
mean effective pressure	mep
mean horizontal candlepower	mhcp
meacycle	mHz
megohm	MΩ
melting point, -temperature	mp, T_m
meter	m
meter-kilogram	m-kg
metre	m
mho	spell out
microsmpere	μa or mu a
microfarad	μf
microinch	μin.
micrometer (formerly micron)	μm
micromicrofarad	μμf
micromicron	μμ
micron	μ
microvolt	μv
microwatt	μw or mu w
mile	spell out
miles per hour	mph
miles per hour per second	mphps
milli	m
milliampere	ma
milliequivalent	meq
milligram	mg
millihenry	mh
millilambert	mL
milliliter or milliliter	ml
millimeter	mm
millimeter or mercury (pressure)	mm Hg
millimicron	mμ or m mu
million	spell out
million gallons per day	mgd
millivolt	mv
minimum	min
minute	min
minute (angular measure)	′

Abbreviations	Symbols
minute (time) (in astronomical tables)	m
mile	spell out
modal	m
modulus of elasticity	E
molar	M
molar enthalpy	H_m
molar Gibbs Energy	G_m
molar heat capacity	C_m
mole	mol
mole fraction	x
molecular weight	mol wt or M
month	spell out
N	number of elementary particles (e.g., molecules, groups, atoms, electrons)
N_L	Avogadro number (Loschmidt's number)
n	amount of a substance (mole)
n	refractive index
nanometer (formerly millimicron)	nm
National Association of Corrosion Engineers	NACE
National Electrical Code	NEC
newton	N
normal	N
number of elementary particles	N
Occupational Safety and Health Administration	OSHA
ohm	Ω
ohm-centimeter	ohm-cm
oil absorption	O.A.
ounce	oz
once-foot	oz-ft
ounce-inch	oz-in.
outside diameter	OD
osomotic pressure	Π
P	permeability of membranes
p	probability
p	dipole moment
$\mathbf{p}_i$	induced dipolar moment
p	pressure

Abbreviations	Symbols
p	extent of reaction
Paint Testing Manual	PTM
parameter	Q
partition function (system)	Q
parts per billion	ppb
parts per million	ppm
pascal	Pa
peck	pk
penny (pency – new British)	p.
pennyweight	dwt
per	diagonal line in expressions with unit symbols or (see Fundamental Rules)
percent	%
permeability of membranes	P
peso	spell out
pint	pt.
Planck's constant (in $E = h\nu$) (6.62517 +/− 0.00023 x 10^{-27} erg sec)	h
polymolecularity index	Q
potential	spell out
potential difference	spell out
pound	lb
pound-foot	lb-ft
pound-inch	lb-in.
pound sterling	£
pounds-force per square inch	psi
pounds per brake horsepower-hour	lb per bhp-hr
pounds per cubi foot	lb per cut ft
pounds per square foot	psf
pounds per square inch	psi
pounds per square inch absolute	psia
power factor	spell out or pf
pressure	p
probability	p
Q	quantity of electricity, charge
Q	heat
Q	partition function (system)
Q	parameter in the Q–e copolymerize-tion equation

Abbreviations	Symbols
Q, Q	polydispersity, polymolecularity in-dex ($Q = \overline{M_w}/\overline{M_n}$)
q	partition function (particles)
quantity of electricity, charge	Q
quart	qt
quod vide (which see)	q.v.
R	molar gas constant
R	electrical resistance
R_G	radius of gyration
R_n	run number
R_ϑ	Rayleigh ratio
r	radius
r_o	initial molar ratio of reactive groups in polycondensations
radian	spell out
radius	r
radius of gyration	R_G
rate constant	k
Rayleigh ratio	R_ϑ
reactive kilovolt-ampere	kvar
reactive volt-ampere	var
reference(s)	ref
refractive index	n
relaxation time	τ
resistivity	ρ
revolutions per minute	rpm
revolutions per second	rps
rod	spell out
root mean square	rms
S	entropy
S^m	molar entropy
S	solubility coefficient
s	sedimentation coefficient
s	selectivity coefficient in osmotic measurements)
Saybolt Universal seconds	SUS
secant	sec
second	s or sec
second (angular measure)	″
second-foot (see cubic feet per second)	

Abbreviations	Symbols
second (time) (in astronomical tables)	s
Second virial coefficient	A_2
shaft horsepower	shp
shilling	s
sine	sin
sine of the amplitude, an elliptic function	sn
society	Soc.
Soluble	sol
solubility coefficient	S
solubility parameter	δ
solution	soln
specific gravity	sp gr
specific heat	sp ht
specific heat capacity (formerly: specific heat)	c
specific optical rotation	$[\alpha]$
specific volume	sp vol
spherical candle power	scp
square	sq
square centimeter	sq cm or cm^2
square foot	sq ft
square inch	sq in.
square kilometer	sq km or km^2
square meter	sq m or m^2
square micron	sq μ or μ^2
square root of mean square	rms
standard	std
Standard	Stnd.
Standard deviation	σ
Staudinger index	$[\eta]$
stere	s
syndiotactic	st
T	temperature
t	time
t	*trans* conformation
tangent	tan
temperature	T or temp
tensile strength	ts
threodiisotactic	tit
thousand	M
thousand foot-pounds	kip-ft
thousand pound	kip
ton	spell out
ton-mile	spell out
trans conformation	t
trans-tactic	tt
U	voltage
U	internal energy
U^m	molar internal energy
u	excluded volume
ultraviolet	UV
United States	U.S.
V	volume
V	electrical potential
v	rate, rate of reaction
v	specific volume always with an in-dex
vapor pressure	vp
versed sine	vers
versus	vs
volt	v or V
volt-ampere	va
volt-coulomb	spell out
voltage	U
volume	V or vol.
Volume (of a publication)	Vol
W	weight
W	work
w	mass function
watt	w or W
watthour	whr
watts per candle	wpc
week	spell out
weight	W or w
weight concentration*	c
work	y yield
X	degree of polymerization
X	electrical resistance
x	mole fractio y yield
yard	yd
year	yr
Young's	E
Z	collision number
Z	z fraction
z	ionic charge

Abbreviations and Symbols

Abbreviations	Symbols
z	coordination number
z	dissymmetry (light scattering)
z	parameter in excluded volume theory
α	angle, especially angle of rotation in optical activity
α	cubic expandion coefficient [$\alpha = V^{-1}(\partial V/\partial T)_p$]
α	expansion coefficient (as reduced length, e.g., α_L in the chain end-to-end distance or α_R for the radius of gyration)
α	degree of crystallinity (always with an index)
α	electric polarizability of a molecule
[α]	"specific" optical rotation
β	angle
β	coefficient of pressure
β	excluded volume cluster integral
Γ	preferential solvation
γ	angle
γ	surface tension
γ	linear expansion coefficient
δ	loss angle
δ	solubility parameter
δ	chemical shift
ε	linear expansion ($\varepsilon = \Delta l/l_o$)
ε	expectation
ε_r	relative permittivity (dielectric number)
η	dynamic viscosity
[η]	Staudinger index (called J_o in DIN 1342)
Θ	characteristic temperature, especial-ly theta temperature
θ	angle, especially angle of rotation
ϑ	angle, especially valence angle
κ	isothermal compressibility [$\kappa = V^{-1}(\partial V/\partial p)_T$]
κ	enthalpic interaction parameter in solution theory

Abbreviations	Symbols
λ	wavelength
λ	heat conductivity
λ	degree of coupling
μ	chemical potential
μ	moment
μ	permanent dipole moment
ν	mement, with respect to a reference value
ν	frequency
ν	kinetic chain length
ξ	shielding ratio in the theory of random coils
Ξ	partition function
Π	osmotic pressure
ρ	density
σ	mechanical stress (σ_{ii} = normal stress, σ_{ij} = shear stress)
σ	standard deviation
σ	hindrance parameter
τ	relaxation time
τ_i	internal transmittance (transmission factor) (represents the ratio of transmitted to absorbed light)
φ	volume fraction
$\varphi(r)$	potential between two segments separated by a distance r
Φ	constant in the viscosity-molecular-weight relationship
[Φ]	"molar" optical rotation
χ	interaction parameter in solution theory
ψ	entropic interaction parameter in solution theory
ω	angular frequency, angular velocity
Ω	angle
Ω	probability
Ω	skewness of a distribution

*(= weight of solute divided by volume of solvent); IUPAC suggests the symbol ρ for this quantity, which could lead to confusion with the same IUPAC symbol for density.

Notations

The abbreviations for chemicals and polymer were taken from the "Manual of Symbols and Terminology for Physicochemical Quantities and Units," *Pure and Applied Chemistry* **21***1) (1970), but some were added because of generally accepted use.

The ISO (International Standardization Organization) has suggested that all extensive quantities should be described by capital letters and all intensive quantities by lower-case letters. IUPAC doe not follow this recommendation, however, but uses lower-case letters for specific quantities.

The following symbols are used above or after a letter.

Symbols Above Letters

— signifies an average, e.g., $\overline{M}$ is the average molecular weight; more complicated averages are often indicated by $\langle \rangle$, e.g., $\langle R_G^2 \rangle$ is another way of writing $\overline{(R_G^2)}_z$

— stands for a partial quantity, e.g., $\tilde{v}_A$ is the partial specific volume of the compound A; V_A is the volume of A, wherea $\tilde{V}_A^m$xxx is the partial molar volume of A.

Superscripts

°	pure substance or standard state
∞	infinite dilution or infinitely high molecular weight
m	molar quantity (in cases where subscript letters are impractical)
(q)	the q order of a moment (always in parentheses)
‡	activated complex

Subscripts

Initial	State
1	solvent
2	solute
3	additional components (e.g., precipitant, salt, etc.)
am	amorphous
B	brittleness
bd	bond
cr	crystalline
crit	critical
cryst	crystallization
e	equilibrium

Initial	State
E	end group
G	glassy state
i	run number
i	initiation
i	isotactic diads
ii	isotactic triads
Is	heterotactic triads
j	run number
k	run number
m	molar
M	melting process
mon	monomer
n	number average
p	polymerization, especially propagation
pol	polymer
r	general for average
s	syndiotactic diads
ss	syndiotactic triads
st	start reaction
t	termination
tr	transfer
u	monomeric unit
w	weight average
z	z average
Prefixes	
at	atactic
ct	*cis*-tactic
eit	erythrodiisotactic
it	isotactic
st	syndiotactic
tit	threodiisotactic
tt	*trans*-tactic

Square brackets around a letter signify molar concentrations. (IUPAC prescribes the symbol c for molar councentrations, but to date this has consistently been used for the mass/volume unit.)

Angles are always given by °.

Apart from some exceptions, the meter is not used as a unit of length; the units cm and mm derived from it are used. Use of the meter in macromolecular science leads to very impractical units.

Mathematical Signs

Sign	Definition
Operations	
$+$	Addition
$-$	Subtraction
$\times$	Multiplication
$\cdot$	Multiplication
$\div$	Division
$/$	Division
$\circ$	Composition
$\cup$	Union
$\cap$	Intersection
$\pm$	Plus or minus
$\mp$	Minus or plus
Convolution	
$\oplus$	Direct sum, variation
$\ominus$	Various
$\otimes$	Various
$\odot$	Various
$:$	Ratio
$\amalg$	Amalgamation
Relations	
$=$	Equal to
$\neq$	Not equal to
$\approx$	Nearly equal to
$\cong$	Equals approximately, isomorphic
$<$	Less than
$<<$	Much less than
$>$	Greater than
$>>$	Much greater than
$\leq$	Less than or equal to
$\leqslant$	Les than or equal to
$\leqq$	Less than or equal to
$\geq$	Greater than or equal to
$\geqslant$	Grean than or equalt o
$\geqq$	Greater than or equal to
$\equiv$	Equivalent to, congruent to
$\not\equiv$	Not equivalent to, not congruent to
$|$	Divides, divisible by
$\sim$	Similar to, asymptotically equal to
$:=$	Assignment

Sign	Definition
$\in$	A member of
$\subset$	Subset of
$\subseteq$	Subset of or equal to
$\supset$	Superset of
$\supseteq$	Superset of or equal to
$\propto$	Varies as, proportional to
$\doteq$	Approaches a limit, definition
$\rightarrow$	Tends to, maps to
$\leftarrow$	Maps from
$\mapsto$	Maps to
$\hookrightarrow$ or $\hookleftarrow$	Maps into
$\square$	d'Alembertian operator
Σ	Summation
Π	Product
$\int$	Integral
$\oint$	Contour integral
Logic	
$\wedge$	And, conjunction
$\vee$	Or, distunction
$\neg$	Negation
$\Rightarrow$	Implies
$\rightarrow$	Implies
$\Leftrightarrow$	If and only if
$\leftrightarrow$	If and only if
$\exists$	Existential quantifier
$\forall$	Universal quantifier
$\in$	A member o
$\notin$	Not a member of
$\vdash$	Assertion
$\therefore$	Hence, therefore
$\because$	Because
Radial units	
$'$	Minute
$''$	Second
$\circ$	Degree
Constants	
π	pi ($\approx$3.14159265)
e	Base of natural logarithms ($\approx$2.71828183)

Sign	Definition
Geometry	
⊥	Perpendicular
∥	Parallel
∦	Not parallel
∠	Angle
∢	Spherical angle
$\stackrel{v}{=}$	Equal angles
Miscellaneous	
i	Square root of -1
′	Prime
″	Double prime
‴	Triple prime
√	Square root, radical
$\sqrt[3]{\ }$	Cube root
$\sqrt[n]{\ }$	nth root
!	Factorial
!!	Double factorial
∅	Empty set, null set
∞	Infinity

Sign	Definition
∂	Partial differential
Δ	Delta
∇	Nabla, del
∇^2, Δ	Laplacian operator

English–Greek–Latin Numerical Prefixes

English	Greek	Latin
2	bis	di
3	tris	tri
4	tetrakis	tetra
5	pentakis	penta
6	hexakis	hexa
7	heptakis	hepta
8	octakis	octa
9	nonakis	nona
10	decakis	deca

Greek-Russian-English Alphabets

Greek letter		Greek name	English equivalent	Russian letter		English equivalent
A	α	Alpha	(ä)	А	а	(ä)
B	β	Beta	(b)	Б	б	(b)
				В	в	(v)
Γ	γ	Gamma	(g)	Г	г	(g)
Δ	δ	Delta	(d)	Д	д	(d)
E	ε	Epsilon	(e)	Е	е	(ye)
Z	ζ	Zeta	(z)	Ж	ж	(zh)
				З	з	(z)
H	η	Eta	(ā)	И	и	(i, ē)
Θ	θ	Theta	(th)	Й	й	(ē)
I	ι	Iota	(ē)	К	к	(k)
				Л	л	(l)
K	k	Kappa	(k)	М	м	(m)
Λ	λ	Lambda	(l)	Н	н	(n)
				О	о	(ô, o)
M	μ	Mu	(m)	О	о	(ô, o)
				П	п	(p)
N	ν	Nu	(n)	Р	р	(r)
Ξ	ξ	Xi	(ks)	С	с	(s)
				Т	т	(t)
O	o	Omicron	α	У	у	ōō
Π	π	Pi	(P)	Ф	ф	(f)
				Х	х	(kh)
P	ρ	Rho	(r)	Х	х	(kh)
				Ц	ц	(t_s)
Σ	σ	Sigma	(s)	Ч	ч	(ch)
T	τ	Tau	(t)	Ш	ш	(sh)
Υ	υ	Upsilon	(ü, ōō)	Щ	щ	(shch)
				Ъ	ъ	8
Φ	ø	Phi	(f)	Ы	ы	(ë)
X	χ	Chi	(H)	ь	ь	(ë)
Ψ	ψ	Psi	(ps)	Э	э	(e)
				Ю	ю	(ū)
Ω	ω	Omega	(ō)	Я	я	(yä)

English-Greek-Latin Numbers

English	Greek	Latin
1	mono	uni
2	bis	di
3	tris	tri
4	tetrakis	tetra
5	pentakis	penta
6	hexakis	hexa
7	heptakis	hepta
8	octakis	octa
9	nonakis	nona
10	decakis	deca

International Union of Pure and Applied Chemistry: Rules Concerning Numerical Terms Used in Organic Chemical Nomenclature (specifically as prefixes for hydrocarbons)

1	mono-or hen-	10	deca-	100	hecta-	1000	kilia-
2	di- or do-	20	icosa-	200	dicta-	2000	dilia-
3	tri-	30	triaconta-	300	tricta-	3000	trilia-
4	tetra-	40	tetraconta-	400	tetracta	4000	tetralia-
5	penta-	50	pentaconta-	500	pentactra	5000	pentalia-
6	hexa-	60	hexaconta-	600	hexacta	6000	hexalia-
7	hepta-	70	hepaconta-	700	heptacta-	7000	hepalia-
8	octa-	80	octaconta-	800	ocacta-	8000	ocatlia-
9	nona-	90	nonaconta-	900	nonactta-	9000	nonalia-

Source: IUPAC, Commission on Nomenclature of Organic Chemistry (N. Lorzac'h and published in *Pure and Appl. Chem* 58: 1693–1696 (1986))

Elemental Symbols and Atomic Weights

Source: International Union of Pure and Applied Chemistry (IUPAC) 2001Values from the 2001 table *Pure Appl. Chem.*, **75**, 1107–1122 (2003). The values of zinc, krypton, molybdenum and dysprosium have been modified. The *approved name* for element 110 is included, see *Pure Appl. Chem.*, **75**, 1613–1615 (2003). The *proposed name* for element 111 is also included.

A number in parentheses indicates the uncertainty in the last digit of the atomic weight.

List of Elements in Atomic Number Order

At No	Symbol	Name	Atomic Wt	Notes
1	H	Hydrogen	1.00794(7)	1, 2, 3
2	He	Helium	4.002602(2)	1, 2
3	Li	Lithium	[6.941(2)]	1, 2, 3, 4
4	Be	Beryllium	9.012182(3)	
5	B	Boron	10.811(7)	1, 2, 3
6	C	Carbon	12.0107(8)	1, 2
7	N	Nitrogen	14.0067(2)	1, 2
8	O	Oxygen	15.9994(3)	1, 2
9	F	Fluorine	18.9984032(5)	
10	Ne	Neon	20.1797(6)	1, 3
11	Na	Sodium	22.989770(2)	
12	Mg	Magnesium	24.3050(6)	
13	Al	Aluminium	26.981538(2)	
14	Si	Silicon	28.0855(3)	2
15	P	Phosphorus	30.973761(2)	
16	S	Sulfur	32.065(5)	1, 2
17	Cl	Chlorine	35.453(2)	3
18	Ar	Argon	39.948(1)	1, 2
19	K	Potassium	39.0983(1)	1
20	Ca	Calcium	40.078(4)	1
21	Sc	Scandium	44.955910(8)	
22	Ti	Titanium	47.867(1)	
23	V	Vanadium	50.9415(1)	
24	Cr	Chromium	51.9961(6)	
25	Mn	Manganese	54.938049(9)	
26	Fe	Iron	55.845(2)	
27	Co	Cobalt	58.933200(9)	
28	Ni	Nickel	58.6934(2)	
29	Cu	Copper	63.546(3)	2
30	Zn	Zinc	65.409(4)	
31	Ga	Gallium	69.723(1)	
32	Ge	Germanium	72.64(1)	
33	As	Arsenic	74.92160(2)	
34	Se	Selenium	78.96(3)	
35	Br	Bromine	79.904(1)	
36	Kr	Krypton	83.798(2)	1, 3
37	Rb	Rubidium	85.4678(3)	1
38	Sr	Strontium	87.62(1)	1, 2
39	Y	Yttrium	88.90585(2)	
40	Zr	Zirconium	91.224(2)	1
41	Nb	Niobium	92.90638(2)	
42	Mo	Molybdenum	95.94(2)	1
43	Tc	Technetium	[98]	5
44	Ru	Ruthenium	101.07(2)	1
45	Rh	Rhodium	102.90550(2)	
46	Pd	Palladium	106.42(1)	1
47	Ag	Silver	107.8682(2)	1
48	Cd	Cadmium	112.411(8)	1
49	In	Indium	114.818(3)	
50	Sn	Tin	118.710(7)	1
51	Sb	Antimony	121.760(1)	1
52	Te	Tellurium	127.60(3)	1
53	I	Iodine	126.90447(3)	
54	Xe	Xenon	131.293(6)	1, 3
55	Cs	Caesium	132.90545(2)	
56	Ba	Barium	137.327(7)	
57	La	Lanthanum	138.9055(2)	1
58	Ce	Cerium	140.116(1)	1
59	Pr	Praseodymium	140.90765(2)	
60	Nd	Neodymium	144.24(3)	1
61	Pm	Promethium	[145]	5
62	Sm	Samarium	150.36(3)	1
63	Eu	Europium	151.964(1)	1
64	Gd	Gadolinium	157.25(3)	1
65	Tb	Terbium	158.92534(2)	
66	Dy	Dysprosium	162.500(1)	1
67	Ho	Holmium	164.93032(2)	
68	Er	Erbium	167.259(3)	1

At No	Symbol	Name	Atomic Wt	Notes
69	Tm	Thulium	168.93421(2)	
70	Yb	Ytterbium	173.04(3)	1
71	Lu	Lutetium	174.967(1)	1
72	Hf	Hafnium	178.49(2)	
73	Ta	Tantalum	180.9479(1)	
74	W	Tungsten	183.84(1)	
75	Re	Rhenium	186.207(1)	
76	Os	Osmium	190.23(3)	1
77	Ir	Iridium	192.217(3)	
78	Pt	Platinum	195.078(2)	
79	Au	Gold	196.96655(2)	
80	Hg	Mercury	200.59(2)	
81	Tl	Thallium	204.3833(2)	
82	Pb	Lead	207.2(1)	1, 2
83	Bi	Bismuth	208.98038(2)	
84	Po	Polonium	[209]	5
85	At	Astatine	[210]	5
86	Rn	Radon	[222]	5
87	Fr	Francium	[223]	5
88	Ra	Radium	[226]	5
89	Ac	Actinium	[227]	5
90	Th	Thorium	232.0381(1)	1, 5
91	Pa	Protactinium	231.03588(2)	5
92	U	Uranium	238.02891(3)	1, 3, 5
93	Np	Neptunium	[237]	5
94	Pu	Plutonium	[244]	5
95	Am	Americium	[243]	5
96	Cm	Curium	[247]	5
97	Bk	Berkelium	[247]	5
98	Cf	Californium	[251]	5
99	Es	Einsteinium	[252]	5
100	Fm	Fermium	[257]	5
101	Md	Mendelevium	[258]	5
102	No	Nobelium	[259]	5
103	Lr	Lawrencium	[262]	5
104	Rf	Rutherfordium	[261]	5, 6
105	Db	Dubnium	[262]	5, 6
106	Sg	Seaborgium	[266]	5, 6
107	Bh	Bohrium	[264]	5, 6
108	Hs	Hassium	[277]	5, 6
109	Mt	Meitnerium	[268]	5, 6
110	Ds	Darmstadtium	[281]	5, 6
111	Rg	Roentgenium	[272]	5, 6

At No	Symbol	Name	Atomic Wt	Notes
112	Uub	Ununbium	[285]	5, 6
114	Uuq	Ununquadium	[289]	5, 6
116	Uuh	Ununhexium		see Note above
118	Uuo	Ununoctium		see Note above

1. Geological specimens are known in which the element has an isotopic composition outside the limits for normal material. The difference between the atomic weight of the element in such specimens and that given in the Table may exceed the stated uncertainty.
2. Range in isotopic composition of normal terrestrial material prevents a more precise value being given; the tabulated value should be applicable to any normal material.
3. Modified isotopic compositions may be found in commercially available material because it has been subject to an undisclosed or inadvertant isotopic fractionation. Substantial deviations in atomic weight of the element from that given in the Table can occur.
4. Commercially available Li materials have atomic weights that range between 6.939 and 6.996; if a more accurate value is required, it must be determined for the specific material [range quoted for 1995 table 6.94 and 6.99].
5. Element has no stable nuclides. The value enclosed in brackets, e.g. [209], indicates the mass number of the longest-lived isotope of the element. However three such elements (Th, Pa, and U) do have a characteristic terrestrial isotopic composition, and for these an atomic weight is tabulated.
6. The names and symbols for elements 112-118 are under review. The temporary system recommended by J Chatt, *Pure Appl. Chem.*, **51**, 381–384 (1979) is used above. The names of elements 101-109 were agreed in 1997 (See *Pure Appl. Chem.*, 1997, **69**, 2471–2473) and for element 110 in 2003 (see *Pure Appl. Chem.*, 2003, **75**, 1613–1615). The proposed name for element 111 is also included.

List of Elements in Name Order

At No	Symbol	Name	Atomic Wt	Notes
89	Ac	Actinium	[227]	5
13	Al	Aluminium	26.981538(2)	
95	Am	Americium	[243]	5
51	Sb	Antimony	121.760(1)	1

At No	Symbol	Name	Atomic Wt	Notes
18	Ar	Argon	39.948(1)	1, 2
33	As	Arsenic	74.92160(2)	
85	At	Astatine	[210]	5
56	Ba	Barium	137.327(7)	
97	Bk	Berkelium	[247]	5
4	Be	Beryllium	9.012182(3)	
83	Bi	Bismuth	208.98038(2)	
107	Bh	Bohrium	[264]	5, 6
5	B	Boron	10.811(7)	1, 2, 3
35	Br	Bromine	79.904(1)	
48	Cd	Cadmium	112.411(8)	1
55	Cs	Caesium	132.90545(2)	
20	Ca	Calcium	40.078(4)	1
98	Cf	Californium	[251]	5
6	C	Carbon	12.0107(8)	1, 2
58	Ce	Cerium	140.116(1)	1
17	Cl	Chlorine	35.453(2)	3
24	Cr	Chromium	51.9961(6)	
27	Co	Cobalt	58.933200(9)	
29	Cu	Copper	63.546(3)	2
96	Cm	Curium	[247]	5
110	Ds	Darmstadtium	[281]	5, 6
105	Db	Dubnium	[262]	5, 6
66	Dy	Dysprosium	162.500(1)	1
99	Es	Einsteinium	[252]	5
68	Er	Erbium	167.259(3)	1
63	Eu	Europium	151.964(1)	1
100	Fm	Fermium	[257]	5
9	F	Fluorine	18.9984032(5)	
87	Fr	Francium	[223]	5
64	Gd	Gadolinium	157.25(3)	1
31	Ga	Gallium	69.723(1)	
32	Ge	Germanium	72.64(1)	
79	Au	Gold	196.96655(2)	
72	Hf	Hafnium	178.49(2)	
108	Hs	Hassium	[277]	5, 6
2	He	Helium	4.002602(2)	1, 2
67	Ho	Holmium	164.93032(2)	
1	H	Hydrogen	1.00794(7)	1, 2, 3
49	In	Indium	114.818(3)	
53	I	Iodine	126.90447(3)	
77	Ir	Iridium	192.217(3)	
26	Fe	Iron	55.845(2)	

At No	Symbol	Name	Atomic Wt	Notes
36	Kr	Krypton	83.798(2)	1, 3
57	La	Lanthanum	138.9055(2)	1
103	Lr	Lawrencium	[262]	5
82	Pb	Lead	207.2(1)	1, 2
3	Li	Lithium	[6.941(2)]	1, 2, 3, 4
71	Lu	Lutetium	174.967(1)	1
12	Mg	Magnesium	24.3050(6)	
25	Mn	Manganese	54.938049(9)	
109	Mt	Meitnerium	[268]	5, 6
101	Md	Mendelevium	[258]	5
80	Hg	Mercury	200.59(2)	
42	Mo	Molybdenum	95.94(2)	1
60	Nd	Neodymium	144.24(3)	1
10	Ne	Neon	20.1797(6)	1, 3
93	Np	Neptunium	[237]	5
28	Ni	Nickel	58.6934(2)	
41	Nb	Niobium	92.90638(2)	
7	N	Nitrogen	14.0067(2)	1, 2
102	No	Nobelium	[259]	5
76	Os	Osmium	190.23(3)	1
8	O	Oxygen	15.9994(3)	1, 2
46	Pd	Palladium	106.42(1)	1
15	P	Phosphorus	30.973761(2)	
78	Pt	Platinum	195.078(2)	
94	Pu	Plutonium	[244]	5
84	Po	Polonium	[209]	5
19	K	Potassium	39.0983(1)	1
59	Pr	Praseodymium	140.90765(2)	
61	Pm	Promethium	[145]	5
91	Pa	Protactinium	231.03588(2)	5
88	Ra	Radium	[226]	5
86	Rn	Radon	[222]	5
75	Re	Rhenium	186.207(1)	
45	Rh	Rhodium	102.90550(2)	
111	Rg	Roentgenium	[272]	5, 6
37	Rb	Rubidium	85.4678(3)	1
44	Ru	Ruthenium	101.07(2)	1
104	Rf	Rutherfordium	[261]	5, 6
62	Sm	Samarium	150.36(3)	1
21	Sc	Scandium	44.955910(8)	
106	Sg	Seaborgium	[266]	5, 6
34	Se	Selenium	78.96(3)	
14	Si	Silicon	28.0855(3)	2

At No	Symbol	Name	Atomic Wt	Notes
47	Ag	Silver	107.8682(2)	1
11	Na	Sodium	22.989770(2)	
38	Sr	Strontium	87.62(1)	1, 2
16	S	Sulfur	32.065(5)	1, 2
73	Ta	Tantalum	180.9479(1)	
43	Tc	Technetium	[98]	5
52	Te	Tellurium	127.60(3)	1
65	Tb	Terbium	158.92534(2)	
81	Tl	Thallium	204.3833(2)	
90	Th	Thorium	232.0381(1)	1, 5
69	Tm	Thulium	168.93421(2)	
50	Sn	Tin	118.710(7)	1
22	Ti	Titanium	47.867(1)	
74	W	Tungsten	183.84(1)	
112	Uub	Ununbium	[285]	5, 6
116	Uuh	Ununhexium		see Note above
118	Uuo	Ununoctium		see Note above
114	Uuq	Ununquadium	[289]	5, 6
92	U	Uranium	238.02891(3)	1, 3, 5
23	V	Vanadium	50.9415(1)	
54	Xe	Xenon	131.293(6)	1, 3
70	Yb	Ytterbium	173.04(3)	1
39	Y	Yttrium	88.90585(2)	
30	Zn	Zinc	65.409(4)	
40	Zr	Zirconium	91.224(2)	1

Pronounciation Symbols and Abbreviations

ə	Banana, collide, abut
ˈə, ˌə	Humdrum, abut
ᵊ	Immediately preceding \l\, \n\, \m\, \ŋ\, as in battle, mitten, eaten, and sometimes open \ˈō-pᵊm\, lock and key \-ᵊ ŋ-\; immediately following \l\, \m\, \r\, as often in French table, prisme, titre
ər	further, merger, bird
ˈə-, ˈə-r	As in two different pronunciations of hurry \ˈhər-ē, \ˈhə-rē\
a	mat, map, mad, gag, snap, patch
ā	day, fade, date, aorta, drape, cape
ä	bother, cot, and, with most American speakers, father, cart
á	father as pronounced by speakers who do not rhyme it with bother; French patte
aú	now, loud, out
b	baby, rib
ch	chin, nature \ˈnā-chər\
d	did, adder
e	bet, bed, peck
ˈē, ˌē	beat, nosebleed, evenly, easy
ē	easy, mealy
f	fifty, cuff
g	go, big, gift
h	hat, ahead
hw	whale as pronounced by those who do not have the same pronunciation for both whale and wail
i	tip, banish, active
ī	site, side, buy, tripe
j	job, gem, edge, join, judge
k	kin, cook, ache
k̲	German ich, Buch; one pronunciation of loch
l	lily, pool
m	murmur, dim, nymph
n	no, own
ⁿ	Indicates that a preceeding vowel or diphthong is pronounced with the nasal passages open, as in French un bon vin blanc \œⁿ-bōⁿvaⁿ-bläⁿ\
ŋ	sing \ˈsiŋ\, singer \ˈsiŋ-ər\, finger \ˈfiŋ-gər\, ink \ˈiŋk\
ō	bone, know, beau
ó	saw, all, gnaw, caught
ōō	fool
o͝o	took
œ	French coeuf, German Hölle
œ̄	French feu, German Höhle
ói	coin, destroy
p	pepper, lip
r	red, car, rarity
s	source, less
sh	as in shy, mission, machine, special (actually, this is a single sound, not two); with a hyphen between, two sounds as in grasshopper \ˈgras-ˌhä-pər\
t	tie, attack, late, later, latter
th	as in thin, ether (actually, this is a single sound, not two); with a hyphen between, two sounds as in knighthood \ˈnīt-ˌh----d\
t̲h̲	then, either, this (actually, this is a single sound, not two)
ü	rule, youth, union \ˈyün-yən\, few \ˈfyü\
ú	pull, wood, book, curable \ˈky ú r-ə-bəl\, fury \ˈfy----r-ē\
ue	German füllen, hübsch
u̲e̲	French rue, German fühlen
v	vivid, give
w	we, away
y	yard, young, cue \ˈkyü\, mute \ˈmyüt\, union \ˈyün-yən\
ʸ	indicates that during the articulation of the sound represented by the preceding character the front of the tongue has substantially the position it has for the articulation of the first sound of yard, as in French digne \dēnʸ\
z	zone, raise
zh	as in vision, azure \ˈa-zhər\ (actually this is a single sound, not two).
\	reversed virgule used in pairs to mark the beginning and end of a transcription: \ˈpen\
ˈ	mark preceding a syllable with primary (strongest) stress: \ˈpen-mən-ˌship\
ˌ	mark preceding a syllable with secondary (medium) stress: \ˈpen-mən-ˌship\
-	mark of syllable division

( )	indicate that what is symbolized between is present in some utterances but not in others: *factory* \ ▎fak-t(ə-)rē
÷	indicates that many regard as unacceptable the pronunciation variant immediately following: *cupola* \ ▎kyü-pə-lə, ÷- ▎lō\

Explanatory Notes and Abbreviations

(date)	date that word was first recorded as having been used
[. . .]	etomology and origin(s) of word
{. . .}	usage and/or languages, including French, German, Italian and Spanish
adj	adjective
adv	adverb
B.C.	before Christ
Brit.	Britain, British
C	centigrade, Celsius
c	century
E	English
Eng.	England
F	French, Fahrenheit
Fr.	France
fr.	from
G	German
Gr.	Germany
L	Latin
ME	middle English

n	noun
neut.	neuter
NL	new Latin
OE	old English
OL	old Latin
pl	plural
prp.	present participle
R	Russian
sing.	singular
S	Spanish
U.K.	United Kingdom
v	verb

Source: From *Merriam-Webster's Collegiate© Dictionary*, Eleventh Editioh, ©2004 by Merriam-Webster, incorporated, (www.Merriam-Webster.com). With permission.

Languages

French, German and Spanish translations are enclosed in {--} and preceded by F, G, I and S, respectively; and gender is designated by f-feminine, m-masculine, n-neuter. For example: **Polymer**--{F polymere m} represents the French translation "polymere" of the English word polymer and it is in the masculine case. These translations were obtain from multi-language dictionaries including: *A Glossary of Plastics Terminology in 5 Languages*, 5th Ed., Glenz, W., (ed) Hanser Gardner Publications, Inc., Cinicinnati, 2001. By permission).

R

r \är\ Symbol for radius of a circle or sphere, or radial coordinate in cylindrical and spherical coordinate systems, or product–moment correlation coefficient in statistics (Witte RSS, Witte JS, Smith GS (2003) Statistics. Wiley, New York; Box GE, Hunter WG, Hunter JS (2005) Statistics for experimenters: innovation, and discovery, 2nd edn. Wiley, New York).

R In organic chemistry, symbol for a general attached group or radical, frequently an aliphatic or aromatic hydrocarbon group, that may take on any of various specific identities. Abbreviation for Roentgen. Abbreviation for generalized of multiple correlation coefficient, the square root of the Coefficient of Determination. (°R) Abbreviation for Degree of Rankine, See ▶ Rankine Temperature. Symbol for electrical resistance.

Rabinowitsch Correction n The correction factor derived by B. Rabinowitsch (1929) applied to the Newtonian shear rate at the wall of a circular tube (including capillary) through which a non-Newtonian liquid is flowing, gives the true shear rate at the wall. For pseudoplastic liquids such as paints and some polymer melts the correction is always an increase. If the fluid obeys the *Power Law* it reduces to a simple correction factor $(3n + 1)/4n$, where n is the flow-behavior index of the liquid (Munson BR, Young DF, Okiishi TH (2005) Fundamentals of fluid mechanics. Wiley, New York; Harper CA (ed) (2002) Handbook of plastics, elastomers and composites, 4th edn. McGraw-Hill, New York; Patton TC (1964) Paint flow and pigment dispersion. Interscience Publishers, New York).

Racemic \rā-sē-mik\ *adj* (1892) Of, relating to, or constituting a compound or mixture that is composed of equal parts of dextrorotatory (D-) and levorotatory (L-) forms of the same compound and is not optically active (Morrison RT, Boyd RN (1992) Organic chemistry, 6th edn. Prentice-Hall, Englewood Cliffs, NJ).

Rack n A warp-knitting measure consisting of 480 courses. Tricot fabric quality is judged by the number of inches per rack.

Racked Stitch n A knitting stitch that produces a herringbone effect with a ribbed back. It is employed in sweaters for decorative purposes or to form the edge of garments. The racked stitch is a variation of the half-cardigan stitch; it is created when one set of needles is displaced in relation to the other set.

Racking n A term referring to the side-to-side movement of the needles of the needle bed of a knitting machine. Racking results in inclined stitches and reduced elasticity.

Rad \rad\ n [radiation absorbed dose] (1918) (1) A deprecated, but still widely used, unit of energy absorbed by a material, including living matter, from exposure to ionizing radiation. 1 rad = 0.01 gray (Gy) = 0.01 J/kg. The abbreviation for the SI unit of plane angular measure, the *radian*, the angle intercepting a circular arc of length equal to its radius (= $360o/2\pi$ = $57.3°$) (Lide DR (ed) (2004) CRC handbook of chemistry and physics. CRC Press, Boca Raton, FL; (2003) College physics. McGraw-Hill Science/Engineering/Math, New York).

Radiance \rā-dē-ən(t)s\ n (1601) Quotient of the radiant intensity in a given direction of an infinitesimal surface element containing the point under consideration divided by the area of the orthogonal projection of this surface element on a plane perpendicular to the given direction (Holst GC (2003) Electro-optical imaging system performance. J. D. C. Publishing, New York; Klocek P (ed) (1991) Handbook of infrared optical materials. Marcel Dekker, New York).

Radiant Energy n (ca. 1890) That form of energy consisting of the electromagnetic spectrum which travels at 115.890 km (186,500 miles/s) through a vacuum, reducing this speed in denser media (air, water, glass, etc.). The nature of radiant energy is described by its wavelength or frequency although it also behaves as distinct quanta ("corpuscular theory"). The various types of energy may be transformed into other forms of energy (electrical, chemical, mechanical, atomic, thermal, and radiant) but the energy itself cannot be destroyed (Serway RA, Faugh JS, Bennett CV (2005) College physics. Thomas, New York).

Radiant Flux n (1917) Radiant power; radiant energy emitted from, transferred to, or received through a surface per time internal.

Radiant Heat n (1794) Heat transmitted by radiation as contrasted with that transmitted by conduction or convection.

Radiant Heat Baking n Curing treatment in which heat is transferred to the paint surface mainly by radiation from a hot surface, e.g., electric lamps or gas heated panels. This is also known as infrared drying or infrared baking.

Radiant Heat Drying See ▶ Radiant Heat Baking.

Radiant Heating *n* (1794) The net transfer of heat from a hotter body to a cooler one by (usually infrared) radiation. Radiant transfer is one of the three basic mechanisms of Heat Transfer requiring no contact or fluid between the bodies. The net rate is proportional to the differences between the fourth powers of the absolute temperatures of the hotter and cooler bodies $\left(T_1^4 - T_2^4\right)$, and depends also on the thermal "color" of the bodies (their emissivities), their geometries, and their positioning relative to each other. The principal use of radiant heating in the plastics industry is in ▶ Sheet Thermoforming. See ▶ Infrared Drying (or IR Drying) (Strong AB (2000) Plastics materials and processing. Prentice-Hall, Columbus, OH).

Radiation \ ▮rā-dē- ▮ā-shən\ *n* (15c) (1) Emission or transfer of energy in the form of electromagnetic waves or particles. (2) The electromagnetic waves or particles. NOTE — In general, nuclear radiations and radio waves are not considered in this vocabulary, only optical radiations, that is, electromagnetic radiations (photons) of wavelengths between the region of transmission to X-ray (1 nm) and the region of transition to radio waves (1 mm) (Serway RA, Faugh JS, Bennett CV (2005) College physics. Thomas, New York; Weast RC (ed) Handbook of chemistry and physics, 52nd edn. CRC Press, Boca Raton, FL).

Radiation Degradation *n* Breakdown of a plastic caused by too long exposure to radiation, or to radiation of too high energy levels, or both. The radiation may be X-ray, electron, gamma, or neutron beams. The mechanism is ionization and chain scission (Mark JE (ed) Physical properties of polymers handbook. Springer, New York; Zaiko GE (ed) (1995) Degradation and stabilization of polymers. Nova Science Publishers, New York; Dissado LA, Fothergill CJ (eds) (1992) Electrical degradation and breakdown of polymers. Institution of Electrical Engineering (IEE), London).

Radiation Compatibility *n* The ability of a plastic to maintain its properties when exposed to X-ray, gamma, electron, or other ionizing radiation.

Radiation Crosslinking *n* The formation of chemical links between polymer chains through the action of high-energy radiation, commonly gamma radiation from a cobalt-60 source of electrons from an electron gun. The treatment has improved the modulus and raised the use temperature of polyethylene wire coatings and some polymer films. Exposure must be accurately controlled if the crosslinking is to be achieved without degrading the resin (Mark JE (ed) (1996) Physical properties of polymers handbook. Springer, New York).

Radiation Formula, Planck's *n* The emissive power of a black body at wavelength λ may be written

$$E\lambda = \frac{c_1 \lambda^{-5}}{e^{c_2/\lambda T} - 1}$$

where c_1 and c_2 are constants with c_1 being 3.7403×10^{10} mW μm^4 per square centimeter or 3.7403×10^{-12} W cm^2, c_2 being 14,384 micron degrees and T the absolute temperature (Giambattista A, Richardson R, Richardson RC, Richardson B (2003) College physics. McGraw-Hill Science/Engineering/Math, New York; Freir GD (1965) University physics. Appleton-Century-Crofts, New York).

Radiation Polymerization *n* A polymerization reaction initiated by exposure to radiation such as ultraviolet or gamma rays rather than by means of a chemical initiator (Odian GC (2004) Principles of polymerization. Wiley, New York; Lenz RW (1967) Organic chemistry of synthetic high polymers. Interscience Publishers, New York).

Radical \ ▮ra-di-kəl\ *n* [ME, fr. LL *radicalis*, fr. L *radic-*, radix root] (1641) A group of atoms, normally part of a molecule that may replace a single atom (frequently H in organic compounds) and remain unchanged during reactions of the compound. Some examples are the *ethyl* radical, $-C_2H-$, the *acetate* radical, CH_3COO-, and the *phenyl* radical, $-C_6H_5$. Many chemical-reaction mechanisms postulate the transitory existence of unattached ("free") radicals as intermediates, which, because of their charge, are extremely reactive. ▶ Free Radicals play important roles in addition polymerizations. A few free radicals are known that are sufficiently stable to permit their identification ad quantitative determination as chemical entities (Morrison RT, Boyd RN (1992) Organic chemistry, 6th edn. Prentice-Hall, Englewood Cliffs, NJ; Smith MB, March J (2001) Advanced organic chemistry, 5th edn. Wiley, New York; Odian GC (2004) Principles of polymerization. Wiley, New York).

Radical Polymerization *n* A complex mechanism of initiation, propagation, and termination of which the propagation and termination steps are typically very fast. *Also called Free-radical Polymerization* (Odian GC (2004) Principles of polymerization. Wiley, New York).

Radioactive Nuclides *n* Atoms that disintegrate by emission of corpuscular or electromagnetic radiations. The rays most commonly emitted are alpha or beta or gamma rays. The three classes are: *Primary*, which have half-life times exceeding 10^8 years. These may be alpha-emitters or beta-emitters. *Secondary*, which are formed in radioactive transformations starting with U^{238}, U^{235}, or TH^{232}. *Induced*, having geologically short lifetimes and formed by induced nuclear reactions occurring in nature. All these reactions result in transmutation (Serway RA, Faugh JS, Bennett CV (2005) College physics. Thomas, New York; Weast RC (ed) Handbook of chemistry and physics, 52nd edn. CRC Press, Boca Raton, FL).

Radioactive Tracer *n* A chemical compound or other material in which one or more of the ordinary atoms have been replaced by their radioactive isotopes. Carbon-14, tritium (hydrogen-3), and iodine-131 are among the isotopes that have been used in this way. The tracer isotopes have been useful in elucidating chemical-reaction mechanisms and in tracking human and animal metabolisms.

Radioactivity \rā-dē- ō-ak- ti-və-tē\ *n* [ISV] (1899) The decay or decomposition of nuclei of atoms (Serway RA, Faugh JS, Bennett CV (2005) College physics. Thomas, New York).

Radio Frequency *n* (1915) (RF) A frequency of electromagnetic radiation within the broad range of radio and radar transmission, e.g., from about 300 kHz to 20 GHz (Serway RA, Faugh JS, Bennett CV (2005) College physics. Thomas, New York).

Radio-Frequency Drying *n* Use of radio-frequency electromagnetic radiation for drying textiles. The application of RF to wet goods results in the selective heating of the water, which has a partial polarity, because the molecule must do work to align in the RF field causing heat generation within the water droplets. Nonpolar materials, i.e., fabrics, are unaffected. RF drying in very uniform and energy efficient when airflow patterns through the dryer are properly designed and controlled.

Radio-Frequency Heating See ▶ Dielectric Heating.

Radio-Frequency Preheating *n* (RF preheating) A method of preheating used for thermosetting molding materials to facilitate the molding operation or shorten the molding cycle. The frequencies most commonly used are near 20 or 40 MHz.

Radio-Frequency Welding *n* (dielectric welding, high-frequency welding). See ▶ Dielectric Heat Sealing.

Radius of Gyration *n* The radial distance from a given axis at which the mass of a body could be concentrated without altering the rotational inertia of the body about that axis (Giambattista A, Richardson R, Richardson RC, Richardson B (2003) College physics. McGraw-Hill Science/Engineering/Math, New York). For a polymer molecule, a parameter characterizing the size of a polymer random coil. It is defined as

$$R_G^2 = \Sigma m s_i^2 / \Sigma m = \Sigma s_i^2 / n$$

where the polymer chain consists of n segments, each of mass m, located at distance s from the center of gravity of the coil. The radius of gyration is the second moment of mass distribution. The mean square value of the unperturbed radius of gyration, $<R_G^2>_0$, is related to the unperturbed end to end distance, $<r^2>_0$, by the following equation.

$$<R_G^2>_0 = <r^2>_0/6$$

If the polymer is in contact with a solvent other than a theta solvent, the mean square value of the radius of gyration is given by the following equation

$$<R_G^2> = \alpha_R <R_G^2>_0$$

(Mark JE (ed) (1996) Physical properties of polymers handbook. Springer, New York; Elias HG (1977) Macromolecules, vol 1–2. Plenum Press, New York; Miller ML (1966) Structure of polymers. Reinhold Publishing, New York).

Rafaelite *n* Hard, natural asphaltum mined in the Argentine. It is characterized by a high melting point, and good opacity, but limited solubility in drying oils. Special heating treatment is necessary to obtain stable solutions (Usmani AM (1997) Asphalt science and technology. Marcel Dekker, New York).

RAFT *n* Reversible addition–fragmentation transfer. Odian G (2004) Principles of polymerization. Wiley-Intersciene, New York.

Rag Rolled Finish See ▶ Rag Rolling.

Rag Rolling *n* (1) Process of forming in a scumble or glaze over a painted ground, a textured or variegated pattern by rolling a rag or washleather over the surface. (2) Printing a pattern on a dry painted surface by means of a rag or washleather or special paint roller charged with color. The rag is crumpled or screwed up in the form of a rough roller to produce the pattern.

Railroading *n* The horizontal, rather than vertical, application of a wallcovering or the use, horizontally, of an upholstery fabric ((2000) Complete textile glossary. Celanese Corporation, Three Park Avenue, New York).

Railroad Tracks *n* In coated fabric, depressions in surface of a definite pattern as indicated by the name ((2000) Complete textile glossary. Celanese Corporation, Three Park Avenue, New York).

Rails \rā(ə)l\ *n* [ME *raile*, fr. MF *reille* ruler, bar, fr. L *regula* ruler, fr. *regere* to keep straight, direct, rule] (14c) The metal bars on which the spindles of a downtwister are mounted (Vincenti R (ed) (1994)

Elsevier's textile dictionary. Elsevier Science and Technology Books, New York).

Rain Spotting *n* Particular case of weather spotting caused by rain.

Raised Grain *n* Condition of wood surfaces where fibers from the wood structure have become prominent due to wetting with water or materials containing water. Prominence of the harder portions of the grain of wood when the softer portions have suffered shrinkage ((1978) Paint/coatings dictionary. Compiled by Definitions Committee of the Federation of Societies for Coatings Technology; Laurie AP (1967) Painter's methods and materials. Dover Publications, New York; Gross WF (1970) Application manual for paint and protective coatings. MGraw-Hill, New York).

Raising See Lifting and Napping.

RAL *n* Abbreviation for Reichsnormen Ausschuss for Lieferbedingungen, issues of German Colors (Farbregisters RAL 840).

Ram \ˈram\ *n* [ME, fr. OE *ramm*; akin to OHGr *ram*] (before 12c) In compression and matched-die molding, the press member that enters the cavity block and exerts pressure on the molding compound, designated by its position in the assembly as the top force or bottom force. In older injection machines predating the development of screw injection, the plunger that forced the feed pellets through the annulus between cylinder and torpedo, and that also accomplished, in most such machines, the injection of melt into the mold. Very few of these machines are still (1993) in commercial use. The piston of a melt accumulator such as may be used in blow molding large objects, or in special injection-molding techniques. The piston or plunger of a ram extruder.

RAM *n* (1957) Acronymic abbreviation for Random-Access Memory, computer memory for storing and working with programs and data, and erasable by the operator. Compare ROM ((2002) Microsoft computer dictionary, Microsoft Press; Merriam-Webster's Collegiate Dictionary (2004), 11th edn. Merriam-Webster, Springfield, MA).

Raman Spectroscopy (Raman Effect, and Normal Raman Scattering) When light is scattered from a molecule most photons are elastically scattered. The scattered photons have the same energy (frequency) and, therefore, wavelength, as the incident photons. However, a small fraction of light (approximately 1 in 107 photons) is scattered at optical frequencies different from, and usually lower than, the frequency of the incident photons. The process leading to this inelastic scatter is the termed the Raman effect. Raman scattering can occur with a change in vibrational, rotational, or electronic energy of a molecule. Chemists are concerned primarily with the vibrational Raman effect. We will use the term Raman effect to mean vibrational Raman effect only. The difference in energy between the incident photon and the Raman scattered photon is equal to the energy of a vibration of the scattering molecule (Smith E, Dent G (2004) Modern Raman spectroscopy. Wiley, New York). Raman spectroscopy determines the maximum theoretical extents to which plastics may be drawn when high tensile modulus fibers are made.

Ramie \ˈrā-mē, ˈra-\ *n* [Malay *rami*] (1832) A natural vegetable fiber obtained from the stems of the hemp *Boehmeria nivea*, used as a reinforcement.

Ramped Temperature *n* The programmed heating of a sample at a closely controlled linear rate.

Ramsden Circle (Ramsden Disc, Eyepoint) *n* The circular spot of light formed at that distance above the eyepiece where the chief image forming rays cross the back focal plane of the eyepiece. The objective back focal plane is in conjugate focus in this same plane. In visual microscopy, the point where the lens of the eye is placed.

Ram Travel *n* The distance the injection ram (or screw) moves in filling the mold in injection or transfer molding.

Random Chain, Coil, Flight, Walk *n* Simple model that relates the average size of the molecule's cloud changes with molecular weight. It states that the average square distance traveled by a molecule (r_o^2), increases linearly with the number (n) of links, and $r_o^2 = n\,a^2$, where a is the length of the link (Mark JE (ed) (1996) Physical properties of polymers handbook. Springer, New York).

Random Copolymers *n* A copolymer consisting of alternating segments of two different monomeric units of random lengths, including single molecules. A random copolymer usually results from the copolymerization of two monomers in the presence of a free-radical initiator, for example the so-produced rubbery copolymer of ethylene and propylene.

Random-Sheared Carpet *n* A pile carpet with a textured face produced by shearing some of the loops and leaving others intact (Complete textile glossary. Celanese Corporation, Three Park Avenue, New York, NY).

Range \ˈrānj\ *n* {*often attributive*} [ME, row of persons, fr. MF *renge*, fr. OF *rengier* to range] (14c) (R) In sampling of product dimensions and properties, the difference between the largest and smallest values in the sample. Range charts for small samples have long been used in quality-control work. The range is simple to calculate and is almost as efficient as the ▶ Standard Deviation in samples of two to five items.

Rankine Scale of Temperature \ˈraŋ-kən\ [William J. M. *Rankine* † 1872 Scottish engineer & physicist] (ca. 1926) The absolute Fahrenheit scale,

$$°F + 459.59 = °R; \text{ thus } 0°F \text{ Rankin} = 459.69°F$$

(Merriam-Webster's Collegiate Dictionary (2004), 11th edn. Merriam-Webster, Springfield, MA).

Rankine Temperature *n* The absolute temperature scale, now deprecated, derived from the Fahrenheit scale, having its zero at $-459.67°F$. To convert Rankine to SI's Kelvin (K), multiply by 5/9.

Raoult's Law *n* The quantitative relationship between vapor-pressure lowering and concentration in an ideal solution is stated in Raoult's Law: the partial vapor pressure of a component in solution is equal to the mole fraction of that component times its vapor pressure when pure at a temperature;

$$P_1 = X_1 P°_1$$

Where P_1 and $P°_1$ are the vapor pressure of the solution and the pure solvent, respectively, X_1 is the mole fraction of solvent, $X_1 = 1-X_2$ and $P_1 = (X_1-X_2)P°_1$, and,

$$X_2 = \frac{P°_1 - P_1}{P°_1}$$

this means that the fractional vapor–pressure lowering is equal to the mole fraction of the solute; also, the total pressure of the system is equal to the sum of the partial pressures,

$$P_s = P_1 + P_2 + P_2 \ldots P_n.$$

(Atkins P, de Palua J (2009) Physical Chemistry. W. H. Freeman & Company, New York).

Rapier Looms *n* Looms in which either a double or single rapier (thin metallic shaft with a yarn gripping device) carries the filament through the shed. In a single rapier machine, the yarn is carried completely across the fabric by the rapier. In the double machine, the yarn is passed from one rapier to the other in the middle of the shed (Complete textile glossary. Celanese Corporation, Three Park Avenue, New York, NY). Also see ▶ Weft Insertion.

Rapseed Oil *n* (1816) Obtained from the seeds of the species, *Brassica*. It is a nondrying oil, the main constituent acid being erucic acid, with smaller amounts of oleic and linoleic acids. Sp gr, 0.915/15°C; iodine value, 92; saponification value, 173. *Known also as Colza Oil.*

RAPRA *n* Rubber and Plastics Research Association, since 1985 renamed RAPRA Technology, Ltd., consultants for the plastics and rubber industry and an industry-supported organization headquartered at Shawbury, Shrewsbury, England, www.rapra.com. Rapra offers a wide range of consultancy services for a whole spectrum of clients within the plastics and rubber industry and works alongside the client offering guidance and support as an integral part of the product design and development process. RAPRA boasts the necessary skills and experience to undertake entire product design and development programs and using knowledge built up over many years within the industry, has an understanding of how polymer materials perform in service.

Raschel Knitting See ▶ Knitting (1).

Rate Constants *n* The constant of the reaction rate of a reaction. This constant is a function of temperature that follows the Arrhenius equation

Rate of Shear In rheology, rate of shear is often used interchangeably with velocity gradient. See ▶ Shear Rate.

Rate-Process Theory *n* A general theory, derived from statistical mechanics, applicable to both chemical reactions and creep phenomena in plastics. For the latter, the theory relates time-to-rupture to stress and the reciprocal of the absolute temperature. An equation recommended for a pipe in ASTM D 2837 is,

$$\log t = A_0 + A_i/T + (A_2/T) \cdot \log S$$

where the A_i are empirical coefficients, different for each plastic, t = the time to failure under sustained hoop stress S, and T = the absolute temperature (Carley JF (ed) (1993) Whittington's dictionary of plastics. Technomic Publishing).

Rate Law The dependence of the rate of a reaction upon the concentrations of reactants (Levenspiel O (1998) Chemical reaction engineering. Wiley, New York).

Rate of Polymerization In chain polymerization, the rate at which the monomer (M), of concentration [M], is converted to polymer, $-d[M]/dt$. Since the rate of propagation occurs hundreds of times more frequently than initiation, rates of polymerization and propagation are the same and the symbol R_p is used for both (Odian GC (2004) Principles of polymerization. Wiley, New York; Connors KA (1990) Chemial kinetics. Wiley, New York; Solomon DH (1969) Kinetics and mechanisms of polymerization series, vol 2 – Ring opening and vol 3 – step growth. Marcel Dekker, New York). For a steady-state chain polymerization reaction, R_p is defined by the following equation

$$R_p = k_P[M](R_i/2k_t)^{1/2}$$

Where k_p is the rate constant, R_i is the rate of initiation, and k_t is the termination rate constant. In free radical polymerization initiated by thermal decomposition of initiator I with rate constant k_d and initiator efficiency f, the rate of polymerization is given by

$$R_p = k_P[M](f\,k_d[I]/2k_t)^{1/2}$$

(Elias HG (1977) Macromolecules, vol 1–2. Plenum Press, New York; and Lenz RW (1967) Organic chemistry of synthetic high polymers. Interscience Publishers, New York).

Rate-Determining Step *n* The slowest step in a chemical reaction (Masel RI (2001) Chemical Kinetics, 1st edn. John Wiley & Sons, Inc., New York; Russell JB (1980) General chemistry. McGraw-Hill, New York).

Ratine \ ra-tə- nā\ *n* [F *ratiné*] (ca. 1914) (1) A plain-weave, loosely constructed fabric having a rough, spongy texture which is imparted by the use of nubby plied yarns. It is made from worsted, cotton, or other yarns. (2) A variant of spiral yarns in which the outer yarn is fed more freely to form loops that kink back on themselves and are held in place by a third binder yarn that is added in a second twisting operation (Complete textile glossary. Celanese Corporation, Three Park Avenue, New York, NY).

Ravel \ ra-vəl\ *n* [D *rafelen*, fr. *rafel* loose thread] (1582) A type of comb or rail with projecting teeth for separating and guiding warp ends.

Raveling *n* (1658) The process of undoing or separating the weave or knit of a fabric.

Raw Fiber *n* A textile fiber in its natural state, such as silk "in the gum" and cotton as it comes from the bale.

Raw Sienna \-sē- e-nə\ *n* [I *terra di Siena*, literally, Siena earth, fr. *Siena*, Italy] (1787) Naturally occurring iron oxide, limonite type ore. Color range, yellow to light brown, 40–70% Fe_2O_3. Somewhat translucent when dispersed in oil. General uses, tinting colors, artists' colors, and stains. See ▶ Iron Oxides, Natural.

Raw Umber See ▶ Umber.

Rayleigh Ratio \ rā-lē\ *n* [John Welsh S. *Rayleigh*] R⟩ = KcM_2 as $c \rightarrow o$ where K is the optical constant, c is a dilute concentration factor for a mixture of a homologous series of polymers of different molecular weights M_2 corresponds to weight-average molecular weight $\overline{M_w}$; other derivations of the equation $\overline{R_\theta} = \kappa c \overline{M}_w$ are used to determine the molecular weights of copolymers (Mark JE (ed) (1996) Physical properties of polymers handbook. Springer, New York; Chu B (1992) Laser Light scattering: Basic Principles and Practice. Academic Press, New York).

Rayleigh Scattering *n* (1937) Scattering of light by small or molecular size particles to render the effect selective so that different colors are deflected through different angles, scattering = $1/\lambda^4$ where λ = wavelength of light. This relationship explains why earth's sky is blue as viewed in direct sunlight because blue has a lower wavelength compared to other wavelengths of the visible spectrum (Flory PJ (1953) Principles of polymer science. Cornell University Press, Ithaca, NY; and Elias HG (1977) Macromolecules, vol 1–2. Plenum Press, New York).

Raymond Roller Mill *n* A type of mill used for the dry-grinding of pigments or similar materials. The pre-crushed crudes or solids are fed into a circular grinding chamber in the base of the mill and by mechanical means are forced between rolls and a ring. Grinding of the pigment crude is obtained by the centrifugal force and pressure of the revolving roll exerted against the ring. A stream of compressed air passes through the mill and carries out the pigment particles which are fine enough to "float" on it; the larger particles drop back into the mill grinding chamber for further grinding.

Rayon \ rā- än\ *n* [irreg. from ^{2}*ray*] (1924) The definition established by the Federal Commission in 1951 is: "Generic name for a manufactured fiber composed of regenerated cellulose, as well as manufactured fibers composed of regenerated cellulose in which constituents have replaced not more than 15% of the hydrogens of the hydroxyl groups." Prior to that date, going back to 1924 when the name rayon was first used (inspired by its sheen invoking the brilliance of a ray of sunlight), the term was used for all man-made fibers derived from cellulose, including cellulose acetate and cellulose triacetate. Rayon is the oldest of the synthetic fibers, having been produced commercially since 1855. All methods of producing rayon are based on treating fibrous forms of cellulose to make them soluble, extruding the solution through the tiny orifices of a spinneret then converting the filaments into solid cellulose. Most rayon fibers are produced from the intermediate *Viscose*. See also ▶ Cuprammonium Rayon.

Rayon Acetate *n* Generic name for the regenerated manmade fibers of cellulose from cellulose acetate.

Rayon Fiber *n* A manufactured fiber composed of regenerated cellulose, as well as manufactured fibers composed of regenerated cellulose in which substituents have replaced not more than 15% of the hydrogens of the hydroxyl groups (FTC definition). Rayon fibers include yarns and fibers made by the viscose process, the cuprammonium process, and the now obsolete nitrocellulose and saponified acetate processes. Generally, in the manufacture of rayon, cellulose derived from wood pulp, cotton linters, or other vegetable matter is dissolved into a viscose spinning solution. The solution is extruded into an acid–salt coagulating bath and drawn into continuous filaments. Groups of these filaments may be made in the form of yarns or cut into staple. CHARACTERISTICS: Rayon yarns are made in a wide range of types in regard to size, physical characteristics, strength, elongation, luster, handle, suppleness, etc. They may be white or solution dyed. Strength is regulated by the process itself and the structure of the yarn (Also see ▶ Polynosic Fiber). Luster is reduced by including delustering materials, such as titanium dioxide pigments, in the fiber when it is extruded. The suppleness of the yarn is controlled by the number of filaments in the yarn, the denier or gauge of the individual filaments or fibers, and the fiber cross-section. END USES: Rayon is used in draperies, bedspreads, upholstery, blanket, dish towels, curtains, throw rugs, tire cord, industrial products, sport shirts, slacks, suitings, dress goods, and linings and in blends with other fibers to enhance functional and aesthetic qualities, e.g., with polyester in permanent-press fabrics.

Re (N_{Re}) An alternate, older symbol for *Reynolds Number*.

Reactant \rē-ˈakt-tənt\ *n* (ca. 1920) A substance consumed in a chemical reaction.

Reaction Injection Molding *n* (RIM) This term is usually applied to the process of injection molding of urethane reactants in which the two primary constituents, isocyanate and polyol, are pumped by a metering device into a mixing head from which the intimately mixed reactants are quickly injected into a closed mold. The injection pressure is much lower than in conventional injection molding of molten plastics, enabling the use of inexpensive, light-weight molds. However, the mixing head is a high-pressure impingement mixer in which pressures may reach 14–21 MPa. One mixing head may be used to feed up to ten separate molding presses. One of the largest-volume applications of RIM is the production of exterior automotive parts such as body panels and bumpers. Furniture is another big use. The term *Liquid Injection Molding* (LIM) is usually applied to the similar process of molding other thermosetting resins such as polyesters, epoxies, silicones, alkyds, and diallyl phthalate resins. The terms *liquid reaction molding (LRM)* and *high-pressure injection molding (HPIM)* have sometimes been used for either or both processes. If reinforcing fibers are included in the reaction mix, the process is called *reinforced reaction injection molding (RRIM)*. *Structural reaction injection molding (SRIM)* is a variation in which there is some foaming of the polyurethane in the core of the molding, with a solid skin on the outside. This technique reduces part weight with little loss of stiffness or strength and is widely used in the commodity-furniture industry.

Reaction Order *n* The exponent on a concentration term in a simple rate law (order with respect to one component); or the sum of all such exponents (overall order).

Reaction Rate *n* The time rate of change of concentration (or, sometimes, quantity) of a reactant or product in a reaction.

Reaction Spinning See Spinning (2).

Reactive Pigments *n* Those pigments which react with the vehicle, as in the formation of zinc and lead soaps with drying oils; also, pigments such as red lead which react with acids formed at metal surface to prevent rust.

Reactive Plasticizer See ▶ Plasticizer, Polymerizable.

Reactive Processing *n* A molding or extrusion operation in which chemical reactions are carried out. Extruders, mainly specialized twin-screw machines, have successfully carried out partial and complete polymerizations on a large scale. Transfer and compression molding of thermosets have always been reactive processes, but see ▶ Reaction Injection Molding.

Reactive Resins *n* (1) Resins of phenol-formaldehyde type which are believed to react with drying oils on heating, more particularly those of conjugated types. (2) Resins capable of crosslinking with themselves or other resins. (3) Resins with a high acid number.

Ready-Mixed Aluminum Paint See ▶ Aluminum Mixing Varnish.

Reagent Resistance *n* (chemical resistance) The ability of a plastic to withstand exposure to acids, alkalis, oxidants, and solvents.

Realgar \rē-ˈal-ˌgär, -gər\ *n* [ME, fr. ML, fr. Catalan, fr. Arabic *rahj al-ghār* powder of the mine] (15c) As_2S_2. Arsenic disulfide mineral, which was used as a pigment under the name, "arsenic orange." Its chemical and physical properties are similar to those of orpiment. In modern times, it is not used as a pigment because of its toxicity.

Ream \ˈrēm\ *n* [ME *reme*, fr. MF *raime*, fr. Arabic *rizmah*, literally, bundle] (14c) (1) Layers of inhomogeneous material parallel to the surface in a transparent or translucent plastic article. (2) A quantity of paper, 472–500 sheets, depending on the type of paper.

Reaming *n* Further plying of a two-ply yarn with a singles yarn. Reaming is not the same as plying three singles yarns in one operation.

Reciprocating-Screw Injection Molding *n* In this process the screw serves to both plasticate the feedstock and inject the melt into the mold. During part of the cycle, the screw rotates rapidly, moving backward as it accumulates a charge (*shot*) of melt in the forward and of the cylinder. A limit switch stops the rotation and two hydraulic rams, one on either side, push the screw forward, forcing the melt into the mold and holding melt pressure unit the gates freeze. The rams are withdrawn, screw rotation recommences, the mold opens to eject the parts, the mold closes, and a new cycle begins. Most new injection machines sold today (1993) are of this type.

Reciprocating-Screw Molding Machines *n* A combination plasticating and injection unit in which an extrusion device with a reciprocating screw is used to plasticate the plastic.

Reclaim \ri-ˈklām\ *vt* [ME *reclamen*, fr. MF *reclamer* to call back, fr. L *reclamare* to cry out against, fr. *re-* + *clamare* to cry out] (14c) (recycle) To salvage plastics from discarded products such as milk and soda bottles, automobiles, and packaging films.

Reclaiming or Recycling *n* The recovery, recycling, and reuse of scrap materials.

Reconstituted Fibers *n* Fibers made from recovered waste polymer or blends of virgin polymer and recovered waste polymer.

Reconstituted Oils *n* Drying oils which are made by the reesterification of selected fatty acids, previously derived from a natural drying oil.

Recovery See ▶ Elastic Recovery.

Recreational Surfaces *n* Manufactured surfaces providing consistent properties, durability, and special characteristics as needed for the specific application. Included are artificial turf, pool decks, indoor–outdoor carpeting, tennis court surfaces, etc. Most types of constructions (knit, woven, tufted, and nonwoven), and most polymer types find use in this market. The polyolefins are particularly prominent in these applications.

Recycle \(ˌ)rē-ˈsī-kəl\ *n,v* (1926) (1) (regrind) In a processing plant, to recover trim scrap and faulty parts by granulating them and blending the ground material with virgin feed. (2) To ▶ Reclaim.

Recycled Plastic *n* (reclaimed plastic) A plastic prepared from discarded articles that have been cleaned and ground (ISO). This material may or may not be reformulated by the addition of stabilizers, plasticizers, fillers, pigment, etc.

Recycling *n* (1) (regrind) In a processing plant, to recover trim scrap and faulty parts by granulating them and blending the ground material with virgin feed. (2) To ▶ Reclaim.

Red (Magenta) Inks The calcium salt of an azo pigment, it has generally good transparency; good resistance to bleeding and baking; fair lightfastness and alkali resistance.

Red Cinnabar See ▶ Mercuric Sulfide.

Red Iron Oxide See ▶ Iron Oxides, Synthetic.

Red Lake C *n* Family of organic acid azo pigments prepared by coupling the diazonium salt of ortho-chloro-meta-toluidine-para-sulfonic acid with β-naphthol.

Red Lake C Pigments *n* Family of organic acid azo pigments prepared by coupling the diazonium salt of ortho-chloro-meta-toluidine-para-sulfonic acid with β-naphthol.

Red Lake D *n* Strong red lake, sensitive to cobalt driers, and made by coupling diazotized anthranilic acid with β-naphthol.

Red Lake P *n* Made by coupling diazotized p-nitro aniline-*o*-sulfonic acid with β-naphthol.

Red Lead *n* (15c) Pb_3O_4. Pigment Red 105 (77578). Bright red to orange-red tetroxide; excellent opacity with good properties as a primary constituent of anticorrosive primers for iron and steel. Density, 8.9 g/cm^3 (74.2 lb/gal); O.A., 7–9. *Also known as Minium*. See ▶ Orange Mineral.

Red Lead (Nonsetting) *n* Special type of red lead containing a minimum amount of free or reactive litharge and a high content of lead peroxide. The nonsetting types are used when it is desired to prevent undue thickening of oil mixtures or paints.

Red Ocher *n* (1572) A mixture of hematites; any of a number of natural earths used as red pigments. See ▶ Iron Oxides, Natural.

Red Oil Commercial grade of oleic acid.

Red Oxide See ▶ Iron Oxides, Natural.

Redox, Initiation \ˈrē-ˌdäks-\ *adj* [*red*uction + *ox*idation] (1928) Oxidation or reduction of a compound to generate ionic species for initiation of polymer reactions (Mark JE (ed) (1996) Physical properties of polymers handbook. Springer, New York).

Redox Polymers *n* Polymers which are formed when a redox catalyst enters into an oxidation–reduction reaction.

Redox Reaction *n* An oxidation–reduction, or electron transfer, reaction (Mark JE (ed) (1996) Physical properties of polymers handbook. Springer, New York).

Redtone Ink Blue See ▶ Iron Blue.

Reduce \ri-ˈdüs, -ˈdyüs\ *v* [ME, to lead back, fr. L *reducere*, fr. *re-* + *ducere* to lead] (14c) To add a solvent or thinner to a coating, varnish, resin, latex, or emulsion for the purpose of lowering its viscosity and/or nonvolatile content.

Reduced Viscosity *n* (1) (IUPAC: viscosity number) Reduced viscosity is the fluid viscosity increase per unit of polymer solute concentration. Mathematically, it is defined by the following equation:

$$\eta_{reduced} = \eta_{specific}/c$$

where $\eta_{reduced}$ is the reduced viscosity, $\eta_{specific}$ is the specific viscosity, and c is the concentration of the polymer in solution. Reduced viscosity is a measure of the specific capacity of the polymer to elevate viscosity. See ▶ Dilute-Solution Viscosity. (2) The viscosity of a coating or vehicle when thinned (reduced) to a specified percent solids with specified solvents. Often used as a portion of the specifications between buyer and seller (Collins EA, Bares J, Billmeyer FW, Jr (1973) Experiments in polymer science. Wiley-Interscience, New York).

Reducer *n* A volatile compound which is employed to bring coatings to the proper consistency. *Also called Thinner.*

Reducers *n* Varnishes, solvents, oils, or waxy or greasy compounds that are employed to reduce tack or consistency of the ink for use on a press. See ▶ Thinner.

Reducing Agent *n* (1885) A species or substance which loses electrons in a reaction.

Reducing Power *n* The strength of a white pigment, i.e., the degree to which it is able to produce a very pale tint when mixed with a defined proportion of colored pigment. The paler the tint produced, the greater the reducing power.

Reductant adj. The electron donating species in an oxidation reduction reaction.

Reduction \ri-ˈdək-shən\ *n* [ME *reduccion* restoration, fr. MF *reduction*, fr. LL & L *reduction-*, *reductio* reduction (in a syllogism), fr. L, restoration, from *reducere*] (1546) (1) Any chemical process that increases the proportion of hydrogen or base-forming elements or radicals in a compound. (2) The gaining of electrons by an atom, ion, or element, thereby reducing the positive valence of that which gained the electron(s) (Whitten KW, Davis RE, Davis E, Peck LM, Stanley GG (2003) General chemistry. Brookes/Cole, New York). The reverse of *Oxidation.*

Reduction Clearing *n* The removal of unabsorbed disperse dye from the surface of polyester at the end of the dyeing or printing process by treatment in a sodium hydroxide/sodium hydrosulfite bath. A surface-active agent may be employed in the process.

Reduction Potential *n* A measure of the tendency of a reduction half-reaction to occur, expressed as the voltage produced by a cell employing the half-reaction at its cathode and using the standard hydrogen electrode as it anode.

Reed \ˈrēd\ *n* [ME *rede*, fr. OE *hrēod*; akin to OHGr *hriot* reed] (before 12c) A comb-like device on a loom that separates the warp yarns and also beats each succeeding filling thread against that already woven. The reed usually consists of a top and bottom rib of wood into which metal strips or wires are set. The space between two adjacent wires is called a dent (or split) and the warp is drawn through the dents. The fineness of the reed is calculated by the number of dents per inch.

Reed Marks *n* A fabric defect consisting of warpwise light and heavy streaks in a woven fabric, caused by bent, unevenly packed, or weak reed wires.

Reel \rē(ə)l\ *n* [ME, fr. OE *hrēol*; akin to ON *hrœll* weaver's reed, Gk *krekein* to weave] (before 12c) (1) A revolving frame on which yarn is wound to form hanks or skeins. (2) The frame on which silk is wound from the cocoon. (3) A linen yarn measure of 72,000 yards. (4) The large wheel in a horizontal warper onto which the warp sections are wound in the indirect system of warping. (5) A spool of large capacity used to wind yarn or wire.

Reeling *n* In silk fiber production, the process of unwinding the cocoon.

Reentrant Mold *n* A mold containing an undercut that tends to impede withdrawal of the molded product. If the undercut is more than slight, the mold will probably be designed with a side draw that retracts from the undercut region as the mold opens and relieves the undercut.

Reference Material *n* (reference standard) In analytical chemistry, a gas mixture, pure liquid, solution, pure solid, or alloy whose composition is certified with a stated, high degree of accuracy. Reference materials are used to calibrate analytical procedures and instruments. They are available from private companies, also from the National Institute of Standards and Technology (the former National Bureau of Standards).

Reference Standard *n* The standard to which a measurement is compared.

Refiner *n* A machine similar to a two-roll mixing mill, operated with rolls very close together to crush undispersed ingredients and hold them in the bite of the rolls for removal and discarding when the mass has passed through. Refiner rolls are shorter and have a much greater diameter than mixing rolls, and are operated at a higher surface speed ratio to provide more grinding effect and classification.

Reflectance \ri-ˈflek-tən(t)s\ *n* (1926) The ratio of the intensity of reflected radiant flux to that of the incident flux. In popular usage, it is considered as the ratio of the intensity of reflected radiant energy to that reflected from a defined reference standard. See ▶ Light Reflectance.

Reflectance, Absolute See ▶ Absolute Reflectance.

Reflectance, Diffuse *n* Reflectance over a wide range of angles. In popular usage, the diffuse reflectance is all of the reflected radiant energy except that of the specular angle. See ▶ Perfect Diffuser and ▶ Specular Reflectance Excluded.

Reflectance, Directional *n* Measurement of radiant flux reflected for specified directions of illumination and viewing, generally measured relative to that from a perfect diffuse reflector similarly illuminated and viewed. In practice, a reference white which approaches a perfect diffuser, or a secondary calibrated reference standard, may be used.

Reflectance Factor *n* Radiance measured relative to the perfect diffuser rather than to the irradiance.

Reflectance, Fresnel *n* Reflectance of radiant energy at the surface separating media of different refractive indices. The magnitude of the light reflected may be calculated from Fresnel's Law.

Reflectance, Hemispherical *n* Ratio of radiant flux reflected into a hemispherical collector to the incident flux. Normally, the portion reflected backward from a material is measured, although the portion reflected in the forward direction may be measured on translucent materials.

Reflectance, Luminous *n* Reflectance derived from radiant flux by evaluating the radiant energy reflected according to its action upon a selective receptor (such as the eye), whose spectral sensitivity is defined by a standard relative luminous efficiency function. In popular usage, the term is used to describe the Y tristimulus value in the CIE System.

Reflectance, Nonspecular *n* Reflectance of radiant flux from a surface at angles other than that of the specular angle, i.e., diffuse reflectance. See ▶ Reflectance, Diffuse.

Reflectance, Specular *n* Reflectance of a beam of radiant energy at an angle equal but opposite to the incident angle; the mirror-like reflectance. The magnitude of the specular reflectance on glossy materials depends on the angle and on the difference in refractive indices between two media at a surface and may be calculated from the Fresnel Law. See ▶ Reflectance, Fresnel.

Reflectance, Total *n* Reflectance of radiant flux reflected at all angles from the surface, thus including both diffuse and specular reflectances.

Reflection \ri-$^{\textbf{I}}$flek-shən\ *n* [ME, alter. of *reflexion*, fr. LL *reflexion-*, *reflexio* act of bending back, fr. L *reflectere*] (14c) Process or phenomenon of the return of radiant energy from a surface.

Reflection–Absorption Spectroscopy *n* Spectroscopic technique that investigates the vibration of molecules on surfaces. Reflection–absorption spectroscopy uses IR light to excite the molecules on the surface. These molecules will absorb only certain fixed frequencies. Hence, the spectrum will show the absorption peaks characteristic of the molecule as well as its method of bonding to the surface.

Reflection Coefficient or Reflectivity *n* The ratio of the light reflected from a surface to the total incident light. The coefficient may refer to diffuse or to specular reflection. In general it varies with the angle of incidence and with the wavelength of the light.

Reflection of Light by a Transparent Medium in Air *n* (Fresnel's formulae) If i is the angle of incidence, r the angle of refraction, n_1 the index of refraction for air (nearly equal to unity), n_2 index of refraction for a medium, then the ratio of the reflected light to the incident light is,

$$R = \frac{1}{2}\left(\frac{\sin^2(i-r)}{\sin^2(i+r)} + \frac{\tan^2(i-r)}{\tan^2(i+r)}\right)$$

If $i = 0$ (normal incidence), and $n_1 = 1$ (approximate for air),

$$R = \left(\frac{n_2 - 1}{n_2 + 1}\right)^2$$

(Weast RC (ed) Handbook of chemistry and physics, 52nd edn. CRC Press, Boca Raton, FL).

Reflectivity (R^∞) \$_{\textbf{I}}$rē-$_{\textbf{I}}$flek-$^{\textbf{I}}$ti-və-tē\ *n* (1) Reflectance which would be attained if a material were completely opaque; reflectance of a layer of material of such thickness that an increase in thickness will not change its reflectance. (2) The reflectance of a film so thick that a further increase in thickness does not change the reflectance.

Reflectometer \$_{\textbf{I}}$rē-$_{\textbf{I}}$flek-$^{\textbf{I}}$tä-mə-ter, ri-\ *n* (1891) An instrument that measures the total luminous flux from a surface and reports it as a percentage of the incident flux on the surface.

Reflectorization *n* Incorporation of ballotini in, for example, road signs or markings to give selective reflection in the general direction of the light source.

Reflex Blue See ▶ Alkali Blue.

Reflex Reflector *n* A reflecting material so designed that a beam of radiant energy returns, or is reflected back, along the line of incidence. Such reflectors are of importance for automobiles, traffic signals and signs. See also ▶ Retroreflective.

Refract \ri-$^{\textbf{I}}$frakt\ *vt* [L *refractus*, pp. of *refringere* to break open, break up, fr. re- + *frangere* to break] (1612) To subject (as a ray of light) to refraction.

Refraction \ri-ˈfrak-shən\ *n* (1603) Deflection of radiant energy from a straight path in one medium to a different path in another medium of different index of refraction.

Refraction at a Spherical Surface *n* If u be the distance of a point source, v the distance of the point image or the intersection of the refracted ray with the axis, n_1 and n_2 the indices of refraction of the first and second medium, and r the radius of curvature of the separating surface,

$$\frac{n_2}{v} + \frac{n_1}{u} = \frac{n_2 - n_1}{r}$$

In the first medium is air the equation becomes,

$$\frac{n}{v} + \frac{1}{u} = \frac{n-1}{r}$$

Refractive Index *n* (1839) The index of refraction of a material is the ratio of the velocity of the light in a vacuum to that of the specimen; it is expressed as a ratio of the sine of the angle of incidence to the sine of the angle of refraction; it is used as a measure of purity, identification and optical design (ASTM D 542). See ▶ Index of Refraction.

Refractive index of polymers/haze value

Abbr.	Polymer	Refractive index
PHFPO	Poly(hexafluoropropylene oxide)	1.3010
	Water	1.33
	Alginic acid, sodium salt	1.3343
	Hydroxypropyl cellulose	1.3370
	Poly(tetrafluoroethylene-co-hexafluoropropylene)	1.3380
FEP	Fluorinated Ethylene Propylene	1.3380
	Poly(pentadecafluorooctyl acrylate)	1.3390
	Poly(tetrafluoro-3-(heptafluoropropoxy)propyl acrylate)	1.3460
	Poly(tetrafluoro-3-(pentafluoroethoxy)propyl acrylate)	1.3480
PTFE	Poly(tetrafluoroethylene)	1.3500
THV	Tetrafluoroethylene hexafluoropropylene vinylidene fluoride	1.3500
	Poly(undecafluorohexyl acrylate)	1.3560

Refractive index of polymers/haze value (Continued)

Abbr.	Polymer	Refractive index
PFA	Perfluoroalkoxy	1.3400
ETFE	Ethylene Tetrafluoroethylene	1.4000
	Poly(nonafluoropentyl acrylate)	1.3600
	Poly(tetrafluoro-3-(trifluoromethoxy)propyl acrylate)	1.3600
	Poly(pentafluorovinyl propionate)	1.3640
	Poly(heptafluorobutyl acrylate)	1.3670
	Poly(trifluorovinyl acetate)	1.3750
	Poly(octafluoropentyl acrylate)	1.3800
	Poly(methyl 3,3,3-trifluoropropyl siloxane)	1.3830
	Poly(pentafluoropropyl acrylate)	1.3850
	Poly(2-heptafluorobutoxy)ethyl acrylate)	1.3900
PCTFE	Poly(chlorotrifluoroethylene)	1.3900
	Poly(2,2,3,4,4-hexafluorobutyl acrylate)	1.3920
	Poly(methyl hydro siloxane)	1.3970
	Poly(methacrylic acid), sodium salt	1.4010
	Poly(dimethyl siloxane)	1.4035
	Poly(trifluoroethyl acrylate)	1.4070
	Poly (2-(1,1,2,2-tetrafluoroethoxy)ethyl acrylate)	1.4120
	Poly(trifluoroisopropyl methacrylate)	1.4177
	Poly(2,2,2-trifluoro-1-methylethyl methacrylate)	1.4185
	Poly(2-trifluoroethoxyethyl acrylate)	1.4190
PVDF	Poly(vinylidene fluoride)	1.4200
ECTFE	Ethylene Chlorotrifluorotheylene	1.4470
	Poly(trifluoroethyl methacrylate)	1.4370
	Poly(methyl octadecyl siloxane)	1.4430
	Poly(methyl hexyl siloxane)	1.4430
	Poly(methyl octyl siloxane)	1.4450
	Poly(isobutyl methacrylate)	1.4470
	Poly(vinyl isobutyl ether)	1.4507
	Poly(methyl hexadecyl siloxane)	1.4510
PEO	Poly(ethylene oxide)	1.4539
	Poly(vinyl ethyl ether)	1.4540
	Poly(methyl tetradecyl siloxane)	1.4550

Refractive index of polymers/haze value (Continued)

Abbr.	Polymer	Refractive index
	Poly(ethylene glycol monomethyl ether)	1.4555
	Poly(vinyl n-butyl ether)	1.4563
PPOX	Poly(propylene oxide)	1.4570
	Poly(3-butoxypropylene oxide)	1.4580
	Poly(3-hexoxypropylene oxide)	1.4590
	Poly(ethylene glycol)	1.4590
	Poly(vinyl n-pentyl ether)	1.4590
	Poly(vinyl n-hexyl ether)	1.4591
	Poly(4-fluoro-2-trifluoromethylstyrene)	1.4600
	Poly(vinyl octyl ether)	1.4613
	Poly(vinyl n-octyl acrylate)	1.4613
	Poly(vinyl 2-ethylhexyl ether)	1.4626
	Poly(vinyl n-decyl ether)	1.4628
	Poly(2-methoxyethyl acrylate)	1.4630
	Poly(acryloxypropyl methyl siloxane)	1.4630
PMP	Poly(4-methyl-1-pentene)	1.4630
	Poly(3-methoxypropylene oxide	1.4630
PtBuMA	Poly(t-butyl methacrylate)	1.4638
	Poly(vinyl n-dodecyl ether)	1.4640
	Poly(3-ethoxypropyl acrylate)	1.4650
	Poly(vinyl propionate)	1.4664
PVAC	Poly(vinyl acetate)	1.4665
	Poly(vinyl propionate)	1.4665
	Poly(vinyl methyl ether)	1.4670
	Poly(ethyl acrylate)	1.4685
	Poly(vinyl methyl ether) (isotactic)	1.4700
	Poly(3-methoxypropyl acrylate)	1.4710
	Poly(1-octadecene)	1.4710
	Poly(2-ethoxyethyl acrylate)	1.4710
PIPA	Poly (isopropyl acrylate)	1.4728
	Poly(1-decene)	1.4730
	Poly(propylene)(atactic)	1.4735
	Poly(lauryl methacrylate)	1.4740
	Poly(vinyl sec-butyl ether) (isotactic)	1.4740
P-nBuA	Poly(n-butyl acrylate)	1.4740
	Poly(dodecyl methacrylate)	1.4740
	Poly(ethylene succinate)	1.4744

Refractive index of polymers/haze value (Continued)

Abbr.	Polymer	Refractive index
	Poly(tetradecyl methacrylate)	1.4746
	Poly(hexadecyl methacrylate)	1.4750
CAB	Cellulose acetate butyrate	1.4750
CA	Cellulose acetate	1.4750
	Poly(vinyl formate)	1.4757
EVA-40% vinyl acetate	Ethylene/vinyl acetate copolymer-40% vinyl acetate	1.4760
	Poly(2-fluoroethyl methacrylate)	1.4768
	Poly(octyl methyl silane)	1.4780
EC	Ethyl cellulose	1.4790
PMA	Poly(methyl acrylate)	1.4793
	Poly(dicyanopropyl siloxane)	1.4800
POM	Poly(oxymethylene) or Polyformaidehyde	1.4800
	Poly(sec-butyl methacrylate)	1.4800
	Poly(dimethylsiloxane-co-alpha-methylstyrene)	1.4800
	Poly(n-hexyl methacrylate)	1.4813
EVA-33% vinyl acetate	Ethylene/vinyl acetate copolymer-33% vinyl acetate	1.4820
PnBuMA	Poly(n-butyl methacrylate)	1.4830
	Poly(ethylidene dimethacrylate)	1.4831
	Poly(2-ethoxyethyl methacrylate)	1.4833
	Poly(n-propyl methacrylate)	1.4840
	Poly(ethylene maleate)	1.4840
EVA-28% vinyl acetate	Ethylene/vinyl acetate copolymer-28% vinylacetate	1.4845
	Poly(ethyl methacrylate)	1.4850
PVB	Poly(vinyl butyral)	1.4850
PVB-11% hydroxl	Poly(vinyl butyral)-11% hydroxl	1.4850
	Poly(3,3,5-trimethylcyclohexyl methacrylate)	1.4850
	Poly(2-nitro-2-methylpropyl methacrylate)	1.4868
	Poly(dimethylsiloxane-co-diphenylsiloxane)	1.4880
	Poly(1,1-diethylpropyl methacrylate)	1.4889
	Poly(triethylcarbinyl methacrylate)	1.4889

Refractive index of polymers/haze value (Continued)

Abbr.	Polymer	Refractive index
PMMA	Poly(methyl methacrylate)	1.4893
	Poly(2-decyl-1,4-butadiene)	1.4899
PP-isotactic	Poly(propylene), isotactic	1.4900
PVB-19% hydroxyl	Poly(vinyl butyral)-19% hydroxyl	1.4900
	Poly(mercaptopropyl methyl siloxane)	1.4900
	Poly(ethyl glycolate methacrylate)	1.4903
	Poly(3-methylcyclohexyl methacrylate)	1.4947
	Poly(cyclohexyl alpha-ethoxyacrylate)	1.4969
MC	Methyl cellulose	1.4970
	Poly(4-methylcyclohexyl methacrylate)	1.4975
	Poly(decamethylene glycol dimethacrylate)	1.4990
PVAL	Poly(vinyl alcohol)	1.5000
PVFM	Poly(vinyl formal)	1.5000
	Poly(2-decyl-1,4-butadiene)	1.4899
PVFM	Poly(vinyl formal)	1.5000
	Poly(2-bromo-4-trifluoromethyl styrene)	1.5000
	Poly(1,2-butadiene)	1.5000
	Poly(sec-butyl alpha-chloroacrylate)	1.5000
	Poly(2-heptyl-1,4-butadiene)	1.5000
	Poly(vinyl methyl ketone)	1.5000
	Poly(ethyl alpha-chloroacrylate)	1.5020
PVFM	Poly(vinyl formal)	1.5020
	Poly(2-isopropyl-1,4-butadiene)	1.5020
	Poly(2-methylcyclohexyl methacrylate)	1.5028
	Poly(bornyl methacrylate)	1.5059
	Poly(2-t-butyl-1,4-butadiene)	1.5060
	Poly(ethylene glycol dimethacrylate)	1.5063
PCHMA	Poly(cyclohexyl methacrylate)	1.5065
	Poly(cyclohexanediol-1,4-dimethacrylate)	1.5067
IIR or PIBI	Butyl rubber(unvulcanized)	1.5080
	Gutta percha b	1.5090
	Poly(tetrahydrofurfuryl methacrylate)	1.5096

Refractive index of polymers/haze value (Continued)

Abbr.	Polymer	Refractive index
PIB	Poly(isobutylene)	1.5100
LDPE	Polyethylene, low density	1.5100
EMA	Ethylene/methacrylic acid ionomer, sodium ion	1.5100
PE	Polyethylene	1.5100
CN	Cellulose nitrate	1.5100
	Polyethylene Ionomer	1.5100
	Polyacetal	1.5100
	Poly(1-methylcyclohexyl methacrylate)	1.5111
	Poly(2-hydroxyethyl methacrylate)	1.5119
	Poly(1-butene)(isotactic)	1.5125
	Poly(vinyl methacrylate)	1.5129
	Poly(vinyl chloroacetate)	1.5130
	Poly(N-butyl methacrylamide)	1.5135
	Gutta percha a	1.5140
	Poly(2-chloroethyl methacrylate)	1.5170
PMCA	Poly(methyl alpha-chloroacrylate)	1.5170
	Poly(2-diethylaminoethyl methacrylate)	1.5174
	Poly(2-chlorocyclohexyl methacrylate)	1.5179
	Poly(1,4-butadiene)(35% cis; 56% trans; 7% 1,2-content)	1.5180
PAN	Poly(acrylonitrile)	1.5187
	Poly(isoprene),cis	1.5191
	Poly(allyl methacrylate)	1.5196
	Poly(methacrylonitrile)	1.5200
	Poly(methyl isopropenyl ketone)	1.5200
	Poly(butadiene-co-acrylonitrile)	1.5200
	Poly(2-ethyl-2-oxazoline)	1.5200
	Poly(1,4-butadiene)(high cis-type)	1.5200
	Poly(N-2-methoxyethyl) methacrylamide	1.5246
	Poly(2,3-dimethylbutadiene) {methyl rubber}	1.5250
	Poly(2-chloro-1-(chloromethyl) ethyl methacrylate)	1.5270
	Poly(1,3-dichloropropyl methacrylate)	1.5270

Refractive index of polymers/haze value (Continued)

Abbr.	Polymer	Refractive index
PAA	Poly(acrylic acid)	1.5270
	Poly(N-vinyl pyrrolidone)	1.5300
NYLON-6	Nylon 6{Poly(caprolactam)}	1.5300
	Poly(butadiene-co-styrene)(30% styrene)block copolymer	1.5300
	Poly(cyclohexyl alpha-chloroacrylate)	1.5320
	Poly(methyl phenyl siloxane)	1.5330
	Poly(2-chloroethyl alpha-chloroacrylate)	1.5330
	Poly(butadiene-co-styrene)(75/25)	1.5350
	Poly(2-aminoethyl methacrylate)	1.5370
	Poly(furfuryl metacrylate)	1.5381
PVC	Poly(vinyl chloride)	1.5390
	Poly(butylmercaptyl methacrylate)	1.5390
	Poly(1-phenyl-n-amyl methacrylate)	1.5396
	Poly(N-methyl methacrylamide)	1.5398
HDPE	Polyethylene, high density	1.5400
	Cellulose	1.5400
	Poly(cyclohexyl alpha-bromoacrylate)	1.5420
	Poly(sec-butyl alpha-bromoacrylate)	1.5420
	Poly(2-bromoethyl methacrylate)	1.5426
	Poly(dihydroabietic acid)	1.5440
	Poly(abietic acid)	1.5460
	Poly(ethylmercaptyl methacrylate)	1.5470
	Poly(N-allyl methacrylamide)	1.5476
	Poly(1-phenylethyl methacrylate)	1.5487
	Poly(2-vinyltetrahydrofuran)	1.5500
	Poly(vinylfuran)	1.5500
	Poly(methyl m-chlorophenylethyl siloxane)	1.5500
	Poly(p-methoxybenzyl methacrylate)	1.5520
	Poly(isopropyl methacrylate)	1.5520
	Poly(p-isopropyl styrene)	1.5540
	Poly(isoprene), chlorinated	1.5540

Refractive index of polymers/haze value (Continued)

Abbr.	Polymer	Refractive index
	Poly(p,p'-xylylenyl dimethacrylate)	1.5559
	Poly(cyclohexyl methyl silane)	1.5570
	Poly(1-phenylallyl methacrylate)	1.5573
	Poly(p-cyclohexylphenyl methacrylate)	1.5575
CR	Poly(chloroprene)	1.5580
	Poly(2-phenylethyl methacrylate)	1.5592
	Poly(methyl m-chlorophenyl siloxane)	1.5600
	Poly{4,4-heptane bis(4-phenyl) carbonate}	1.5602
	Poly{1-(o-chlorophenyl)ethyl methacrylate)}	1.5624
S/MA	Styrene/maleic anhydride copolymer	1.5640
	Poly(1-phenylcyclohexyl methacrylate)	1.5645
NYLON 6,10	Nylon 6,10{Poly(hexamethylene sebacamide)}	1.5650
NYLON 6,6	Nylon 6,6{Poly(hexamethylene adipamide)}	1.5650
NYLON 6(3)	Nylon 6(3)T {Poly(trimethyl hexamethylene terephthalamide)}	1.5660
	Poly(2,2,2'-trimethylhexamethylene terephthalamide)	1.5660
	Poly(methyl alpha-bromoacrylate)	1.5672
	Poly(benzyl methacrylate)	1.5680
	Poly{2-(phenylsulfonyl)ethyl methacrylate}	1.5682
	Poly(m-cresyl methacrylate)	1.5683
SAN	Styrene/acrylonitrile copolymer	1.5700
	Poly(o-methoxyphenol methacrylate)	1.5705
PPhMA	Poly(phenyl methacrylate)	1.5706
	Poly(o-cresyl methacrylate)	1.5707
PDAP	Poly(diallyl phthalate)	1.5720
	Poly(2,3-dibromopropyl methacrylate)	1.5739
	Poly(2,6-dimethyl-p-phenylene oxide)	1.5750

Refractive index of polymers/haze value (Continued)

Abbr.	Polymer	Refractive index
PET	Poly(ethylene terephthalate)	1.5750
PVB	Poly(vinyl benozoate)	1.5775
	Poly{2,2-propane bis[4-(2-methylphenyl)]carbonate}	1.5783
	Poly{1,1-butane bis(4-phenyl) carbonate}	1.5792
	Poly(1,2-diphenylethyl methacrylate)	1.5816
	Poly(o-chlorobenzyl methacrylate)	1.5823
	Poly(m-nitrobenzyl methacrylate)	1.5845
	Poly(oxycarbonyloxy-1,4-phenyleneisopropylidene-1,4-phenylene)	1.5850
	Poly{N-(2-phenylethyl) methacrylamide}	1.5857
	Poly{1,1-cyclohexane bis[4-(2,6-dichlorophenyl)]carbonate}	1.5858
PC	Polycarbonate resin	1.5860
BPA	Bisphenol-A polycarbonate	1.5860
	Poly(4-methoxy-2-methylstyrene)	1.5868
	Poly(o-methyl styrene)	1.5874
PS	Polystyrene	1.5894
	Poly{2,2-propane bis[4-(2-chlorophenyl)]carbonate}	1.5900
	Poly{1,1-cyclohexane bis(4-phenyl)carbonate}	1.5900
	Poly(o-methoxy styrene)	1.5932
	Poly(diphenylmethyl methacrylate)	1.5933
	Poly{1,1-ethane bis(4-phenyl) carbonate}	1.5937
	Poly(propylene sulfide)	1.5960
	Poly(p-bromophenyl methacrylate)	1.5964
	Poly(N-benzyl methacrylamide)	1.5965
	Poly(p-methoxy styrene)	1.5967
MeOS	Poly(4-methoxystyrene)	1.5967
	Poly{1,1-cyclopentane bis(4-phenyl)carbonate}	1.5993
PVDC	Poly(vinylidene chloride)	1.6000
	Poly(o-chlorodiphenylmethyl methacrylate)	1.6040

Refractive index of polymers/haze value (Continued)

Abbr.	Polymer	Refractive index
	Poly{2,2-propane bis[4-(2,6-dichlorophenyl)]carbonate}	1.6056
	Poly(pentachlorophenyl methacrylate)	1.6080
	Poly(2-chlorostyrene)	1.6098
PaMes	Poly(alpha-methylstyrene)	1.6100
	Poly(phenyl alpha-bromoacrylate)	1.6120
	Poly{2,2-propane bis[4-(2,6-dibromophenyl)cabonate]}	1.6147
	Poly(p-divinylbenzene)	1.6150
	Poly(N-vinyl phthalimide)	1.6200
	Poly(2,6-dichlorostyrene)	1.6248
	Poly(chloro-p-xylene)	1.6290
	Poly(beta-naphthyl methacrylate)	1.6298
	Poly(alpha-naphthyl carbinyl methacrylate)	1.6300
PEI-ULTEM	Polyetherimide (880 nm wavelength)	1.630
	Polyetherimide (643.8 nm wavelength)	1.651
	Polyetherimide (587.6 nm wavelength)	1.660
	Polyetherimide (546.1 nm wavelength)	1.668
	Polyetherimide (480 nm wavelength)	1.687
	Poly(phenyl methyl silane)	1.6300
	Poly(sulfone) {Poly[4,4'-isopropylidene diphenoxy di(4-phenylene)sulfone]}	1.6330
PSU	Polysulfone resin	1.6330
	Poly(2-vinylthiophene)	1.6376
Mylar Film	Polyethylene terephthalate (boPET)	1.64–1.67
	Poly(2,6-diphenyl-1,4-phenylene oxide)	1.6400
	Poly(alpha-naphthyl methacrylate)	1.6410
	Poly(p-phenylene ether-sulphone)	1.6500
	Poly{diphenylmethane bis(4-phenyl)carbonate}	1.6539
	Poly(vinyl phenyl sulfide)	1.6568

Refractive index of polymers/haze value (Continued)

Abbr.	Polymer	Refractive index
	Poly(styrene sulfide)	1.6568
	Butylphenol formaldehyde resin	1.6600
	Poly(p-xylylene)	1.6690
PVN	Poly(2-vinylnapthalene)	1.6818
PVK	Poly(N-vinyl carbazole)	1.6830
	Naphthalene-formaldehyde rubber	1.6960
PF	Phenol-formaldehyde resin	1.7000
	Poly(pentabromophenyl methacrylate)	1.7100
MFA	Polytetrafluoroethylene-Perfluoromethylvinylether	Unknown
PEEK1	(amorphous) Polyetheretherketone	1.65–1.71
PEEK2	(crystalline) Polyetheretherketone	1.68–1.77

Refractive Index and Dielectric Constant n The dielectric constant, dependent on wavelength, is simply the square of the complex refractive index in a nonmagnetic medium, one with a relative permeability of unity. The refractive index is used for optics in Fresnel equations and Snell's law while the dielectric constant is used in Maxwell's equations and electronics.

Refractivity \ rē- frak- ti-və-tē\ n (1673) The ▶ Index of Refraction minus 1. *Specific refractivity* is given by the expression $(n - 1)/d$ where n = the index of refraction and d = the density of the material.

Refractometer \ rē- frak- ta-mə-tər\ n [ISV] (ca. 1859) An instrument used to measure the index of refraction of transparent materials, both solid and liquid. Refractive indices are often taken as a guide to the purity of raw materials, such as drying oils. The Abbé design is convenient and is used worldwide. See ▶ Index of Refraction, ▶ Differential Refractometer, and ▶ Abbé Re-Fractometer.

Refractory \ri- frak-t(ə-)rē\ adj [alter. of *refractary*, fr. L *refractarius*, irreg. fr. *refragari* to oppose, fr. re- + fragari (as in *suffragari* to support with one's vote)] (1606) Having a very high melting point.

Refractory Fiber n Oxide or nonoxide, amorphous, or crystalline, manufactured fiber generally used for applications at temperatures greater the 1,063°C in both oxidizing and nonoxidizing atmospheres, i.e., Al_2O_3, ZrO_2, $Al_2O_3SiO_2$.

Regain Standard See ▶ Standard Moisture Regain.

Regenerated Cellulose n A transparent cellulosic plastic made by mixing cellulose expatiate with a dilute sodium hydroxide solution to form a ▶ Viscose extruding the viscose into film, sheeting, or fiber form, then treating the extrudate with acid to effect regeneration. In fiber form, the material is called ▶ Rayon. The term ▶ Cellophane is used for films and sheets.

Register v (14c) When a design or form is printed in parts, as in multiple colors, it is a requirement that all parts match exactly. When they do, they are "in register"; otherwise they are "out of register."

Regression Analysis n (method of least squares) A family of statistical techniques for fitting equations to data based on the principle that the best-fitting parameters are those that minimize the sum of squares of the differences between the original observations and the corresponding equation values. A simple case is fitting a straight line to a set of data (x_i, y_i) to obtain the equation $\hat{y} = a + b \cdot x$. *Multiple linear regression* includes the application of the least-squares principle to any form of relationship between a dependent variable y and several "independent" Factors $x_{1i}, x_{2i}, \ldots x_{ki}$ or explicit functions of the x_{ji}, that is "linear with respect to the constants to be fitted." *Nonlinear regression*, which requires one or another iterative-search method, deals with relationships that are not linear in the constants sought. Powerful programs for all these techniques are now available for personal computers. Regression equations provide the vehicle for estimating the outcomes of future experiments in the systems investigated, with known errors of estimate. They have been a powerful tool in polymer science, plastics processing, compound development, and in estimating future performances of plastics products in service.

Regrind n Thermoplastic waste material such as sprues, runners, excess parison material, sheet trimmings, and rejected parts from molding, extrusion, and thermoforming operations, that has been reclaimed by shredding or granulating. Regrind is usually mixed with virgin compound at a predetermined percentage for remolding, etc.

Regular Block n In the chemical structure of polymers, a block that can be described by only one species of ▶ mer in a single sequential arrangement (IUPAC, slightly modified).

Regular Polymer n A polymer whose molecules can be described by only one species of ▶ mer in a single sequential arrangement (IUPAC).

Regulator In polymerization reactions, a chain-transfer agent used at low concentration to limit the molecular weight of the polymer.

Reichert–Meissl Value *n* The number of ml of N/10 potassium hydroxide solution required to neutralize the free volatile and soluble fatty acids in 5 g of the sample under test.

Reinforced Molding Compound *n* A compound containing reinforcing fibers and supplied by the raw-material producer in the form of ready-to-use material, as distinguished from ▶ Premix.

Reinforced Plastic *n* (RP) A plastic composition in which are embedded fibers that are much stronger and typically much stiffer than the matrix resin. The reinforcements are usually fibers, rovings, fabrics, or mats, or mixed forms of glass, carbon, asbestos, metals, ceramics, paper, sisal, cotton, or nylon. Resins most commonly used are polyesters, phenolics, aminos, silicones, epoxies, and various thermoplastics. The term *reinforced plastic* includes some forms of ▶ Laminate and molded parts in which the reinforcements are not in layered form. When the resin is thermoplastic, the term *reinforced thermoplastic* is often used. Methods of forming reinforced-plastics articles from thermosetting resins are defined under the entries listed below.

Axial winding	Perform
Bag molding	Premix
Centrifugal casting	Prepreg
Ceraplast	Prepreg molding
Contact-pressure molding pulforming	
Fiberfill molding	Pulp molding
Filament winding	Pultrusion
Furan prepreg	Reaction injection
Impregnation molding	Molding
Laminate resin-transfer	Resin-transfer molding
Lap winding	Reverse helical winding
Layup molding	Sheet-molding compound
Low-pressure laminate	Slurry performing

(Pittance JC (ed) (1990) Engineering plastics and composites. SAM International, Materials Park, OH; Carley JF (ed) (1993) Whittington's dictionary of plastics. Technomic Publishing).

Reinforced Thermoplastic *n* (RTP) A reinforced structure in which the bonding resin is a thermoplastic rather than a thermoset. Over the past 2 decades, applications for RTP have grown rapidly, mainly based on nylons, polycarbonates, acetal resins, polystyrene, polypropylene, and ▶ Advanced Resins. The tensile strength and modulus of a thermoplastic can be at least doubled by the addition of glass reinforcement, and creep under load is greatly decreased. Because thermoplastics are remeltable, RTPs are most commonly produced as palletized molding compounds for injection molding.

Reinforcement *n* A strong, inert, fibrous material incorporated in a plastic mass to improve mechanical properties. Typical reinforcements are ▶ Asbestos, ▶ Boron Fiber, ▶ Carbon Fiber, ▶ Ceramic Fiber, ▶ Flock, ▶ Glass-Fiber Reinforcement, ▶ Graphite, ▶ Mica, ▶ Sisal, and ▶ Whiskers, all of which see. Others sometimes used are chopped paper, macerated fabrics, synthetic fibers, and metal wires. To be effective, a reinforcement must form a strong adhesive bond with the matrix resin, to which end adhesion, promoting substances known as coupling agents are often preapplied to the fibers. Reinforcements differ from ▶ Fillers in that they markedly improve modulus and strength, whereas filler do not.

Reinforcement Fabrics See ▶ Geotextiles.

Reinforcing Pigment *n* A pigment that also serves to improve the strength of the finished product. An example is carbon black.

Related Shades *n* Colors of similar tone in the same or different depths.

Relative \ˈre-lə-tiv\ *adj* (15c) Comparative; specifically applied to measurements made relative to a reference standard, which values are dependent on the method of measurement used, as opposed to absolute measurements made in fundamental units and obtainable by different methods.

Relative Density Syn: ▶ Specific Gravity.

Relative Dry Hiding Power *n* Of a paint is the ability of that paint to reduce the contrast of a black and white surface to which it is applied and allowed to dry, It is quantitatively expressed in terms of the proportional spreading rate of paint required to produce the same contrast reduction as obtained with the paint chosen as standard.

Relative Humidity *n* (1820) (RH) The ratio, always expressed in percent, of the quantity of water vapor present in the ambient atmosphere to the quantity that would saturate it at the prevailing temperature. It is also the ratio of the partial pressure of water vapor present to the vapor pressure of water at the prevailing temperature. High relative humidity, which occurs in summer in many manufacturing-plant atmospheres, can cause condensation of water on chilled surfaces such as injection molds, with attendant mold defects. It can also cause blushing problems in the painting of plastics and metals. See also ▶ Relative.

Relative Leveling *n* A measure of the ability of a coating to flow out after application so as to obliterate any surface irregularities, such as brush marks or orange peel or peaks and craters which have been produced by the mechanical process of applying the coating. The term is used with the "0" to "10" sale of leveling ratings of ASTM Method D 2801. Leveling Characteristics of Paints by Draw-down Method.

Relative Standard Deviation *n* (1894) See ▶ Coefficient of Variation.

Relative Viscosity *n* The ratio of the kinematic viscosity of a specified solution of the polymer to the kinematic viscosity of the pure solvent. This can also be written as the ratio of the viscosities of pure solvent (η_{solv}) and a solute in solution (η_{sol}) at a concentration according to the method of calculating the Staudinder Index [η]:

$$\eta_{sol}/\eta_{solv} = \eta_{rel}$$

See also ▶ Dilute-Solution Viscosity (Kamide K, Dobashi T (2000) Physical chemistry of polymer solutions. Elsevier, New York; Pethrick RA, Pethrick RA (eds) (1999) Modern techniques for polymer characterization. Wiley, New York; Staudinger H, Heuer W (1930) A relationship between the viscosity and the molecular weight of polystyrene (German). Ber 63B:222–234; Collins EA, Bares J, Billmeyer FW, Jr (1973) Experiments in polymer science. Wiley-Interscience, New York).

Relaxation \rē- lak- sā-shən, ri- lak-, *esp British* re-lək-\ *n* (1548) A gradual decrease in stress in a structure under sustained constant strain. See ▶ Stress Relaxation.

Relaxation-Map Analysis *n* (RMA) A technique used on the results of a series of ▶ Thermally Stimulated Current experiments in which the TSC data are transformed into relaxation times and plotted versus reciprocal absolute temperature to estimate enthalpy and entropy of activation for the molecular relaxations.

Relaxation Time *n* Of a viscoelastic material under constant strain (specifically, one behaving as a Maxwell element), the time required for the stress to diminish to 1/e(= 0.368) of its initial value. Compare ▶ Retardation Time.

Relaxed Yarn *n* A yarn treated to reduce tension and produce more uniform shrinkage or torque. Relaxation produces more uniform dyeing characteristics in regular filament yarns of nylon or polyester.

Release Agent See Mold Wash and Parting Agent.

Release Paper *n* A layer of paper which can be readily separated from the surface of a plastic article to which it has been applied or against which the plastic article has been formed. The term applies to papers used to protect the surfaces of plastic sheets, to temporary backings for pressure-sensitive adhesives, and to papers used as temporary carriers in film- and foam-casting processes. See also ▶ Transfer Coating.

Releasing Liquid *n* A base for receiving printed matter, which base later can be removed by treatment with a solvent, e.g., water or alcohol.

Relief \ri- lēf\ *n* [ME, fr. MF, fr. OF, fr. *relever* to relieve] (14c) Decoration in which the design is given prominence by cutting away the background.

Relief Angle In an injection or blow mold, the relief angle is the angle between the narrow pinch-off land and the cutaway portion adjacent to the pinch-off land.

Relief Offset *n* Process using a relief plate on an offset press.

Relief Printing *n* (1875) Process utilizing those portions of the matrix which are in relief for taking ink from a supply roller and transferring it to the surface to be printed.

Relset® Process *n* A process of Richen, Inc., for continuous heat-setting of carpet or other heavy yarns. Individual ends are continuously fed into a heat-setting chamber and withdrawn into take-up cans or fed to winders

Reluctance \ri- lək-tən(t)s *n* (1710) That property of a magnetic circuit which determines the total magnetic flux in the circuit when a given magnetomotive force is applied. Unit, the reluctance of 1 cm length and 1 cm^2 cross-section of space taken in a vacuum. Dimensions, [$el\ t^{-2}$]; [$\mu\ l^{-1}$].

Reluctivity *n* Reluctivity or specific reluctance is the reciprocal of magnetic permeability. The reluctivity of empty space is taken as unity. Dimensions, [$el\ t^{-2}$]; [$\mu\ l^{-1}$].

Remover See ▶ Paint and Varnish Remover.

Render \ ren-dər\ *v* ME *rendren*, fr. MF *rendre* to give back, yield, fr. (assumed) Vulgar L *rendere*, alter. of L *reddere*, partly fr. *re-* + *dare* to give & partly fr. *re-* + *-dere* to put] (14c) To cover (brick, wood or stone) with a first coat of plaster.

Rennet Casein *n* A type of casein precipitated from milk to means of rennet, the dried extract of stomach secretions from calves or other ruminants containing the enzyme rennin. See also ▶ Casein and Casein Plastic.

Repact Order *n* (1) An order requiring special packaging, as for export. (2) A small order for a number of items requiring a breakdown of large case.

Repainting *n* Cleaning and recoating (with similar materials) of extensive areas which are being redecorated or on which the existing coatings have deteriorated, or otherwise do not provide adequate protection.

Repeat *n* (15c) (1) The printing length of a plate cylinder as determined by one revolution of the plate cylinder gear. (2) The distance covered by a single unit of a pattern that is duplicated over and over, measured along the length of a fabric.

Repellency \ri-ˈpe-lən(t)sē\ *n* (1747) The ability to resist wetting and staining by oils, water, soils, and other materials.

Representative Element *n* A member of one of the main, or A, groups in the periodic table.

Reprocessed Plastic *n* Thermoplastic material that has been left over from molding, extrusion, or thermoforming, such as sprues and runners, sheet, trim, trim between thermoformed parts, and rejected parts, then molded, extruded, or thermoformed again into useful articles by other than the original processors. See also ▶ Recycled Plastic.

Reprography \ri-prä-grə-fē\ *n* [*repro*duction + -*graphy*] (1956) A term coined in 1963 to cover the arts and sciences involved in the copying and duplicating of information by photographic, mechanical or reprinting in quantities below the commercial printing level ((1996–2004) The American heritage dictionary. Houghton Mifflin Company, New York).

Rescue Orange *n* Special color used for lifebelts and life rafts, etc., to improve visibility at sea.

Resene, β *n* Dammar is a mixture, and one of its constituents is β resene, sometimes known as dammar wax. It is this material which is removed in the so-called dewaxing process, which is an essential operation to make the dammar suitable for use in nitrocellulose lacquers.

Resenes *n* (1) Alkali-resistant constituents of certain natural resins, e.g., dammar, and characterized by the presence of oxygen. The resenes are not, however, acids, aldehydes, esters, ketones, or lactones. (2) As applied to naval stores, those constituents of rosin which cannot be saponified with alcoholic alkali, but which contain carbon, hydrogen, and oxygen in the molecule.

Reserve Dyeing See ▶ Dyeing.

Residual Monomer *n* The unpolymerized monomer that remains incorporated in a polymer after the polymerization reaction has been completed (Odian GC (2004) Principles of polymerization. Wiley, New York; Harris DC (2002) Quantitative chemical analysis. W.H. Freeman, New York).

Residual Shrinkage *n* A term describing the amount of shrinkage remaining in a fabric after finishing, expressed as a percentage of the dimensions before finishing.

Residual Solvent *n* Solvent, usually polymerization solvent, remaining in an unfinished resin, or a pelletized resin ready for market, or solvent remaining in a solvent-cast film after drying. Either is usually expressed as a weight percent.

Residual Strain *n* Strain remaining in a part that has been chilled while undergoing plastic deformation or immediately thereafter. Often, if the part is reheated, some or nearly all of the strain may be recovered. Usually associated with ▶ Residual Stress.

Residual Stress *n* (frozen-in stress) Stress remaining in a part that has been chilled quickly during or after molding, extrusion, or forming. It remains because there was too little time for the stress to relax while the material was soft. Over time, high residual stress an cause parts to warp and shrink. It can be relieved and rendered harmless by annealing residually stressed parts while restraining them in fixtures.

Residual Tack *n* Tackiness remaining in a film which, although set, does not reach the really tack-free stage. See ▶ Tack.

Resilience \ri-ˈzil-yən(t)s\ *n* (1824) (1) The degree to which a body can quickly resume its original shape after removal of a deforming stress. When the body is a standard test specimen, the resilience, expressed as the percentage recovery from a stated maximum strain, may be attributed to the material from which the specimen was made. ASTM Tests D 926 and D 945 (section 09.01) describes compression and shear tests for resilience of rubber and foam rubber. (2) The fractional return, to an impacting body, of the energy with which it strikes a resilient test specimen. ASTM D 1054 details a pendulum-rebound test, while D 2632 and D 3574 describe drop-weight-rebound tests, all employing this principle and all in section 09.01.

Resiliency \-yən(t)-sē\ *n* (ca. 1836) Ability of a fiber or fabric to spring back when crushed or wrinkled.

Resin \ˈre-zᵊn\ *n* [ME, fr. MF *resine*, fr. L *resina*; akin to Gk *rhētinē* pine resin] (14c) The term *resin* is defined by ASTM D 883 as a solid or pseudosolid material, often of high molecular weight, that exhibits a tendency to flow when subjected to stress, usually has a softening or melting range, and usually fractures conchoidally. A note appended to this definition explains that in a broad sense, the term is used to designate any polymer that is a basic material for plastics. However, common uses of the term in the plastics industry do not always conform to this definition. The term is also used for uncured fluid thermosetting materials, some chemically modified natural resins, and often synonymously with the terms *plastic* and *polymer* ((2001) Paint: pigment, drying oils, polymers, resins, naval stores, cellulosics esters, and ink vehicles, vol 3. American Society for Testing and Material, West Conshohocken, Pennsylvania; Flick EW (1991) Industrial synthetic resins handbook. Williams Andrews Publishing/Noyes, New York).

Resinamines *n* Resinous basic products obtained from low temperature tar.

Resin Applicator In filament winding, a device that deposits the liquid resin onto the reinforcement band.

Resinates, Metallic *n* Rosin in which part or all of the rosin aids have been chemically reacted with those metals which give soaps or salts which are water insoluble. Limed rosin, zinc-treated rosin, and the resinates of lead, cobalt, copper, and manganese, are of the greatest importance.

Resin Content The percentage, by weight or volume, or resin in a laminate or filled or reinforced thermoplastic molding.

Resin Emulsion Paint *n* A water paint consisting of a water emulsion of an oil-modified alkyd or other resin; when dry, leaves a tough film of resin.

Resin, Natural *n* When certain trees and other plants are wounded, either by natural accident or by tapping, they exude liquids that harden or partially harden upon exposure to air to form resinous or balsam-like products. Deposits of such secretions undergo chemical changes – oxidation and polymerization – when buried for long periods, forming solid or semisolid products that are soluble in oils and organic liquids but insoluble in water. Such water-insoluble resins are generally known as the natural resins, sometimes called *varnish resins*. Those that are water-soluble are known as gums or essential oils. Examples of natural resins are accroides (acaroid resin), amber, Canada balsam, Congo copal, dammar, elemi, Kauri copal, and sandarac. They are used in varnishes and lacquers, also as modifiers for plastics (Langenheim JH (2003) Plant resins: chemistry, evolution ecology and ethnobotany. Timber Press, Portland, OR).

Resinography See ▶ Polymerography.

Resinoid \ˈre-zᵊn-ˌȯid\ *n* (1880) Any of the class of thermosetting synthetic resins, either in their initial temporarily fusible state or in their final infusible state. See also ▶ Novolak and ▶ Thermosetting.

Resin Oil See ▶ Tall Oil.

Resinous Timber *n* Wood from certain trees which contain resinous material in the cells. The resins have high-solvent power for many paints and paint media. This resin frequently exudes through paint films applied to such wood, especially in sunny positions.

Resin Picket See ▶ Pitch Pocket.

Resin Pocket *n* An apparent accumulation of excess resin in a small localized area between laminations in a laminated-plastic article, visible on a cut edge or a molded surface.

Resin-Rich Area *n* A region in a reinforced-plastic article in which there is an objectionable excess of resin.

Resin-Starved Area *n* (dry spot) An area of a reinforced-plastic article that has an insufficient amount of resin to wet out the reinforcement completely, evidenced by low gloss, dry spots, or exposure of fiber. The condition may be caused by improper wetting or impregnation, or by excessive molding pressure.

Resin Streak *n* A long, narrow surface imperfection on the surface of a laminated plastic caused by local excess of resin.

Resin, Synthetic (Synthetic Polymer) A Resin that has been produced from simple materials, or intermediates made from such chemicals, by either addition or condensation polymerization. Of the commercial plastics, all but the cellulosics are based on synthetic resins. Among commercial elastomers, only natural rubber, or refined from the sap of the *Hevea* tree, is natural.

Resin, Synthetic *n* Originally, a member of a group of synthetic substances which resemble and share some of the properties of natural resins, but now used for materials which bear little resemblance to natural resins. The term is generally understood to mean a member of the heterogeneous group of compounds produced from simpler compounds by condensation and/or polymerization. Chemically modified natural polymers, such as cellulose acetate and hardened casein, are not considered to be synthetic resins. (Flick EW (1991) Industrial synthetic resins handbook. Williams Andrews Publishing/Noyes, New York).

Resin Transfer *n* Resin transfer molding is a variation of Matched-Metal-Die molding in which, after placing the preformed reinforcement in the heated mold, premixed, quick-curing resin is injected while the mold is closing or after it has closed.

Resin Transfer Molding *n* (RTM) A variation of Matched-Metal-Die Molding in which, after placing the preformed reinforcement in the heated mold, premixed, quick-curing resin is injected while the mold is closing or after it has closed. The technique has been used to make body parts for specialty cars and aircraft components.

Resin Treated *n* Usually a term descriptive of a textile material that has received an external resin application for stiffening or an internal fiber treatment (especially of cellulosics) to give wrinkle resistance or permanent press characteristics.

Resist \ri-╹zist\ *n* (1836) (1) In preparing zinc printing plates, a material, such as wax, that covers the areas of the plate that are not to be etched by acid (*acid resist*); or, in electroplating, a material that covers areas that are not to be plated. (2) In photolithography and microlithography, widely used in making solid-state electronic devices and printed circuits, a thin film, applied over a substrate, whose solubility in a developer solvent is reduced (negative resist) or enhanced (positive resist) by exposure to UV or other radiation. In polymeric resists, the mechanism of solubility reduction is crosslinking and of solubility enhancement, chemical reaction or chain scission (Leach RH, Pierce RJ, Hickman EP, Mackenzie MJ, Smith HG (1993) Printing ink manual, 5th edn. Blueprint, New York; Leach RH, Pierce RJ (eds) (1976) Printing ink handbook. National Association of Printing Ink Manufacturers, Woodbridge, New Jersey).

Resistance \ri-╹zis-tən(t)s\ *n* (14c) A property of conductors depending on their dimensions, material and temperature which determines the current produced by a given difference of potential. The practical unit of resistance, the *ohm* is that resistance through which a difference of potential of 1 V will produce a current of 1 ampere. The *international ohm* is the resistance offered to an unvarying current by a column of mercury at 0°C, 14.4521 g in mass, of constant cross-sectional area and 106.300 cm in length, sometimes called the legal ohm. Dimensions, $[e^{-1} l^{-1} t]$; $[\mu l \, l^{-1}]$.

Resistance Heating *n* Heating by means of dissipation of electrical energy in resistive circuit elements. The rate of heating (watts, W) is given by the quotient of the square of the voltage drop across the heater (V^2) divided by its at-temperature resistance (Ω), also by the product of the voltage drop and the current (V·A).

Resistance of a Conductor *n* At 0°C, of length *l*, cross-section *s* and specific resistance ρ,

$$R_o = \rho \frac{1}{s}$$

The resistivity may be expressed as ohm-centimeter when *R* is in ohms, *l* in centimeter and *s* in square centimeter, $R_t = R_o(1 + \rho l)$ (Lide DR (ed) (2004) CRC handbook of chemistry and physics. CRC Press, Boca Raton, FL; Giambattista A, Richardson R, Richardson RC, Richardson B (2003) College physics. McGraw-Hill Science, New York).

Resistance of Conductors in Series and Parallel *n* The total resistance of any number of resistances joined in series is the sum of the separate resistances. The total resistance of conductors in parallel whose separate resistances are $r_1, r_2, r_3 \ldots r_n$ is given by the formula

$$\frac{1}{R} = \frac{1}{r_1} + \frac{1}{r_2} + \frac{1}{r_3} \cdots + \frac{1}{r_n}$$

Where *R* is the total resistance. For two terms this becomes,

$$R = \frac{r_1 r_2}{r_1 + r_2}$$

(Lide DR (ed) (2004) CRC handbook of chemistry and physics. CRC Press, Boca Raton, FL; Giambattista A, Richardson R, Richardson RC, Richardson B (2003) College physics. McGraw-Hill Science/Engineering/Math, New York).

Resistance, Specific (Resistivity) A proportionality factor characteristic of different substances equal to the *resistance* that a centimeter cube of the substance offers to the passage of electricity, the current being perpendicular to two parallel faces. It is defined by the expression:

$$R = \rho \frac{l}{A}$$

where *R* is the resistance of a uniform conductor, *l* is its length, *A* is its cross-sectional area, and ρ is its resistivity. Resistivity is usually expressed in ohm-centimeters (Giambattista A, Richardson R, Richardson RC, Richardson B (2003) College physics. McGraw-Hill Science/Engineering/Math, New York).

Resistance Thermometer *n* A temperature-measuring device consisting of an encapsulated, fine coil of platinum wire whose resistance increases substantially and nearly linearly with rising temperature. See ▶ Thermistor. The change in resistance is sensed and converted to a temperature reading. Resistance thermometers have found considerable use in plastics processing.

Resistance Welding See ▶ Thermoband Welding.

Resist Dyeing See Dyeing, Reserve Dyeing.

Resistivity \ri-╻zis-╹ti-və-tē\ *n* (1885) (specific resistance) (1) When this word stands alone, it usually means ▶ Volume Resistivity. But also see ▶ Surface Resistivity. (2) The electrical resistance of a body of unit length and unit cross-sectional area or unit weight. Reciprocal of conductivity.

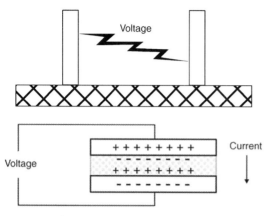

Measurement of volume resistivity

Resist Printing See ▶ Printing.

Resist Technique *n* Taping a fine line effect over a surface, and then blocking out the rest of the area and spraying into the opening left by the tape.

Resite *n* A phenol-formaldehyde resin in the final state of the curing process. In this stage it is insoluble in alcohol and acetone, and infusible. Alternate term for C-stage. See ▶ C-Stage; also called ▶ Resitol and ▶ Resol.

Resitol *n* A phenol-formaldehyde resin in the transition state of the curing process. Under heating it softens to rubberlike consistency but without melting. It swells when it is immersed in alcohol or acetone, but does not dissolve. Alternate term for B-stage. See ▶ Resite, ▶ Resol, and ▶ Resolite.

Resol *n* A flusible, soluble, phenolic resin containing sufficient reactive methylol groups to enable the resin to become infusible on further reaction. Alternate term for A-stage. See ▶ Novolak.

Resolite Syn: a resin at its B-Stage.

Resolving Power *n* The resolving power of a telescope or microscope is indicated by the minimum separation of two objects for which they appear distinct and separate when viewed through the instrument.

Resonance \▪re-zən-ən(t)s, ▪rəz-nən(t)s\ *n* (15c) In chemistry, the periodic cycling of electrons from one atom of a molecule or ion to another atom of the same molecule or ion. Thus, given atoms remain in a fixed spatial arrangement with their electrons oscillating between atoms so as to satisfy two (or more) possible structural formulas. Resonance was first conceived to account for the outstanding stability of the benzene ring and it took almost a century for researchers to prove its reality (Smith MB, March J (2001) Advanced organic chemistry, 5th edn. Wiley, New York; Morrison RT, Boyd RN (1992) Organic chemistry, 6th edn. Prentice-Hall, Englewood Cliffs, NJ). A large-amplitude vibration of a mechanical or electrical system caused by a relatively small periodic stimulus applied at the natural frequency of the system. It is a phenomenon to be avoided in most structures. That is done by designing the natural frequency and the first few harmonics to be very different from the loading frequencies expected during use (Lide DR (ed) (2004) CRC handbook of chemistry and physics. CRC Press, Boca Raton, FL).

Resonance (Chemical) *n* The moving of electrons from one atom of a molecule or ion to another atom of that molecule or ion. It is simply the oriented movement of the bonds between atoms (Smith MB, March J (2001) Advanced organic chemistry, 5th edn. Wiley, New York; Morrison RT, Boyd RN (1992) Organic chemistry, 6th edn. Prentice-Hall, Englewood Cliffs, NJ).

Resonance Hybrid *n* A structure which cannot be represented by a single valence-bond Lewis structure but rather is shown as a combination, or average, of two or more structures.

Resorcinol \-▪ól, -▪ōl\ *n* (1881) (resorcin, *m*-dihydroxybenzene, 3-hydroxyphenol) $C_6H_6O_2$. A white, needlelike, water-soluble solid, a benzene derivative (1,3-substituted with two hydroxy group) originally obtained from certain resins, now usually synthesized. When reacted with formaldehyde, produces resins suitable for cold-setting adhesives

Resorcinol modified Phenolic Resins *n* Thermosetting polymers of phenol, formaldehyde, and resorcinol having good heat and creep resistance and dimensional stability.

Resorcinol Monobenzoate *n* A white, crystalline solid used as an ultraviolet screener in plastics. It is particularly useful in applications requiring a high degree of transparency, and can be used with cellulosics, vinyls, and certain polyesters

Resorcinol Test Used to detect the presence of phthalic acid in phthalate plasticizers or alkyd resins. It involves heating the product under test with concentrated sulfuric acid and resorcinol, and pouring the resulting mixture into excess of an alkaline solution. If phthalic acid is present, the fluorescein produced yields its characteristic yellow-green color.

Resox *n* A contraction (and reversal) of the term "exudation-reduction." A *redox catalyst* is one entering into an oxidation–reduction reaction. A polymerization initiator comprising a mixture of a peroxide and a reducing agent is called a *redox initiator*. Polymers formed by such reactions are sometimes called *redox polymers*.

Restitution, Coefficient of *n* For two bodies on impact. The ratio of the difference in velocity, after impact to the difference before impact.

Restraint Systems *n* An end use for textile fibers; restraint systems are devices such as air bags, seat belts, and shoulder harnesses for passenger protection in automobile, trucks, airplanes, etc.

Restricted Gate *n* A small orifice between runner and cavity in an injection or transfer mold. Such a gate freezes (or cures) quickly when the cavity has filled and flow stops, preventing packing of the region near the gate in the cavity. When the piece is ejected, this gate breaks cleanly, simplifying separation of the runner from the molded item. It may be so small (see ▶ Pinpoint Gate) that its tiny stub need not be removed from the piece.

Restrictor Bar Syn: ▶ Choker Bar.

Restrictor Ring *n* A ring-shaped part protruding from the torpedo surface within the cylinder of a plunger-type injection machine. It was claimed that the ring provided higher injection pressure on the melt, higher rate of filling the cavities, and improved weld strength.

Retainer Plate *n* In injection molding, a plate that reinforces the cavity block against the injection pressure, and also serves as an anchor for the cavities, ejector pins, guide pins, and bushings. The retainer plate is usually cored for circulating water or steam.

Retaining Pin *n* A pin in an injection or transfer mold on which an insert is placed and located prior to molding. The term is sometimes used to mean *guide pin* or *dowel pin*.

Retardation \ ▪rē- ▪tär- ▪dā-shən, ri-\ *n* (1) To reduce the rate at which a polymer is formed. Retardation occurs as a result of a reaction between the chain-initiating species and the inhibitor or retarder. (2) The actual distance of one of the doubly refracted rays behind the other as they emerge from an anisotropic particle. It depends on the difference in the two refractive indices, $n_2 - n_1$, and the thickness.

Retardation Time *n* Of a stressed viscoelastic material (specifically, one behaving like a Voigt element), the time required after release of stress for the strain to decrease to $1/e$ (= 0.368) of its original value.

Retarder (1) Generally, a component added to a composition to slow down a chemical or physical change. A slowly evaporating solvent may be added to a paint, varnish or lacquer to delay the set of the film after application and so improve the application properties, or to give a better film, e.g., one with improved flow, or a retarder may be added to plaster to retard its setting. (2) In flexography, gravure and heatset printing, high boiling solvent added to ink to slow the evaporation rate. (3) In letterpress and offset. See ▶ Antioxidant. (4) A material which in small amount added to a rubber compound retards vulcanization or slows down the activity of the accelerator. Specifically, phthalic anhydride and salicylic acid are retarders. The most valuable retarder is one which slows the vulcanization at processing and early curing temperatures, but does not affect or may even activate the rate of cure at full curing temperatures. Syn: ▶ Inhibitor. (5) A chemical that, when added to the dyebath, decreases the rate of dyeing but does not affect the final exhaustion ((1998) Kirk-Othmer encyclopedia of chemical technology. Wiley-Interscience, New York; Salamone JC (ed) (1996) Polymeric materials encyclopedia. CRC Press, Boca Raton, FL; Carley JF (ed) (1993) Whittington's dictionary of plastics. Technomic Publishing).

Retarders *n* (1) In flexography, gravure, and heatset printing – high boiling solvents added to ink to slow the evaporation rate. (2) In letterpress and offset. See ▶ Antioxidant.

Reticulated Polyurethane Foam *n* A urethane foam of extremely low density, characterized by a three-dimensional skeletal structure of strands with dew or no membranes between the strands, and containing up to 97+% of void space. Such foams are made by treating an open cell foam structure with a dilute aqueous sodium hydroxide solution under controlled conditions so that the thin membranes are dissolved, leaving the strands substantially unaffected. Ultrasonic vibrating is sometimes used to assist the solution process. These foams are used in filters for air conditioners, automobile carburetors, and air-cleaning systems; and in acoustical panels, humidifiers, and various household products.

Reticulation \ri-ˌti-kyə-ˈlā-shən\ n (1671) The crepe-like appearance of the surface of the colloid on collotype plates.

Retina \ˈre-tᵊn-ə, ˈret-nə\ n [ME rethina, fr. ML retina, prob. fr. L rete net] (14c) The light-sensitive layer and highly organized neuron structure within the choroids, and lining the posterior two thirds of the eyeball. This is the part of the human eye that is responsible for detecting color.

Retinal \ˈre-tᵊn-əl, ˈret-nəl\ adj (1838) Of or pertaining to the retina.

Retort Pine Tar See ▶ Pine Tar, Retort.

Retouching See ▶ Spotting In.

Retrogradation n (ca. 1545) (a) A change of starch pastes from low to high consistency on aging.

Retroreflective n (ca. 1965) A retroreflective surface or device is one which reflects in incident ray of light back along the direction of incidence with only limited spread, without requiring the incident ray to be normal to the surface. Such a surface or device differs from both a perfect diffuser and a perfect mirror. It may utilize small triple-reflection cube corners, small transparent spheroids, multiple hemispheric indentations, a multiple lens sheet backed by a reflective surface, or some simple derive. See ▶ Reflex Reflector.

Return Pin n In injection molding, any of the set of pins that return the ejector mechanism to its molding position.

Reverse-Flighted Screw n A short section of a screw in a twin-screw extruder in which the flights have a helical direction opposite to that of the main screw sections, thus opposing the forward flow of the heat-softened plastic. The purpose is to improve mixing and dispersion of compound ingredients such as pigments, and sometimes to heat the melt just before it enters a vacuum-extraction zone. Reverse-flight sections have rarely been used in a single-screw machines.

Reverse Helical Winding n In filament winding, a pattern in which a continuous helix is laid down, reversing direction at each of the polar ends. It differs from biaxial compact, or sequential winding in that the fibers cross each other at definite equators, the number depending on the helical lead, with the minimum number of crossovers being three.

Reverse Impact Test n A test for sheet material in which one side of the specimen is struck by a pendulum or falling object and the reverse side is inspected for damage. See ▶ Impact Test.

Reverse Printing n (1) Printing on the underside of a transparent film. (2) Design in which the copy is "dropped out" and the background is printed.

Reverse-Roll Coating n A method of coating wherein the coating material is premetered between two rolls, one of which deposits the coating on a substrate. The thickness of the coating is controlled by the gap between the rollers and also by the speed of rotation of the coating roll.

Reversible Addition–Fragmentation Chain Transfer Polymerization n Reversible addition–fragmentation chain transfer polymerization or RAFT polymerization agent (Odian, 2004) is one kind of controlled radical polymerizations. The RAFT agent can be considered a "catalytic" polymerization agent because an additional free radical initiator must be used in conjunction with a RAFT agent. RAFT was discovered by the CSIRO in 1998. CSIRO is the Commonwealth Scientific and Industrial Research Organization is the national government body for scientific research in Australia. The RAFT agent, catalysis a relatively new method for the synthesis of living radical polymers that may be more versatile than atom transfer radical polymerization (ATRP) or Nitroxide-Mediated Polymerization| nitroxide-mediated polymerization (NMP). RAFT polymerization uses thiocarbonylthio compounds, such as dithioesters, dithiocarbamates, trithiocarbonates, and xanthates in order to mediate the polymerization via a reversible chain-transfer process. With respect to its exceptional effectiveness and the wide range of applicable monomers as well as solvents, RAFT polymerization has developed into an extremely versatile polymerization technique. Indeed, the molecular weight of the polymer can be easily predetermined and the molecular weight distribution can be controlled fairly well. Controlled molecular weight polymers with narrow polydispersity indexes are desirable because their properties (solubility, etc.) are thus controllable. Moreover, RAFT polymerization is used to design polymers of complex architectures such as linear block, comb-like, star, brush polymers and dendrimers. Addition-fragmentation chain transfer process was first reported in the early 1970s (Moad, 2007). However, the technique was irreversible so the transfer reagents could not be selected to control radical polymerization at this time. For the first few years RAFT was used to help synthesize end-functionalized polymers. Scientists began to realize the potential of RAFT in controlled radical polymerization in the 1980s (Cacioli et al.). Macromonomers were known as reversible chain transfer agents during this time, but had limited applications on controlled radical polymerization. In 1995, a key step in the "degenerate" reversible chain transfer step for chain equilibration was brought to attention. The essential feature is that the product of chain transfer is also a chain transfer agent with similar activity to the precursor transfer agent (Matyjaszewkki et al.). RAFT polymerization today is mainly carried out by thiocarbonylthio chain transfer agents. It was first reported by Thang et al. in 1998 (Chiefari). RAFT is one of the most versatile methods of controlled radical polymerization because it is tolerant of a very wide range of functionality in the monomer and solvent, including aqueous solutions. RAFT polymerization has also been effectively carried out over a wide temperature range.

A RAFT polymerization system consists of: initiator, monomer, chain transfer agent, solvent, and temperature. RAFT polymerization can be performed by simply adding a chosen quantity of an appropriate RAFT agent (thiocarbonylthio compounds) to a conventional free radical polymerization. Usually the same monomers, initiators, solvents and temperatures can be used. Because of the low concentration of the RAFT agent in the system, the concentration of the initiator is usually lower than in conventional radical polymerization. Radical initiators such as Azobisisobutyronitrile(AIBN) and 4,4′-Azobis(4-cyanovaleric acid)(ACVA) are widely used as the initiator in RAFT. RAFT polymerization is known for its compatibility with a wide range of monomers compared to other controlled radical polymerizations. These monomers include (meth)acrylates, (meth)acrylamides, acrylonitrile, styrene and derivatives, butadiene, vinyl acetate, and N-vinylpyrrolidone.

RAFT agents (also called chain-transfer agents) must be thiocarbonylthio compounds where the Z and R groups perform different functions. The Z group primarily controls the ease with which radical species add to the C=S bond. The R group plays an important role – it must be a good homolytic leaving group which is able to initiate new polymer chains. RAFT Mechanism

RAFT is a type of living polymerization involving a conventional radical polymerization in the presence of a reversible chain transfer reagent (Cowie and Arrighi, 2008). Like other living polymerizations, there is no termination step in the RAFT process, leading to very large molecular weight polymers. It is an very versatile method to form low polydispersity polymer from monomers capable of radical polymerization (Moad, 2004). The reaction is usually done with a dithioester (Chiefari). The dithio compound must have a good hemolytic leaving group, R, whose radical must be capable of initiating a polymerization reaction (Cowie and Arrighi, 2008). There are four steps in raft polymerization: initiation, addition–fragmentation, reinitiation, and equilibration. Initiation: The reaction is started by radical initiators such as AIBN. In this step, the initiator (I) reacts with a monomer unit to create a radical species which starts an active polymerizing chain. Addition–fragmentation: The active chain (P_n) reacts with the dithioester, that leaves the homolytic leaving group (R). This is a reversible step, with an intermediate species capable of losing either the leaving group (R) or the active species (P_n). Reinitiation: The leaving group radical then reacts with another monomer species, starting another active polymer chain. This active chain (P_m) is then able to go through the addition–fragmentation or equilibration steps. Equilibration: This is the fundamental step in the RAFT process (Cowie and Arrighi, 2008) which traps the majority of the active propagating species into the dormant thiocarbonyl compound. This limits the possibility of chain termination through. Active polymer chains ($P_{m>}$ and P_n) are in an equilibrium between the active and dormant stages. While one polymer chain is in the dormant stage (bound to the thiocarbonyl compound), the other is active in polymerization. By controlling the concentration of initiator and capping agent (dithioester), it is possible to produce controlled molecular weight with low polydispersities. In RAFT polymerization, the concentration on the active species is kept low relative to the dormant species by controlling the amount of initiator and capping agent. This in turn will limit termination steps such as radical combination and disproportionation, increasing the polymer length. Many high molecular weight polymers with low PDIs have been synthesized using RAFT polymerization (Chiefari et al.). As seen in the figure, poly(methyl methacrylate) and polyacrylic acid have been synthesized using AIBN as the initiator and various dithioester compounds.

RAFT polymerization has successfully synthesized a wide range of polymers with controlled molecular weight and low polydispersities (between 1.05 and 1.4 for many monomers). Some monomers capable of polymerizing by RAFT include styrenes, acrylates, acrylamides, and many vinyl monomers. Additionally, the RAFT process allows the synthesis of polymers with specific macromolecular architectures such as block, gradient, statistical, comb/brush, star, hyperbranched, and network copolymers.

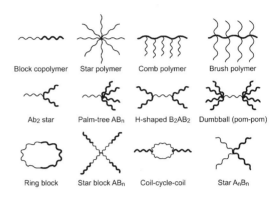

These properties make RAFT useful in many types of polymer synthesis (Perrier and Takolpuckdee, 2005). Block Copolymers: Because RAFT is a form of living radical polymerization, it is ideal for synthesis of block copolymers. For example, in the copolymerization of two monomers (A and B) allowing A to polymerize via RAFT will exhaust the monomer in solution without termination. After monomer A is fully reacted, the addition of monomer B will result in a block copolymer. One requirement for maintaining a narrow polydispersity in this type of copolymer is to have a chain transfer agent with a high transfer constant to the subsequent monomer (monomer B in the example) (Perrier and takolpuckdee, 2005). Multiblock copolymers have also been reported by using difunctional R groups or symmetrical trithiocarbonates with difunctional Z groups. Star Copolymers: Using a multifuntional CTA can result in the formation of a star copolymer. RAFT differs from other forms of LRPs because the core of the copolymer can be introduced by functionalization of either the R group or the Z group. While utilizing the R group results in similar structures found using ATRP or NMP, the use of the Z group makes RAFT unique. When the Z group is used, the reactive polymeric arms are detached from the core while they grow and react back into the core for the chain-transfer reaction (Perrier and Takolpuckdee, 2005). Controlled Grafting onto Polymeric Surfaces: Producing grafted polymers onto a polymer bead via noncontrolled radical polymerization results in a broad molecular weight distribution and high polydispersity. By employing RAFT polymerization, grafting from these microspheres becomes a one-step process. As an additional benefit the grafted polymer would have a RAFT end group, providing the potential for reinitiating the chains to form block copolymer shells (Barner).

Reversible Bonded Fabric *n* A bonded structure in which two face fabrics are bonded together so that the two sides may be used interchangeable. There are limitations to the fabrics that may be used because of increased fabric stiffness resulting from bonding.

Reversible Colloid *n* See ▶ Colloid, Reversible.

Reversible Reaction *n* One which can be caused to proceed in either direction by suitable variation in the conditions of temperature, volume, pressure, or of the quantities of reacting substances.

Reversion *n* A reduction in the modulus and tensile strength of a rubber vulcanizate as a result of prolonged thermal ageing. It occurs in sulfur vulcanized rubbers containing polysulphide crosslinks that have been overcured or during exposures to temperatures above 150°C. Reversion is due to a breakdown in the network structure, probably as a result of crosslink forming cyclic structures. Reversion can be also defined as the recombination of the hydrolysis products of polysaccharides. The softening of vulcanized rubber (usually natural but sometimes synthetic rubber) when heated too long or at too high a temperature. Reversion is evidenced by increase in extensibility, decrease in tensile strength, and lowering of the stress to produce a given elongation. Extreme reversion usually results in tackiness (Harper CA (ed) (2002) Handbook of plastics, elastomers and composites, 4th edn. McGraw-Hill, New York).

Revolving Spinning Ring *n* A driven ring that rotates in the direction of the traveler on a ring spinning frame. Since both the ring and the yarn package turn when this ring system is used, productivity is increased.

Rewind *n* In the coil coating industry, the apparatus used to recoil the strip after it is coated.

Reworked Material *n* A thermoplastic from a processor's own production that has been reground, pelletized, or solvated, after having been previously processed by molding, extrusion, etc. NOTE – In many specifications the use of reworked material is limited to clean plastic that meets the requirements specified for the virgin material, and yields a product essentially equal in quality to one made from only virgin material (ASTM D 883). See also ▶ Regrind.

Reynolds Number (Re, N_{Re}) \ˈre-nᵊldz-\ *n* [Osborne Reynolds † 1912 English physicist] (1910) The Reynolds number Re is the ratio of kinetic energy to viscous energy; the equation is,

$$Re = 2rds/vd = 2rs/vk$$

where
s = speed of the fluid, cm/s,
r = radius of tube, cm
d = fluid density
vd = dynamic viscosity, Poise (dyne-s/cm^2)
vk = kinetic viscosity in Stokes (vd/d)
Re = Reynolds number; small Re means that the fluid's viscosity is dominant, large Re (10,000) means viscosity is negligible and the kinetic or inertia effects rule; when Re is high, turbulent, cavitations and chaos describe the flow, when Re is low, laminar flow dominates.
(2) A dimensionless ratio that relates to frictional pressure drop in fluid flow in pips, defined by $D·V·\rho/\mu$ for gasses and Newtonian liquids, where D is the inside pipe diameter, V is the average velocity of the fluid (= flow rate/cross-section), ρ is the fluid density, and μ is the mass-based viscosity [Pa = kg/(m·s) in SI units]. When N_{Re} = 2,100–4,000, the character of the flow ranges from streamline (laminar) to turbulent. For Non-Newtonian liquids the criterion for flow transition by the Reynolds number is redefined. For a power–law liquid it becomes $D^n·V^{2-n}·\rho/(g_c·K·8^{n-1})$, where n = the flow-behavior index of the liquid, K = its temperature-dependent consistency index, and g_c = the proportionality constant in Newton's law of momentum change. See ▶ Power Law. The Reynolds number is also applicable to other flow geometries, such as packed beds, with appropriate modifications (Perry RH, Green DW (1997) Perry's chemical engineer's handbook, 7th edn. McGraw-Hill, New York).

R-Factor *n* See ▶ Thermal Resistance.

RF Curing *n* Hastening the final crosslinking of thermosetting compounds and laminates by application of radio-frequency energy.

RF Heating Syn: ▶ Dielectric Heating.

RF Shielding *n* Enclosing equipment that generates radio-frequency radiation in a conductive housing that prevents the radiation from being broadcast and causing interference in nearby electronic devices that operate in the same frequency range. Required by law for all RF-heating equipment except that operating at 27.12 MHz, the frequency band reserved for industrial use.

RH *n* Abbreviation for ▶ Relative Humidity.

Rhe *n* In the now deprecated cgs system, the rhe was the unit of fluidity, equal to 1 poise^{-1} [10 (Pa·s)$^{-1}$]. (2) The cgs unit of measurement of fluidity. A material has a fluidity of one rhe when a shearing force of 1 dyne/cm^2 induces a rate of shear of 1 cm/s.

Rheogoniometer See ▶ Weissenberg Rheogoniometer.
Rheological \re-ˈä-lə-jē-\ *n* [ISV] (1929) Refers to rheology.
Rheological Properties *n* The properties of viscous substances including polymers that deal with deformation and flow. Includes viscosity and flow rate measurements.
Rheology \re-ˈä-lə-jē\ *n* [ISV] (1929) Viscosity refer to how thick a liquid is or how easily it flows. A viscometer measures the resistance to flow of a rotating probe in a liquid in its of dyne.cm/s^2 (Gooch JW (1997) Analysis and deformulation of polymeric materials. Plenum Press, New York). Viscosity is the study of deformation and flow of matter. Rheology, derived from the Greek *rheos* meaning "something flowing," was proposed by Bingham in 1929 for the rapidly growing science of flow and deformation properties of materials, including liquids, solids, and even powders, in terms of stress, strain rates, and time. The following plot demonstrates the behavior of pure liquids (usually Newtonian), dispersions (pseudoplastic and dilatant). (ASTM, D5165-93 (2004) West Conshohoken, Pennsylvania; Goodwin JW, Goodwin J, Hughes RW (2000) Rheology for chemists. Royal Society of Chemistry; Munson BR, Young DF, Okiishi TH (2005) Fundamentals of fluid mechanics. Wiley, New York). See ▶ Dilatancy, ▶ Nonnewtonian, ▶ Thixotropy, ▶ Viscoelasticity, ▶ Viscosity, ▶ Bingham Body, and ▶ Yield Value.

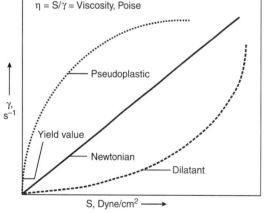

Rheology curves of liquids and dispersions

Rheomalaxis Some fluids demonstrate shear-thinning due to a break-down or "degradation of structure," (Goodwin JW, Goodwin J, Hughes RW (2000) Rheology for chemists. Royal Society of Chemistry), and this phenomenon is known as *rheomalaxis* and an example is the shearing of gypsum paste, whereas an example of thixiotropic phenomenon is shearing of paint (Patton TC (1979) Paint flow and pigment dispersion. Wiley, New York; ▶ Varnish.
Rheometer \rē-ˈa-mə-tər\ *n* [ISV] (ca. 1859) (Plastometer) An instrument for measuring the flow behavior of high-viscosity materials such as molten thermoplastics, rubbers, pastes, and cements. The most widely used principle is that of the Capillary Rheometer of which a variety of makes and models are in daily use. Instruments for measuring the flow properties of less viscous fluids, e.g., dilute polymer solutions, are called ▶ Viscometers but the terms *rheometer* and *viscometer* are often used interchangeably. Currently (1993), computerized, on-line capillary rheometers linked to a single control station can simultaneously monitor melt viscosity in ten or more extruders in a resin-finishing plant.
Rheometer, Foam *n* A special type of capillary rheometer developed by W. Kostyrzewski to measure the density and viscosity of foaming polyurethane immediately after mixing. Creaming time is also measured. It consists of a three-section, vertical acrylic tube, adapted at the bottom to a reaction–injection–molding mixer (RIM machine). The bottom chamber is large enough to contain the mix charge delivered by the RIM machine, and the mixture rises through an instrumented and temperature-controlled "capillary" as the foam forms and expands. The instruments are connected to an analog-to-digital converter and computer. The foam passes up through the capillary into a larger-diameter collection chamber and raises a floating disk that can be weighted to adjust the pressure on the foam as it rises. Flow rate is inferred from foam density; rate of change of density, and shear rate at the capillary wall are computed from flow rate. Shear stress is inferred from the pressure–transducer readings.
Rheopexy *n* (1) The inverse of ▶ Thixotropy. The viscosity of a rheopectic material increases with time under an applied constant stress, approaching a constant value. When the stress is removed or reduced, the viscosity diminishes toward its original value. (2) A special form of thixotropic gel which possesses the property of solidifying more rapidly when sheared (stirred) very slowly than when at complete rest. It should not be confused with dilatancy. The equilibrium state of the former is a solid gel, while the latter is a liquid.
Rhodamine \ˈrō-də-ˌmēn\ *n* [ISV] (1888) Any of a class of organic dyes that exhibit bright orange to red fluorescent colors when viewed under ultraviolet light. An interesting characteristic is that in vinyl

compound, the color shade is dependent on the degree of fusion. Thus, they can be used as indicators of fusion completeness.

Rhodamine B n Pigment Violet 10 (45170). Brilliant, strong dyestuff made by condensing diethyl-m-aminophenol and phthalic anhydride together, in the presence of zinc chloride.

Rhodamine Reds n A class of clean, blue shade organic reds possessing good light-fastness; often called magenta in process printing.

Rhodamine 6G Pigment Red 81 (45160). Xanthene type dyestuff made by condensing monoethyl-m-aminophenol with phthalic anhydride, and converting the product into the ethyl ester.

Rhodester n Cellulose acetate. Manufactured by Soc. Rhone Poulane, France.

Rhodia n Cellulose $2\frac{1}{2}$ acetate. Manufactured by Soc. Rhodiaceta, France.

Rhodiaceta-Nylon n Nylon 66. Soc. Rhodiaceta, France.

Rhoduline Blue 66 n A bright, greenish blue pigment dyestuff, made by condensing o-chlorobenzaldehyde with dimethylaniline, which is then oxidized. *Also known as Setoglaucine or Permanent Peacock Blue.*

Rhoduline Heliotrope n Aniline dye used in colored lacquers. Soluble in lacquer solvents, fast to light, and is not affected by exposure to air.

Rhombohedron \ˌräm-bō-ˈhē-drən\ n [NL] (1836) A crystal having six faces each of which is a rhombus.

Rhovil n Poly(vinyl chloride). Manufactured by Soc. Rhovil, France.

Rhus Vernicifera n $C_{22}H_{31}O_3$. Chinese tree which delivers a milky type of liquid. When the water is removed, a dark oily liquid is obtained, which yields a tough flexible film on atmospheric oxidation. The chief constituent of the film-forming material is an unsaturated phenol–acid–urushiol. The liquid is used for the preparation of Japanese lacquer.

Ribbing n In the coil coating industry, longitudinal streaks that do not flow out on painted strip.

Ribbon n \ˈri-bən\ n [ME *riban*, fr. MF *riban*, *ruban*] (14c) Narrow fabric made in several widths and a variety of weaves and used as a trimming.

Ribbon Blender n A type of low-intensity mixer comprising helical, ribbon-shaped blades supported by spokes from the horizontal shaft, with the blade edges fitting fairly closely to the lower half of the U-shaped shell of the mixer. The shell is usually jacketed for temperature control and the shaft, spokes and blades may also be cored for circulation of heat-transfer liquid (usually water). They are used in the plastics industry mostly for cooling dry blends such as PVC extrusion and calendering compounds discharged from high-intensity mixers before storage and/or shipment.

Ribbon Straw n Cellulose $2\frac{1}{2}$ acetate (artificial straw). Manufactured by British Celanese, Great Britain.

Rib Knit n A double-knit fabric in which the wales or vertical rows of stitches intermesh alternately on the face and the back. In other words, odd wales intermesh on one side of the cloth and even wales on the other. Ribknit fabrics of this type have good elasticity, especially in the width.

Ribs Raised ridge in the finish caused by heavy brush marks which were not sanded or rubbed from underneath coats before finishing coats were applied.

Ricinoleic Acid \ˌrī-sən-ō-ˈlē-ik-\ n [L *ricin*us + E *oleic acid*] (1848) Chief acid of castor oil, almost unique by reason of the presence of a hydroxyl group in the fatty acid chain. To exhibit drying properties, the ester must be dehydrated. Bp, 250°C/15 mm Hg; mp, 17°C; sp gr, 0.945. See ▶ DCO

Ricinus Oil Syn: ▶ Castor Oil.

Rickrack \ˈrik-ˌrak\ n [reduplication of ^{4}*rack*] (1884) Flat braid in zigzag formation. It is produced by applying different tensions to individual threads during manufacture.

Ridge Forming n A variation of ▶ Sheet Thermoforming.

Ridgy Beam n A beam of yarn on which the ends are not evenly distributed across the barrel, causing a profile of peaks (ridges) and valleys. A ridgy beam can give poor removal characteristics.

Ridgy Cloth See ▶ Baggy Cloth.

Rigidity \rə-ˈji-də-tē\ n (1624) The ability of a structure to resist deformation under load. It is a function of both the material's modulus of elasticity and, often more critically, of the geometry of the structure. In a loaded beam, whatever the load distribution or type of beam supports, the maximum deflection is inversely proportional to the product, $E \cdot I$, of the material's elastic modulus and the moment of inertia of the beam's cross-section about its neutral axis. See also

▶ Section Modulus. The term *rigidity* is often applied loosely to materials themselves without reference to a particular structure when what the speaker actually has in mind is the elastic modulus.

Rigidity, Modulus of *n* The slope of the linear portion of the stress–strain graph of a specimen tested in shear. See ▶ Shear Modulus.

Rigid Plastic *n* For the purposes of general classification, a plastic that has a modulus of elasticity ether in flexure or in tension greater than 700 MPa (100 kpsi) at 23°C and 50% relative humidity when tested in accordance with ASTM methods D 747, D 790, D 638, or D 882 (ASTM D 883). This simple ASTM criterion has not always been adequate, especially with respect to vinyls whose impact strengths and other properties can vary widely while elastic modulus remains fairly constant. Vinyls are classified as rigid if their moduli are 1.4 GPa or higher, semi-rigid from 0.4 to 1.4 GPa, and flexible below 0.4 GPa.

Rigid PVC *n* See ▶ Rigid Plastic and ▶ Polyvinyl Chloride.

Rigid Thermoplastic Polyurethanes *n* rigid thermoplastic polyurethanes are not chemically crosslinked. They have high abrasion resistance, good retention of properties at low temperatures, but poor heat resistance, weatherability, and resistance to solvents. Rigid thermoplastic polyurethanes are flammable and can release toxic substances. Processed by injection molding and extrusion. Also called rigid thermoplastic urethanes, nonelastomeric thermoplastic polyurethanes.

Rigid Thermoplastic Urethanes See ▶ Rigid Thermoplastic Polyurethanes.

Rigidsol *n* A coined term for a plastisol that forms an article of very high Durometer hardness. Such hardness is obtained by means of compounding techniques that permit the use of relatively small amounts of plasticizer, and/or by the incorporation of monomers that serve as diluents at room temperature but crosslink or polymerize upon heating.

Rilsan *n* Nylon 11. Manufactured by Acquitaine-Organico, France.

RIM Abbreviation for ▶ Reaction Injection Molding.

RIM Molders *n* Molders for reaction injection molding (RIM). These types of molders are lightly constructed machines and consists of preformer, clamp, metering unit, and mixhead. The preformer types include thermoforming, fiber-directed, knitting, braiding, or a combination of two or more (Mark HF, Kroschwitz JI (eds) Encyclopedia of polymer science and technology: resins, plastics, rubbers and fibers. Wiley-Interscience, New York).

Ring *n* (1) A narrow band around hosiery appearing different from the rest of the hose. Principal causes: variations in yarn size, dye, absorption, or luster. (2) The device that carries the traveler up and down the package in ring spinning. Also see ▶ Ring Spinning and ▶ Revolving Spinning Ring.

Ring *n* [ME, fr. OE *hring*; akin to OHGr *hring* ring] (before 12c) A polymeric structure resulting from the reaction of one end of a molecule with the other end, forming a ring structure that may be likened to a snake biting its own tail. The stability of ring molecules formed from carbon–carbon chains is generally greatest in five- to six-membered rings, and least in 9- to 11-membered rings. The probability of ring formation decreases rapidly as the length of the molecule increases. Thus the presence of a few small rings in the polymer is usually insignificant.

Ring and Ball Method See ▶ Ball and Ring Method.

Ringer *n* On a section beam, ringer is a term used for one or more filaments that have left the parent end; as the beam revolves, the filaments continue to unwind, wrapping around the beam (hence the word "ringer"). The severity of a ringer is dependent upon the number of filaments contained therein at the time the filaments break. In slashing, the term ringer is often used when an end breaks on the slasher can, adheres to the can, and continues to wrap around it. This condition should not be confused with ringers on the section beam ((2000) Complete textile glossary. Celanese Corporation, Three Park Avenue, New York).

Ring Gate *n* A gate for molding tubular objects in which the top of the cavity is encircled by a thick runner connected to the cavity through a thin web (the ring gate) along its entire circumference. This promotes filling the mold without objectionable weld lines.

Ring-Opening Polymerization *n* Ring-opening polymerization, or olefin metathesis reaction, proceeds by the facile cleavage and reformation of the double bond.

Ring-Opening Reactivity *n* A polymerization in which cyclic monomer is converted into a polymer which does not contain rings. The monomer rings are opened up and stretched out in the polymer chain.

Ring-Spinning *n* A system of spinning using a ring-and-traveler takeup wherein the drafting of the roving and twisting and winding of the yarn onto the bobbin proceed simultaneously and continuously. Ring frames are suitable for spinning all counts up to 150s, and they usually give a stronger yarn and are more productive than mule spinning frames. The latest innovation in ring spinning involves the use of a revolving ring (Also see ▶ Revolving Spinning Ring) to increase productivity. Ring spinning equipment is also widely used to take-up manufactured filament yarns and insert producer-twist at extrusion.

Ring Spinning Frame See ▶ Spinning Frame.

Ring Strain *n* The extra strain associated with the cyclic structure of a compound, as compared with a similar acyclic compound.

Ring Tensil Test See ▶ Split-Disk Method.

Rinmann's Green See ▶ Cobalt Green.

Ripening *n* Hydrolysis of cellulose acetate after acetylation to obtain the desired acetyl value. This is generally accomplished by heat and agitation of the acid cellulose acetate solution under controlled conditions of time, temperature, and acidity. Rapid ripening is accomplished by using increased temperature for the reaction.

Rip Out See ▶ Pick-Out Mark.

Ripped Selvage See ▶ Cut Selvage.

Ripple Finish *n* An intentional, uniformly wrinkled finish which is obtained usually by baking. Syn: ▶ Crinkle Finish, ▶ Wrinkle Finish, and ▶ Suede Finish.

Riser *n* In textile fabric designing, a colored or darkened square on the design paper which indicates that the warp end is over the filling pick at that point. The opposite of riser is sinker.

Riveling, Rivellilng See ▶ Wrinkling.

RMA *n* Abbreviation for ▶ Relaxation-Map Analysis.

Rocker *n* A colloquial term for a blown container that is defective by reason of a bulged or deformed bottom causing the container to rock when placed upright on a flat surface.

Rocker Hardness Tester *n* An instrument made up of two circular stainless runners connected through traverse bars spaced 25 mm apart. Rocking frequency is calibrated by weights on a glass standard. It works on the principle that a "soft" surface dampens the oscillations of a rocking wheel more than a "hard" one. Reference ASTM D 2134.

Rock Salt Moss See ▶ Carrageen.

Rockwell A See ▶ Rockwell Hardness.

Rockwell E See ▶ Rockwell Hardness.

Rockwell Hardness *n* The Rockwell hardness test method consists of indenting the test material with a diamond cone or hardened steel ball indenter. The indenter is forced into the test material under a preliminary minor load $F0$, usually 10 kgf. When equilibrium has been reached, an indicating device, which follows the movements of the indenter and so responds to changes in depth of penetration of the indenter is set to a datum position. While the preliminary minor load is still applied an additional major load is applied with resulting increase in penetration. When equilibrium has again been reach, the additional major load is removed but the preliminary minor load is still maintained. Removal of the additional major load allows a partial recovery, so reducing the depth of penetration. The permanent increase in depth of penetration, resulting from the application and removal of the additional major load is used to calculate the Rockwell hardness number.

$$HR = E - e$$

The deprecated unit kilogram–force (kgf) or kilopond (kp) is defined as the force exerted by 1 kg of mass in standard Earth gravity. Although the gravitational pull of the Earth varies as a function of position on earth, it is here defined as exactly 9.80665 m/s. So 1 kgf is by definition equal to 9.80665 newtons. The kilogram–force has never been a part of the International System of Units (SI), which was introduced in 1960. The SI unit of force is the newton. Typical scales for calculating Rockwell Hardness are listed in the following table.

Rockwell Hardness Scales

Scale	Indenter	Minor load F0 kgf	Major load F1 kgf	Total load F kgf	Value of E
A	Diamond cone	10	50	60	100
B	1/16″ steel ball	10	90	100	130
C	Diamond cone	10	140	150	100
D	Diamond cone	10	90	100	100
E	1/8″ steel ball	10	90	100	130
F	1/16″ steel ball	10	50	60	130
G	1/16″ steel ball	10	140	150	130
H	1/8″ steel ball	10	50	60	130
K	1/8″ steel ball	10	140	150	130
L	1/4″ steel ball	10	50	60	130
M	1/4″ steel ball	10	90	100	130
P	1/4″ steel ball	10	140	150	130
R	1/2″ steel ball	10	50	60	130
S	1/2″ steel ball	10	90	100	130
V	1/2″ steel ball	10	140	150	130

$F0$ = preliminary minor load in kgf
$F1$ = additional major load in kgf
F = total load in kgf
e = permanent increase in depth of penetration due to major load F1 measured in units of 0.002 mm
E = a constant depending on form of indenter: 100 units for diamond indenter, 130 units for steel ball indenter
D = diameter of steel ball
HR = Rockwell hardness number
(Strong AB (2000) Plastics materials and processing. Prentice-Hall, Columbus, OH; ww.gordonengland.co.uk/hardness/rockwell.ht).

Rockwell K See ▶ Rockwell Hardness.
Rockwell M See ▶ Rockwell Hardness.
Rockwell R See ▶ Rockwell Hardness.
Rococo \rə-ˈkō-(ˌ)kō, rō-kə-ˈkō\ *n* (1840) Extremely ornate eighteenth century style featuring scrolls, shells, and fanciful forms. This term is often used in connection with furniture and fashionable fabric materials.
Rodent Resistance *n* The ability of a plastic to withstand or repel attacks by rodents. Some plastics require additives to prevent rodents from chewing objects such as cable insulation. One such additive is Tetramethyl-Thiuram Disulfide.
Rod Mill *n* A grinding machine consisting of a cylindrical, horizontal shell containing a number of free steel rods and rotated about its axis at a speed such that each rod is lifted almost to the top of the cylinder before it falls to impact the charge of material being ground. Compare ▶ Ball Mill. **Roentgen** \ˈrent-gən, ˈrənt-, -jən, -shən\ *adj* [ISV, fr. Wilhelm *Röntgen*] (1896) (R) The former international unit, now being phased out, of quantity or dose of X-rays or gamma rays, a measure of the ionization induced by these rays, equal to 2.58×10^{-4} coulomb per kilogram (C/kg).
Roll Bending *n* In calendering of sheet, the practice of applying a bending moment to the ends of the calender rolls that opposes the bending caused by the pressure forces as the plastic is squeezed between the rolls. The object is to produce sheet whose thickness varies minimally across its width, See also ▶ Crown and ▶ Roll Crossing.
Roll Crossing *n* In calendering, a method of compensating for the slight, but significant, bending apart of the rolls that occur under working pressure, in which the roll axes are set slightly askew of each other. The effect is to make the thickness of the calendered sheet more nearly uniform across the sheet's width.
Rolled Ends *n* (1) On a section beam, rolled ends are adjacent ends that do not unwind parallel to each other. Rolled ends can be caused by such factors as uneven tension, ridgy beams, and static. (2) The ends can also roll behind the hook reed in slashing and can tangle with each other, resulting in broken ends and ends doubling.
Rolled Selvage *n* A curled selvage.
Roller *n* (13c) Cylinder covered with lamb's wood, felt, foamed plastics or other materials, used for applying paint. See ▶ Paint Roller.
Roller Application See ▶ Roller Coating.
Roller Card Generally, any type of card in which rollers do the carding. Usually this refers to a woolen card with a main cylinder and four to seven stripper rolls and worker rolls working in pairs.

Roller Cleaner See ▶ Brush or ▶ Roller Cleaner.
Roller Coating *n* Process by which a film is applied mechanically to sheet metal. The machine consists of a series of horizontal cylindrical rollers. The coating is picked up by one of the first rollers rotating in a trough containing the coating, is transferred to subsequent rollers until evenly distributed over the last roller, and thence to the flat surface of mild steel, tinned iron, or other suitable sheets. Application of paint by means of a hand-operated roller to wall surfaces, etc. The process of coating substrates with liquid resins, solutions, or dispersions by contacting the substrate with a roller on which the fluid material has been spread. The process is often used to apply a contrasting color or raised lettering or markings. See also ▶ Gravure Coating and ▶ Reverse-Roll Coating ((1976) Printing ink handbook. National Association of Printing Ink Manufacturers).
Roller Coating Enamel *n* Special type of enamel, designed for application by a roller coating machine. See ▶ Roller Coating.
Roller Mill *n* Mill involving one or more rolls; three and five rolls are commonly used. The rolls, which revolve at different speeds, are adjustable as to clearance between the milling surfaces, making it possible to obtain various fineness of pigment dispersion.
Roller Printing See ▶ Printing.
Rollformer In the coil coating industry, a machine that bends continuous strip into various shapes by a series of revolving metal wheels or discs.
Roll Forming See ▶ Cold Forming.
Roll Goods *n* Fabric rolled up on a core after it has been produced. It is described in terms of weight and width of the roll and length of the material on the roll.
Rolling Up *n* The application of ink to a plate by means of a roller.
Roll Lapping *n* A condition in which groups of fibers attach themselves to the drafting rolls instead of following the normal path through the drafting system. These fibers cause the trailing fibers to wind around the rolls and to bread the end down completely. Cleaning of the rolls is required to remove the accumulated fiber.
Roll-Leaf Stamping See ▶ Hot Stamping.
Roll Mill *n* An apparatus for mixing a plastic material with compounding ingredients, comprising two horizontal rolls placed close together. The rolls turn at speeds that differ by about 10% to produce a shearing action in the material being compounded. Mixing plows and slitting knives are sometimes used to work the stock across the rolls, thus improving the uniformity of additive distribution in the compound. See also ▶ Refiner.

Roll of Wallcovering *n* A bolt, consisting of 36 square feet of wallcovering, of which 30 square feet are estimated as usable. Bolts come in single, double, and triple rolls.

Roll-Out Ink spread for test or sampling purposes by use of a hand roller.

ROM \ ▌räm\ *n* (1966) Acronymic abbreviation for Read-Only Memory, the permanent part of computer memory not alterable by the user ((2002) Microsoft computer dictionary. Microsoft Press; Merriam-Webster's Collegiate Dictionary (2004), 11th edn. Merriam-Webster, Springfield, MA).

Roof \ ▌rüf\ *n* [ME, fr. OE *hrōf*, akin to ON *hrōf* roof of a boathouse and perhaps to Old Church Slavonic *stropŭ* roof] (before 12c) The cover of a building, including the roofing and all other material and construction (such as supporting members) necessary to carry and maintain it on the walls or uprights.

Roof Cladding See ▶ Roofing.

Roofing *n* (1) Any material used as a roof covering (such as shingles, slate, sheet metal, or tile) to make it wind- and waterproof, and often provide thermal insulation. (2) A roof. Also called ▶ Roof Cladding.

Room Temperature *n* A temperature in the range of 20–30°C (68–85°F).

Room Temperature Setting/Adhesive Setting *n* See ▶ Adhesive. A resin obtained as a residue in the distillation of crude turpentine from the sap of the pine tree (gum rosin) or from an extract of the stumps and other parts of the tree (wood rosin).

Root Diameter *n* The smaller (*minor*) diameter of an extruder screw at any point along its length. In most industrial extruder screws, the increase in root diameter from the feed section to the metering section is the most important factor creating the ▶ Compression Ratio.

R.O.P. Colors *n* An abbreviation for RUN OF PRESS COLORS; a term used to describe the colored inks generally accepted as "standard" colors for newspaper printing.

Rope \ ▌rōp\ *n* [ME, fr. OE *rāp*; akin to OHGr *reif* hoop] (before 12c) A heavy, strong cord, made from either natural or manufactured fibers or from wire, in a wide range of diameters. Yarns are twisted together to form strands. These strands are then twisted together in the opposite direction to form the rope. The fact that the twist directions alternate at different stages of rope assembly assures that the rope will be twist-stable and will not kink during use, also called "cord." A fabric in a process without weft tension, thus having the appearance of a thick rope (Vincenti R (ed) (1994) Elsevier's textile dictionary. Elsevier Science and Technology Books, New York).

Rope Mark *n* A fabric defect consisting of long, irregular, longitudinal markings on dyed or finished goods. A principal cause is abrasion while wet processing the fabric in rope form. Rope marks are often related to overloading of the fabric during wet processing (Vincenti R (ed) (1994) Elsevier's textile dictionary. Elsevier Science and Technology Books, New York).

Ropiness See ▶ Ropy Finish.

Ropy Finish *n* Finish in which the brush marks have not flowed out, this being the normal appearance of a paint or varnish having poor leveling properties. A similar appearance may also be produced in a paint, which normally has good leveling properties, by continuing to brush the paint after the film has begun to set.

Rosin \ ▌rä-z°n, ▌rȯ-, *dial* ▌rȯ-zəm\ *n* [ME, mod. of MF *resine* resin] (13c) An important "natural" resin, obtained from pine trees of several species by collection of exudates from cuts through the bark or by extraction from stumps, referred to as wood rosin (Merriam-Webster's Collegiate Dictionary (2004), 11th edn. Merriam-Webster, Springfield, MA). Rosin is brittle and friable, and is used mostly for sizing paper. Modified rosins are used in printing inks, pressure-sensitive adhesives, and chewing gum. Two general kinds of rosin are commercially available: gum rosin obtained from living trees, and wood rosin obtained from dead wood, such as stumps and knots. Tall oil rosin, a by-product of the paper industry, is a chemically similar material, also known a colophonium and colophony (Flick EW (1991) Industrial synthetic resins handbook. Williams Andrew Publishing/Noyes, New York).

Rosin Acids *n* Predominately monocarboxylic acids with the empirical formula $C_{19}H_{29}COOH$. They are classified into two groups: the abietic type and the pimaric type. Both types and their derivatives are found in wood, gum, and tall oil rosins.

Rosin-Bonded Laminate See Laminate and Reinforced Plastic.

Rosin Essence *n* That portion of rosin distilled below 360°C.

Rosin Esters A large family of resins produced from rosin aids and polyols. See ▶ Ester Gum.

Rosin, Limed or Zinc-Treated *n* Rosin which has been chemically reacted with a calcium or zinc compound, or both, to form a product containing a metallic salt of the rosin acids, along with unreacted rosin. The amount of calcium or zinc contained in the finished products of this category may vary over a wide range. See ▶ Resinates, Metallic.

Rosin, Modified *n* Rosin that has been treated with heat or catalysts, or both, with or without added chemical substances, so as to cause substantial change in the structure of the rosin acids, as isomerization, hydrogenation, dehydrogenation, or polymerization, without substantial effect on the carboxyl group.

Rosin Oil *n* Relatively viscous oily portion of the condensate obtained when rosin is subjected to dry destructive distillation: also used to describe specially compounded oils having a rosin oil base. See ▶ Bloom Oils.

Rosin Pitch *n* Dark, acidic residue obtained from the destructive distillation of rosin.

Rosin, Reclaimed *n* Rosin that has been recovered or reclaimed by any means from waste or deteriorated material, provided that the concentration of rosin acids is not below that normal for rosin, and any residual or contaminating component from the waste material itself or from any article used in the recovery process is not in sufficient quantity to cause the physical or chemical properties of the reclaimed product to differ materially from those of rosin.

Rosin Size *n* Soda rosin soap containing varying amounts of free rosin; used for sizing paper.

Rosin Spirit *n* Relatively light, volatile portion of the condensate obtained in the first stages when rosin is subjected to dry destructive distillation. Syn: ▶ Pinolin.

Rosin Standards *n* The Color Standards for the US Naval Stores Rosin Scale consists of color standards specified in ASTM D509. The US Naval Stores Rosin scale are in the form of colored glasses mounted in metal cubes for direct comparison with the rosin samples. Rosin darker in color than the standard for grade D or FF is graded B. The designation Opaque with the grade letters OP is used to describe rosin which, because of a turbid, cloudy, or nontransparent condition due to occluded moisture, excessive crystallization, or presence of foreign matter other than dirt cannot be accurately graded by comparison with any of the rosin grade standards. The recognized official standards are those developed and issued by the U.S. Department of Agriculture, or similar standards made of Lovibond glass (Lovibond, Salisbury, England, www.tintometer.com). Lovibond PFX195 Automatic Colorimeters are designed to automatically measure color tint. The most widely used method for assessing the color quality of rosins consists of includes 15 color standards varying in color from yellow to reddish orange, each assigned letters - XC (lightest), XB, XA, X, WW, WG, N, M, K, I, H, G, F, E, D (darkest). FF is a special additional grade used for dark wood resins.

Rosin Test (Qualitative) *n* See Halphen-Hicks Test and Lieber-Man-Storch Method.

Rosin Type (Sample) *n* Sample of rosin, or a mold of thermosetting plastic material, used as an unofficial standard in grading rosin. Such sample shall be so selected, sized and surface-finished that it will have the form of an approximate 7/8 in. cube with at least two opposite faces having smooth parallel surfaces and shall have a color when viewed through these faces which matches within rather narrow tolerances the color the corresponding official government standard made of glass.

Rossi-Peakes Tester *n* An instrument for measuring the temperature at which a given amount of a molding powder will flow through a standard orifice in a prescribed period of time and pressure (ASTM Test D 569). The instrument tests a peculiar combination of properties of a thermoplastic molding compound that may relate to its moldability.

Rot \rät\ *v* [ME *roten*, fr. OE *rotian*; akin to OHGr *rōzzēn* to rot] (before 12c) Decomposition in wood by fungi and other microorganisms; reduces its strength, density, and hardness.

Rotameter \rō-tə-ˌmē-tər, ro-ˈta-mə-tər\ *n* [L *rota* + English –*meter*] (1907) A simple flow-measuring device for gases and low-viscosity liquids consisting of a vertical glass tube whose internal diameter increases gradually from bottom to top, graduated on the outside, and consisting a float that reaches an equilibrium position that rises in proportion to the flow rate. By calculation or calibration, the flow rate can be derived from the float position and the pressure and temperature of the fluid at the discharge end. Floats of small rotameters may be simple spheres interchangeable with others of different density to expand the instrument's range.

Rotary Joint *n* A plumbing fixture used with cored extruder screws and chilled rolls that permits circulation of liquid within the screw or roll, without leakage at the point of connection of the inlet and return lines.

Rotary Molding *n* A term sometimes used to denote a type of injection, transfer, compression, or blow molding utilizing several mold cavities mounted on a rotating table or dial. Not to be confused with ▶ Rotational Molding.

Rotary Press *n* High-speed press which passes continuously running webs of paper between an impression cylinder and the curved form on the plate cylinder.

Rotary Screen Printing See ▶ Printing.

Rotary-Vane Feeder *n* A device for conveying and metering dry materials, comprising a cylindrical housing containing a concentric shaft with blades or flutes attached, rotating at a rate selected to feed the material at a desired rate.

Rotating Die See ▶ Oscillating Die.

Rotating Spreader *n* A type of torpedo used in plunger-type injection machines that consisted of a finned torpedo rotated by a shaft extending through the cored injection ram behind it. This was soon superseded by the screw-injection machine.

Rotational Casting *n* The process of forming hollow articles from liquid materials by rotating a mold containing a charge of the material about one or more axes at relatively low speeds until the charge is distributed on the inner walls of the mold by gravity and hardened by heating, cooling, or curing. Rotation about one axis is suitable for cylindrical objects. Rotation about two axes and/or rocking motions are employed for completely closed articles. The process of rotational casting of plastisols comprises placing a measured charge of plastisol in the bottom half of an opened mold, closing the mold, rotating the mold about one or more axes while applying heat until the charge has been distributed and fused against the mold walls, cooling the mold until the deposit has gained strength, opening the mold, and removing the article. See also ▶ Centrifugal Casting and ▶ Rotational Molding.

Rotational Injection Molding *n* A modified injection–molding process applicable to hollow, axisymmetrical articles such as cups and beakers, in which the male half of the mold is rotated during the molding cycle until the material has hardened to a prechosen degree. The rotation produces orientation and increased crystallinity of some polymers, resulting in improved toughness and stress–craze resistance.

Rotational Molding *n* (rotomolding) The preferred term for a variation of ▶ Rotational Casting utilizing dry, finely divided, sinterable powders, such as polyethylene, rather than liquid slurries. The powders are first distributed, then sintered against the heated mold walls, the mold is cooled, and the product stripped from the mold.

Rotational Rheometer *n* An instrument for measuring the viscosity of molten polymers (any many other fluid types) in which the sample is held at a controlled temperature between a stator and a rotor. From the torque on either element and the relative rotational speed, the viscosity can be inferred. The most satisfactory type for melts is the *cone-and-plate* geometry, in which the vertex of the cone almost touches the plate and the specimen is situated between the two elements. This provides a uniform shear rate throughout the specimen. It may be operated in steady rotation or in an oscillatory mode.

Rotational Viscometer *n* An instrument for measuring the viscosity of pourable liquids, slurries, plastisols, and solutions. Most are of the *bob-and-cup* type. In these, the bob is a polished, accurate cylinder that is immersed in the liquid contained in the cup. Either the bob or cup is rotated and the torque on one or the other is measured, as is the rotational velocity. From these and the dimensions, the viscosity can be inferred, either directly by calculation from principles or indirectly by calibration with standards of known viscosity. An instrument widely used in the plastics industry is the ▶ Brookfield Viscometer.

Rotatory Power *n* The power of rotating the plane of polarized light, given in general by θ/l where θ is the total rotation which occurs in a distance l. The *molecular* or *atomic rotary power* is the product of the specific rotatory power by the molecular or atomic weight. Magnetic rotatory power is given by

$$\theta/e\,H\cos\alpha$$

where H the intensity of the magnetic field, and α the angle between the field and the direction of the light.

Roto-Dipping *n* Patented method of pretreating and priming motor bodies in which the body is mounted on a longitudinal axle and is caused to pass through a series of spray zones and dip tanks. At appropriate stages in the process the body is revolved on the axle.

Rotoflex *n* A fatigue or endurance test developed by Goodyear for industrial yarns or cords.

Rotogravure \ rō-tə-grə-ˈvyür\ *n* [G *Rotogravur*, blend of L *rota* wheel and Gr *Photogravur* photogravure] (1913) See ▶ Gravure.

Rotomolding *n* A contraction of the term *rotational molding*, sometimes used indiscriminately for both the processes of ▶ Rotational Casting and ▶ Rotational Molding.

Rotor Spinning See Open-End Spinning.

Rot Resistance *n* The ability of textile materials to resist physical deterioration resulting from the action of bacteria and other destructive agents such as sunlight or sea water.

Rottenstone \ ˈrä-tᵊn-ˌstōn\ *n* (1677) Brown, amorphous, siliceous limestone, similar in nature to pumice stone, but softer in texture. Principal uses are as an abrasive and filter medium. *Also known as Tripoli and Terra Cariosa.*

Rouge \ ˈrüzh, *esp Southern* ˈrüj\ *n* [F, fr. MF, fr. *rouge* red, fr. L *rubeus* reddish] (1753) See ▶ Iron Oxides, Synthetic.

Rough *n* A fabric condition in which the surface resembles sandpaper. Principal causes are the shuttle rebounding in the box, jerky, or loose shuttle tension, an incorrectly timed harness, and wild twist in the filling.

Roughing Pump *n* In high-vacuum work, e.g., vacuum metallizing, a mechanical pump that removes most of the air from the chamber, leaving the remainder to a secondary pump, usually a diffusion pump, capable of reducing the pressure to about 0.13 Pa absolute.

Roughness See ▶ Finish.

Round (Round Color) *n* Description of a full-bodied paint, i.e., one which gives good build.

Rovicella *n* Cellulose (viscose), manufactured by Feldmühle Rorsch., Switzerland.

Roving *n* [⁴*rove*] (1802) (1) A form of fibrous glass comprising from 8 to 120 single filaments or strands gathered together in a bundle. When the strands are twisted together, the term *spun roving* is used. Roving is used in continuous lengths for filament winding, chopped into short lengths for use in reinforced-plastic molding compounds and in sprayup, and woven into skeins or mats for use in laminates. See ▶ Reinforcement. (2) In spun yarn production, an intermediate state between sliver and yarn. Roving is a condensed sliver that has been drafted, twisted, doubled, and redoubled. The product of the first roving operation is sometimes called slubbing. (3) The operation of producing roving (see 0). (4) In the manufacture of composites, continuous strands of parallel filaments.

Roving Frame *n* A general name for all the machines used to produce roving, different types of which are called slubber, intermediate, fine, and jack. Roving frames draft the stock by means of drafting rolls, twist it by means of a flyer, and wind it into a bobbin.

Rows *n* In pile floor covering, the average number of tufts or loops per inch in the warpwise direction.

Royalene *n* Poly(ethylene) or poly(propylene), manufactured by U.S. Rubber Company, U.S.

RP *n* Abbreviation for ▶ Reinforced Plastic.

rpm *n* Abbreviation for Revolutions Per Minute, a convenient industrial unit of rotational speed. One rpm equals 0.10472 radian per second.

RRIM *n* Abbreviation for Reinforced Reaction Injection Molding. See ▶ Reaction Injection Molding.

RTM Abbreviation for Resin-Transfer Molding.

RTP *n* Abbreviation for Reinforced Thermoplastic.

RT 700 *n* Cellulose (viscose). Manufactured by Glanzstoff, Germany.

RTV *n* Abbreviation for Room-Temperature Vulcanizing, a characteristic of some elastomers that do not require heating to cure.

RTV Eastomers *n* Room-temperature vulcanizing (RTV) elastomer is an elastomer that does not require heating to cure.

Ru *n* Measure for surface roughness. 0.025 μm (10^{-6} in.) BSS 1134.

Rub See Abrasion Mark.

Rubazote *n* Natural rubber. Manufactured by Expanded Rubber, Great Britain.

Rubber *n* \ˈrə-bər\ *n* (1536) (1) A highly resilient material, capable of recovering from large deformations quickly; manufactured from the juice of rubber trees as well as other trees and plants. (2) Any of various synthetically produced materials having similar properties; an elastomer.

Rubber Cement *n* (1886) See ▶ Cement, Rubber.

Rubber-Emulsion Paint See ▶ Latex Paint.

Rubber Filament A filament extruded from natural or synthetic rubber and used as the core of some elastic threads.

Rubber Hydrochloride *n* A nonflammable thermoplastic material containing about one third chlorine, obtained by treating a solution of rubber with anhydrous hydrogen chloride (HCI) under pressure at low temperature. The packaging film Pliofilm® is a representative product.

Rubber Latex *n* Colloidal aqueous emulsion of an elastomer.

Rubber Materials See ▶ Elastomers.

Rubber, Natural *n* (India rubber, Caoutchouc) An amorphous polymer consisting essentially of *cis*-1,4-polyisoprene, obtained from the sap (latex) of certain trees and plants, mainly the *Hevea brasiliensis* tree. The material is shipped from tropical plantations in one of two primary forms: latex, usually stabilized and preserved with ammonia and centrifuged to remove part of the water; or sheets made by milling the coagulum from the latex. Natural rubber has very high molecular weight and is usually masticated to reduce the molecular weight and improve processability. A major use is sidewalls of automotive tires.

Rubber-Plate Printing *n* A marking method sometimes employed for intricate parts such as molded terminal blocks. Numbers, instructions, etc., are stamped with conventional rubber stamps or printing plates.

Rubber-Plunger Molding *n* A variation of matched-die molding that employs a deformable rubber plunger and a heated metal concave mold. The process enables the use of high fiber loadings.

Rubber-Seed Oil *n* Oil obtained from the seeds of *Hevea brasiliensis*. Sp gr, 0.927/15°C; saponification value, 190, iodine value, 138.

Rubber, Synthetic See ▶ Elastomer.

Rubber, Synthetic Natural An awkward term sometimes used for synthetic elastomers having the composition of natural rubber, *cis*-1,4-polyisoprene.

Rubber Toughening *n* The practice of compounding into a brittle plastic 5–20% of a rubber in the form of spherical particles, in order to improve the plastic's resistance to impact. The process has been used with both thermoplastics and thermosets, e.g., polystyrene and epoxies. Some users of this term include toughening achieved by copolymerization with elastomer-forming monomers.

Rubber Transition *n* (rubbery transition, gamma transition) See ▶ Glass Transition.

Rubbing *n* Process of leveling and flatting a dried coating film by rubbing it, either wet or dry and usually with a suitable abrasive, to remove nibs and other irregularities and to provide a surface with a suitable key for a subsequent coating. *Also called Flatting Down.* See ▶ Sanding.

Rubbing Flat *n* Process of flatting a coating means to subject it to abrasive action of remove nibs and other irregularities and to provide a surface with a suitable key for a subsequent coating.

Rubbing Oil *n* Neutral, medium–heavy mineral oil used as a lubricant for pumice stone in rubbing varnish or lacquer.

Rubbing Test *n* A procedure for testing the ease with which a lacquer or enamel can be rubbed to a smooth high-gloss finish. Federal Standard 141a Method 6332.

Rubbing Varnish *n* See ▶ Polishing Varnish and ▶ Flatting (Rubbing) Varnish.

Rubine *n* Pertaining to blue shade red, probably derived from the Latin word "rubia" for the madder plant from which madder lake (Alizarin Red), a blue shade red, was originally made. The name of the bluish red colored gem, the ruby, would appear to be similarly derived.

Rubine Reds *n* See Permanent Red 2B.

Rub-Of *n* Ink on printed sheets which, after sufficient drying time, smears or comes off on the fingers when handled.

Rubometer (or Rub Tester) *n* An instrument used for the measurement of rub or scuff resistance of a printed design.

Rule of Mixtures See ▶ Law of Mixtures.

Rule 66 *n* The original air quality control regulation enacted by Los Angeles County in California for the purpose of eliminating smog caused by the photochemical degradation of some organic solvents. The Federal Environmental Protection Agency (EPA) has used Rule 66 as a basis for minimum Federal guidelines on which the various states could pattern their mandated local regulations. See ▶ Photochemically-Reactive Solvent.

Rumbling See ▶ Tumbling.

Run *n* (1) In experimental or test work, performance of the experiment or test at one set of experimental factors. (2) Resin flow down the vertical or sloping surface of a sprayed-up or laid-up reinforced-plastic fabrication before it can be cured (usually unwanted). (3) Narrow downward movement of a paint or varnish film; may be caused by the collection of excess quantities of paint at irregularities in the surface, e.g., cracks, holds, etc., the excess material continuing to flow after the surrounding surface has set. Sometimes called ▶ Tear. (4) See ▶ Running.

Runner *n* In an injection or transfer mold, the feed channel, often branched to serve multiple cavities, and usually of semicircular or trapezoidal cross-section, that connects the sprue with the cavity gates. The term is also used for the plastic piece formed in this channel.

Runner Length *n* In knitting, the number of inches of yarn from a warp to make one rack of fabric.

Runnerless Injection Molding *n* (1) In molding thermoplastics, a process in which the runners are insulated from the cavities and kept hot, so that the molded parts are ejected with only small gates attached. See also ▶ Hot-Runner Mold. (2) In thermoset molding, the same as ▶ Cold-Runner Injection Molding.

Runner System *n* This term is sometimes used for the entire moldfeeding system, including sprue, runners, and gates, in injection or transfer molding.

Running *v* Process by which solubility of a resin in an oil is achieved. Fossil resins such as Congo are "run" by heating the resin alone to elevated temperatures in incremental steps, thereby distilling off unwanted volatile components and altering the resin so that it becomes compatible with additions of oil. Resin so treated are said to have been "run."

Run-of-the-Mill *n* See ▶ Mill Run.

Run-Proof A knitted construction in which the loops are locked to prevent runs.

Run-Resistant *n* A type of knitting stitch that reduces runs.

Rupture \ˈrəp(t)-shər\ *n* [ME *ruptur*, fr. MF or L; MF *rupture*, fr. L *ruptura* fracture, fr. *ruptus*, pp of *rumpere* to break] (15c) (fracture) Failure by cracking or tearing of, or in, a test specimen or working part.

Rupture Disk *n* A thin metal disk, contained in a *rupture-disk fitting*, that, when subjected to a known high fluid pressure, will tear, permitting outflow of fluid and rapid relief of pressure. Rupture disks are now used routinely on extruders and gear pumps to protect both personnel and equipment against the dangers of excessive melt pressure.

Rust \ˈrəst\ *n* [ME, fr. OE *rūst*; akin to OE *rēad* red] (before 12c) The reddish, brittle coating formed on iron or its alloys resulting from exposure to humid atmosphere or chemical attach. NOTE — Not to be confused with white rust (Baboian R (2002) Corrosion engineer's handbook, 3rd edn. NACE International – The Corrosion Society, Houston, TX; Uhlig HH (1948, 1971, 2000) Corrosion and corrosion control. Wiley, New York).

Rust Grade Scale *n* In evaluating the resistance to rusting, the linear, numerical rust grade scale is an exponential function of the area of rust so that slight amounts of first rusting have the greatest effect on lowering the rust grade. There is also a European rust scale and an ISO rust scale (*ISO 4624*).

Rustic Siding See ▶ Drop Siding.

Rusting *v* (13c) Corrosion on the surface of iron or ferrous metals resulting in the formation of products consisting largely of hydrous ferric oxide.

Rust-Inhibiting Paint See Anti-Corrosion Paint (Hare CH (2001) Paint film degradation – mechanisms and control. Steel Structures Paint Council, Pittsburg, Pennsylvania; Baboian R (2002) Corrosion engineer's handbook, 3rd edn. NACE International – The Corrosion Society, Houston, TX; Weismantal GF (1981) Paint handbook. McGraw-Hill, New York; Keane JD (1970) Surface preparation: new trends in anti-corrosion coatings. International Congress, Milan, Italy).

Rust-Inhibitive Washes See ▶ Chemical Conversion Coating.

Rust Resistance *n* The ability of a coating to protect the substrate of iron or its alloys from rusting. See ▶ Rust.

Rutile \ˈrü-ˌtēl\ *n* [Gr *Rutil*, fr. L *rutilus*] (1803) One of the crystalline forms of ▶ Titanium Dioxide. Used to designate a crystal structure of titanium dioxide with a sp gr of 4.2 and a refractive index of 2.69. The rutile type is characterized by higher opacity, greater density, and greater inertness as compared to the anatase type ((2002) McGraw-Hill dictionary of geology and mineralogy. McGraw-Hill, New York; Hurlbut CS, Jr. (1966) Mineralogy. Wiley, New York).

Ruvea *n* Nylon 66 (artificial straw). Manufactured by DuPont, U.S.

R-Value See ▶ Thermal Resistance.

Rydberg Formula *n* A formula, similar to that of Balmer, for expressing the wave-numbers (v) of the lines in a spectral series:

$$v = R\frac{1}{(n+a)^2} - \frac{1}{(m+b)^2}$$

where n and m are integers and $m > n$, a, and b are constants for a particular series, and R is the *Rydberg constant*, 109667.8 cm^{-1} for hydrogen.

Ryton *n* Poly(thio-1,4-phenylene), manufactured by Phillips Petroleum, U.S.

S

s *n* \ˈes\ (1) Abbreviation for SI's basic time unit, the second. (2) Symbol for sample Standard Deviation.

S *n* (1) Abbreviation for the SI unit of conductance, the Siemens. (2) Chemical symbol for the element sulfur. (3) Symbol for Entropy.

Sa (0.1,2,−1/2,3) Grades of cleanliness of blasted steel, Swedish Standard SIS 05 50 00, ASTM D-2200.

σ *n* Symbol for Standard Deviation (2).

Sable Pencil \ˌsā-bəl ˈpen(t)-səl\ *n* Fine pointed brushes of sable hair, set in quills, varying in size from lark's, through crow's, to duck's and goose's, or set in a metal ferrule. Used for decorative detail. Because of the high cost of sable, squirrel hair may be used instead.

Sable Writer \-ˈrī-tər\ *n* A brush like a sable pencil, but longer, with a flat edge or a point, used for sign writing.

SABRA *n* Abbreviation or Surface Activation Beneath Reactive Adhesives, a method of bonding plastics, such as polyolefins and polytetrafluoroethylene, that are normally unreceptive to adhesives without pretreatment. The method consists of mechanical abrasion of the surfaces to be joined to roughen their outer layers scission of bonds with creation of free radicals, and further reaction with primers in the liquid, vapor, or gaseous phase. An adhesive such as an epoxy is then applied.

SACMA *n* Acronymic abbreviation for Suppliers Of Advanced Composite Materials Association.

Sacrificial Protection \ˌsa-krə-ˈfi-shəl prə-ˈtek-shən\ *n* (1) The use of a metallic coating, such as a zinc-rich paint, to protect steel. In the presence of an electrolyte, such as salt water, a galvanic cell is set up and the metallic coating corrodes instead of the steel. (2) Zinc or aluminum anode.

SAE *n* Abbreviation for Society Of Automotive Engineers.

Safety Factor \ˈsāf-tē\ *n* See ▶ Factor of Safety.

Safety Inks Printing inks which will change color or bleed when ink eradicator or water is applied to prints. Also inks which will erase easily. Usually, mixtures of a drying-oil ink with an aqueous solution of a water-soluble color.

Safflower Oil \ˈsa-ˌflaú(-ə)r-\ *n* (ca. 1857) Semidrying to drying oil, obtained from *Carthamus tinctorius*, a native of India. It is now available from seed grown in the United States. Its main constituent acid is linoleic, with a small quantity of linolenic acid. It dries and bodies more slowly than linseed oil. Sp gr, 0.925/15°C; iodine value, 139; saponification value, 189. Its drying characteristics lie between those of linseed and soybean oils. One of its main advantages for paint and varnishes is its extremely low after-yellowing due to its very low linolenic acid content.

Saflex *n* Poly(vinyl acetal), manufactured by Monsanto, US.

Sag \ˈsag\ *v* [ME *saggen*, prob. of Scand origin; akin to Sw *sacka* to sag] (14c) (1) In blow molding, the local reduction in parison diameter, or *necking down*, caused by gravity. It is usually greatest on the portion nearest the die, and increases as the parison grows longer. (2) In thermoforming, sag is the downward bulge in the heat-softened sheet. See ▶ Sagging.

Sagging *n* Downward movement of a paint film between the times of application and setting, resulting in an uneven coating having a thick lower edge. The resulting sag is usually restricted to a local area of a vertical surface and may have the characteristic appearance of a draped curtain, hence the synonymous term, "curtaining."

SAIB Abbreviation for ▶ Sucrose Acetate Isobutyrate.

Sailcloth \ˈsā(ə)l-ˌklóth\ *n* (13c) Any heavy, strongly made woven canvas of cotton, linen, jute, polyester, nylon, aramid, etc., that is used for sails. Laminated fabrics are also finding use in this market. Sailcloth is used for apparel, particularly sportswear.

Salicylic Acid \ˌsa-lə-ˈsi-lik-\ *n* [ISV, fr. *salicyl* (the group HOC_6H_4CO)] (1840) $C_7H_6O_3$. A crystalline phenolic acid used especially in the form of salts and other derivatives as an analgesic and antipyretic.

Salloped Selvage *n* A fabric defect consisting of an abrupt, narrow place along the selvage. Principal cause is the failure of the clip on the tenter frame to engage or hold the fabric.

Sa!t \ˈsólt\ *n* [ME, fr. OE *sealt*; akin to OHGr *salz* salt, Lithuanian *saldus* sweet, L *sal* salt, Gk *hals* salt, sea] (before 12c) (1) In inorganic chemistry, an ionic substance formed by the reaction of an acid with a base. Polybasic acids can form acid salts and, when dissolved in water, generally yield acidic solutions, i.e., their pH is less than 7. Similarly, polyacidic bases can form basic salt whose solutions are generally basic. (2) By analogy with the above, some organic reaction products of

Jan W. Gooch, *Encyclopedic Dictionary of Polymers*, DOI 10.1007/978-1-4419-6247-8,
© Springer Science+Business Media LLC 2011

diacids and diamines that have ionic character are called salts. (Whitten KW, Davis RE, Davis E, Peck LM, Stanley GG (2003) General chemistry. Brookes/Cole, New York)

Salt Spray Test *n* A test applied to metal finishes to determine their anti-corrosive properties, involving the spraying of common salt (sodium chloride) solution on the surface of a coated steel panel.

Sample Mean See ▶ Arithmetic Mean.

SAN *n* Copolymer from styrene and acrylonitrile. Abbreviation for ▶ Styrene Acrylonitrile Polymer.

SAN Copolymer See ▶ Styrene Acrylonitrile Copolymer.

Sand \ █sand\ *n* Used as a filter medium in fiber manufacture, particularly used in spinning packs for nylon or polyester production.

Sandarac \ █san-də- █rak\ *n* [L *sandaraca* red coloring, fr. Gk *sandarakē* realgar, red pigment fr. realgar] (1543) Natural resin which exudes from *Callitris quadrivalvis*, found in North Africa.

Sandblast \- █blast\ *n* (1871) To use sand, flint or similar abrasive propelled by an air blast, or metal, masonry, concrete, etc., to remove dirt, rust, or paint, or to decorate the surface with a rough texture.

Sand-Dry *n* Descriptive of a stage in drying of a paint film at which sand will not adhere to the surface.

Sanderson-Milner Zeta Space Color Difference Equation *n* A color difference equation designed to fit the spacing of the Munsell System and based on Munsell value functions as described in the Adams Chromatic Value Color Difference Equation:

$$\Delta E = [(\Delta \xi_1)^2 + (\Delta \xi_2)^2 + (\Delta \xi_3)^2]^{1/2}$$

where

$$\xi_1 = (V_X - V_Y)(9.37 + 0.79 \cos \theta)$$
$$\xi_2 = kV_Y$$
$$\xi_3 = (V_Y - V_Y)(3.33 + 0.87 \sin \theta)$$

$$\theta = \text{angle calculated from } \tan \theta = \frac{0.4(V_Z - V_Y)}{V_X - V_Y}$$

k = constant ranging from 1 to 5 depending on observation conditions such as fineness of line dividing the two colors being compared. Using a constant k of 2, the average magnitude of ΔE is one-sixth that of an NIST Unit. (Johnson SF (2001) History of light and colour measurement: A science in the shadows. Taylor & Francis, UK; Colour physics for industry, 2nd edn. McDonald, Roderick (eds) Society of Dyers and Colourists, West Yorkshire, England, 1997; Billmeyer FW, Saltzman M (1966) Principles of color technology. Wiley, New York)

Sand Finish *n* Rough finish plaster wall.

Sand Grinder *n* A type of pigment dispersing equipment consisting of pumping pigment/vehicle slurry (the mill base) through a cylindrical bank of sand which is being subjected to intense agitation. During passage upward through the agitated sand zone, the mill base is caught and dispersed between the sand particles; a strong shearing action which effects the dispersion of the pigment into the vehicle results. On emerging from the active sand zone, the dispersed mill base overflows through an exit screen sized to permit a free flow-through of the pigment dispersion while holding back the sand particles.

Sand Grinding *n* Process of dispersing pigments in the sand grinder.

Sanding *n* A finishing process employing abrasive belts or disks, sometimes used on thermosetting-resin parts to remove heavy flash or projections, or to produce radii or bevels that cannot be formed during molding.

Sanding Sealer *n* Specially hard first coat that has the property of sealing or filling, but not obscuring, the grain of wood. The surface is then suitable for sanding before the application of subsequent coats.

Sanding Surfacer *n* A heavily pigmented finishing material used for building the surface to a smooth condition. It is sanded after drying.

Sand Mill *n* An apparatus used for preparing pigment dispersions, consisting of a vertical cylinder with a centrally mounted agitator shaft on which are mounted several flat, annular disk impellers. The mill is charged with coarse natural sand or high-silica ceramic beads as the grinding medium. The pigment slurry is pumped into the bottom of the mill and becomes mixed with the grinding medium. As the mixture is forced upward through the mill, it passes through several zones of agitation and finally flows through a screen at the top that retains the sand or beads but allows the much smaller pigment particles to pass.

Sandpaper \- █pā-pər\ *n* (1825) (1) A tough paper which is coated with an abrasive material such as silica, garnet, silicon carbide, or aluminum oxide; used for smoothing and polishing; graded by a grit numbering system according to which the highest grit numbers (360–600) are used for fine polishing, and the lowest grit numbers (16–40) are used for coarse smoothing. Alternatively, sandpaper may be designated by the "0 grade" system, according to which "very fine" includes grades from 10/0 to 6/0, "fine" from 5/0 to 3/0, "medium", 2/0, 1/0, ½; "coarse", 1, 1½, and 2; "very coarse," 2½, 3, and 3½, and 4. (2) See ▶ Coated Abrasive. The general term of the Coated Abrasives Manufacturer's Institute for materials of this type.

Sandwich Blend \-ˈwich-\ *n* A method of preparing fiber mixtures by layering them horizontally in alternating layers with all elements in the proper proportion.

Sandwich Heating *n* (two-sided heating) A method of heating a plastic sheet for thermoforming in which the sheet is positioned between two radiant heaters.

Sandwich Structure *n* A term often employed for a laminate comprising at least three layers; for example, a cellular-plastic or honeycomb core sandwiched between two layers of glass-reinforced laminate. See also ▶ Laminate.

Sanforized® \ˈsan-fə-ˌrīzd\ {*trademark*} A trademark of Cluett, Peabody & Co., Inc., denoting a controlled standard of shrinkage performance. Fabrics bearing this trademark will not shrink more than 1% because they have been subjected to a method of compressive shrinkage involving feeding the fabric between a stretched blanket and a heated shoe. When the blanket is allowed to retract, the cloth is physically forced to comply.

Sanfor-Set® {*trademark*} A trademark of Cluett, Peabody & Co., Inc., denoting a controlled standard of shrinkage performance originally developed for denims. Fabrics bearing this trademark will not shrink under homewash, tumble-dry conditions because they have been subjected to a liquid ammonia treatment and compressive shrinkage.

SAN Resin See ▶ Styrene Acrylonitrile Copolymer.

Saponification \sə-ˌpä-nə-ˌfī\ *v* [F *saponifier*, fr. L *sapon-, sapo*] (1821) (1) Alkaline hydrolysis of fats whereby a soap is formed; more generally, the hydrolysis of an ester by an alkali with the formation of an alcohol and a salt of the acid portion. (Morrison RT, Boyd RN (1992) Organic chemistry, 6th edn. Prentice Hall, Englewood Cliffs, NJ) (2) The decomposition of the medium of a paint or varnish film by alkali and moisture in a substrate, e.g., new concrete or fresh plaster. Saponified paint may become sticky and discolored. In severe cases, the film may be completely liquefied by saponification. Loss of adhesion may occur as a saponified layer develops next to the substrate. (Wicks ZN, Jones FN, Pappas SP (1999) Organic coatings science and technology, 2nd edn. Wiley-Interscience, New York)

Saponification Value *n* A measure of the alkali reactive groups in oils and fatty acids which is expressed as the number of milligrams of potassium hydroxide that react with 1 g of sample. (See ASTM, www.astm.org) Also known as *Koettstorfer Number*.

Sapwood \-ˌwu̇d\ *n* (1791) The layers of wood next to the bark, usually lighter in color than the heartwood and 1/2–3 in. or more wide. It is relatively involved in the life processes of the tree. Under most conditions, sapwood is more susceptible to decay than heartwood.

Saran \sə-ˈran\ *n* [fr. *Saran*, a trademark] (1940) (1) Generic name for thermoplastics consisting of polymers of ▶ Vinylidene Chloride (PVDC) or copolymers of same with lesser amounts of other unsaturated compounds. (2) Saran® is also Dow Chemical's trade name for its copolymer films. PVDC's very low permeability to gases and vapors makes it an excellent barrier material for packaging and food wrapping. (3) Generic name for fibers from polymers with not less than 80% vinylidene chloride. (Strong AB (2000) Plastics: Materials and processing. Prentice Hall, New York; Pittance JC (ed) (1990) Engineering plastics and composites. SAM International, Materials Park, OH)

Saran Fiber *n* Generic name for a manufactured fiber in which the fiber-forming substance is any long-chain synthetic polymer composed of at least 80% by weight of vinylidene chloride units ($-CH_2CCl_2-$) (Federal Trade Commission).

Sardine Oils \sär-ˈdēn\ *n* Fish, oils, mainly of Japanese or American origin, which, when adequately refined, are useful drying oils. Sp gr, 0.930/15°C; iodine value, 187; saponification value, 190.

SASE *n* Acronym for stress at specified elongation; the stress experienced by a yarn or cord at a given elongation.

Sash \ˈsash\ *n* {plural sash *also* sashes} [prob. mod. of F *châssis* chassis (taken as plural] (1681) Any framework of a window; may be movable or fixed; may slide in a vertical plane (as in a double-hung window) or may be pivoted as in a casement window; a pivoted sash is also called a ventilator. Also known as *Window Sash*.

Sash Tool *n* A round brush made in various sizes, bound with metal or string and used for painting sashes, frames and other small areas.

Sateen \sa-ˈtēn, sə-\ *n* [alter. of *satin*] (ca. 1878) A cotton fabric made in a satin weave.

Satin Finish \ˈsa-tᵊn-\ *n* A finish having gloss in the general range overlapping eggshell and semigloss, depending on manufacturer's or customer's specifications. See ▶ Gloss. Syn: ▶ Silk.

Satin Weave *n* (ca. 1883) One of the basic weaves, plain, satin, and twill. The face of the fabric consists almost completely of warp or filling floats produced in the repeat of the weave. The points of intersection are distributed evenly and widely separated as possible. Satin-weave fabric has a characteristic smooth, lustrous surface and has a considerably greater number of yarns in the set of threads, either warp or filling, that forms the face than in the other set.

Satin White *n* Filler made by precipitating together a mixture of calcium sulfate and aluminate. The ingredients used for its preparation are calcium hydroxide and basic aluminum sulfate. It is used chiefly in aqueous pigmented coatings for paper, and also as a base for lake pigments.

Saturated Compound *n* A compound in which all of the valences of the elements are satisfied; in organic chemistry, having no double or triple bonds, i.e., incapable of absorbing substances by addition, such as hydrogen or iodine. (Morrison RT, Boyd RN (1992) Organic chemistry, 6th edn. Prentice Hall, Englewood Cliffs, NJ)

Saturated Hydrocarbon *n* A hydrocarbon containing no multiple bonds. (Morrison RT, Boyd RN (1992) Organic chemistry, 6th edn. Prentice Hall, Englewood Cliffs, NJ)

Saturated Polyester See ▶ Polyester, ▶ Saturated. (Morrison RT, Boyd RN (1992) Organic chemistry, 6th edn. Prentice Hall, Englewood Cliffs, NJ)

Saturated Solution A solution which is or can be at equilibrium with excess solute. (Whitten KW, Davis RE, Davis E, Peck LM, Stanley GG (2003) General chemistry. Brookes/Cole, New York)

Saturation *n* (1) The attribute of color perception that expresses the degree of departure from the gray of the same lightness. All grays have zero saturation. (2) The maximum intensity or purity of a color. If the color is as brilliant as possible, it is at saturation; if the color is subdued or grayed, it is dull, weak, and low in intensity. (3) The upper limit concentration of a solute in a solvent, i.e., no more solute can be dissolved at a fixed temperature and pressure. (Colour physics for industry, 2nd edn. McDonald, Roderick (eds) Society of Dyers and Colourists, West Yorkshire, England, 1997)

Saturation Bonding See ▶ Bonding (2).

Saturation Value *n* The maximum amount of dye that can be absorbed by a textile fiber under defined conditions. (Kadolph SJJ, Langford AL (2001) Textiles. Pearson Education, New York)

Saturator *n* A machine designed to impregnate paper, fabrics, and the like with resins. The web to be saturated is conveyed by rollers through a pan containing a solution of the resin, then through metering devices such as squeeze rolls, scraper blades, or suction elements that control the amount of resin retained.

Saunderson Correction *n* Equation for calculating a correction for the Fresnel reflectances at the surface of a dielectric material for both the incident light (k_1) and exiting light (k_2). It is related to the internal reflectance, R_i, and to the total reflectance. R_T, by the following equation,

$$R_T = k_1 + \frac{(1-k_1)(1-k_2)R_i}{1-k_2 R_i}$$

(The equation was popularized by Saunderson in his work on computer color matching but actually dates back many years, at least to Walsh in the 1920s). (Moller KD (2003) Optics. Springer, New York)

Saxony \sak-s(ə-)nē\ *n* {*plural* -nies, *often capitalized*} [*Saxony*, Germany] (1842) (1) A high-grade fabric for coats, made from Saxony Merino wool. (2) A soft woolen with fancy yarn effects, used in sport-coat fabric. (3) A highly twisted worsted knitting yarn. (4) A term describing a cut-pile carpet having highly twisted, evenly sheared, medium-length pile yarns.

SAXS Abbreviation for ▶ Small-Angle X-Ray Scattering.

Saybolt Universal Viscosity *n* Efflux time in seconds for 60 mL or fluid to flow through a calibrated orifice under specified conditions.

Saybolt Viscosity *n* (SSU or Seconds Saybolt Universal) The time in seconds required to fill at 60-cm^3 flask with an oil specimen preheated to 38, 54, or 99°C and draining through a standard orifice. The measurement is convertible to kinematic viscosity by the equation: $v = 0.00222\ t - 1.83/t$, where v = kinematic viscosity, cm^2/s and t = Saybolt seconds. Several ASTM tests (www.astm.org) make use of the Saybolt viscometer. (Paint and coating testing manual (Gardner-Sward handbook) MNL 17, 14th edn. ASTM, Conshohocken, PA, 1995)

Sb *n* Chemical symbol for the element antimony (Latin: stibnium).

SB Abbreviation for ▶ Copolymers of ▶ Styrene and ▶ Butadiene.

S.B.P. Special boiling point spirits. A range of solvent obtained from petroleum.

SBR Elastomer from styrene and butadiene. Abbreviation for ▶ Styrene Butadiene Rubber.

SBS Abbreviation for Styrene-Butadiene-Styrene block terpolymer.

Scale \skā(ə)l\ *n* [ME, fr. MF *escale*, of Gr origin; akin to OE *scealu* shell, husk] (14c) (1) An adherent oxide coating that is thicker than the superficial film referred to as tarnish. (2) A deposit formed from solution directly in place upon a confining surface. NOTE – Scale is a deposit that usually will retain its physical shape when mechanical means are used to remove it from the surface on which it is deposited. Scale which may or may not adhere to the underlying surface is usually crystalline and dense, frequently laminated, and occasionally columnar in structure. (3) The oxide formed on the surface of the metal during heating. (4) Rust occurring in thin layers.

Scale of Segregation In mixing, the average distance between regions of the same component. See also ▶ Intensity of Segregation.

Scaler *n* (1568) A hand-cleaning chisel.

Scaling (1) Process of forming scale with or without acid fumes; sometimes refers to the spontaneous detachment of scale. (2) See ▶ Flaking.

Scaling Resistance See ▶ Flaking Resistance.

Scanning Electron Microscope *n* (1953) (SEM) An electron microscope that uses electrons reflected from the sample surface to form images. The surface must first be made opaque and reflective, usually by sputtering onto it a very thin layer of gold. The microscope has more depth of field than the transmission-type instrument and can produce very sharp and insightful pictures. See ▶ Electron Micrograph.

Scarf Joint \ˈskärf ˈjóint\ *n* {*plural* scarfs} [ME *skarf*, prob. of Scand origin; akin to ON *skarfr* scarf] (15c) A joint made by cutting away congruent acute-angular segments on two pieces to be joined, then bonding the cut surfaces to make the adherends coplanar. See also ▶ Butt Joint, ▶ Lap Joint and Joint, ▶ Scarf.

Scarlet Chrome \ˈskär-lət ˈkrōm\ Mixed pigment containing lead chromate, molybdate and sulfate. See ▶ Molybdate Orange.

Scarlet Lake $C_{18}H_{14}N_2O_7S_2Ba$. Lake, Acid Red 26 (16150). Pigment made by precipitating Scarlet 2R, which is the product of coupling diazotized *m*-xylidine with 2-naphthol-3,6disulfonic acid, on gloss white or hydrate of alumina.

Scatter, Coefficient of The rate of increase of reflectance with thickness (weight per area) at infinitesimal thickness of material over an ideally black backing, $S = (dR_o/dX)x \rightarrow o$.

Scattering Diffusion or redirection of radiant energy encountering particles of different refractive index; scattering occurs at any such interface, at the surface, or inside a medium containing particles.

Scattering Coefficient, Kubelka-Munk Multiple (diffuse) scattering coefficient for a unit thickness and concentration of scattering material in a medium of different refractive index as used in the Kubelka-Munk Equation. It is the rate of increase of reflectance of a layer over black as thickness is increased. Hence, the assumption is made that all of the scattering is in the backward or reverse direction from that of the incident diffuse light.

Scattering Coefficient, Mie Single scattering coefficient of a particle in a medium of different refractive index, expressed as the ratio between scattering cross section and geometric cross section of the particle. It should properly be called scattering efficiency, but in popular usage, the term "scattering coefficient" is common. See ▶ Mie Theory. (Kamide K, Dobashi T (2000) Physical chemistry of polymer solutions. Elsevier, New York; Saleh BEA, Teich MC (1991) Fundamentals of photonics. Wiley, New York)

Scattering Loss That part of transmitted energy lost due to roughness of reflecting surface.

Scattering, Multiple Diffusion or redirection of radiant energy which results from the observation of the reflected energy after multiple reflections from one particle to others, as opposed to the observation of single scattering. See ▶ Kubelka-Munk Theory.

Scattering, Single Diffusion or redirection of radiant energy by a single particle spaced sufficiently distant from any other particle that the observed reflected energy from each particle has not impinged on any other particle. See ▶ Mie Theory. (Kamide K, Dobashi T (2000) Physical chemistry of polymer solutions. Elsevier, New York)

SCE See ▶ Specular Reflectance Excluded.

Schaffer's Acid Acid used in the manufacture of dyestuffs. Chemically, it is 2-naphthol-6-sulfonic acid.

Schappe A yarn from partly degummed silk waste.

Scheiber One of the original dehydrated castor oils.

Schiff Base \ˈshifs-\ A reaction intermediate compound characteristic of reactions involving amines in general; formula is $RNH–CO–N = CH_2$; often used in formation of ureaformaldehyde polymers. (Smith MB, March J (2001) Advanced organic chemistry, 5th edn. Wiley, New York)

Schlieren Optics \ˈshlir-ən ˈäp-tiks\ An optical system used in an ultracentrifuge for the determination of the solute concentration gradient at any point in the cell. A parallel beam of light from a slit passes through the cell and an image of the slit is obtained which is deflected by different amounts along its length by radically directed refractive index gradient in the cell. The image is focused on a suitable phase plate. An image of this is photograph to give a two-dimensional plot in which the abscissa corresponds to the distance from the end of the cell and the ordinate is proportional to the refractive index gradient and hence to the concentration.

Schnitzer's Green See ▶ Chromium Oxide Green.

Schotten-Baumann Reaction Consists of reactions diamines or dials with acid chlorides of difunctional carboxylic or sulfonic acids to yield high polymers in which degrees of polymerization of well over 100 are easily attainable; e.g., step-growth polymerization in the condensation polymerization of terephthaloyl chloride with ethylene glycol to form poly(ethylene terephthalate). (Odian GC (2004) Principles of polymerization. Wiley, New York)

Schweinfurt Green \ˈshvīn-ˌfúrt\ Another name for Paris green. Chemically, cupric acetoarsenite. *Also known as Emerald Green.*

SCI See ▶ Specular Reflectance Included.

Scintillometer \ˌsin-tᵊl-ˈä-mə-tər\ *n* [L *scintilla* + ISV *-o-* + *-meter*] (1877) (scintillation counter) An instrument used to detect the presence of and measure the concentration of beta-emitting radioactive isotopes. When the emitted beta particle is intercepted by a molecule of the scintillant material in solution, a tiny flash of light is given off. By counting the flashes over a period of hours one obtains an estimate of concentration of the known beta emitter.

Scleroscope An instrument for measuring impact resilience by dropping a ram with a flattened-cone tip from a specified height onto the specimen, then noting the height of rebound.

Scorching A term denoting premature vulcanization of a rubber compound, occurring during the mixing operation when the compound is calendered or extruded. Scorch is often controlled or prevented by selection of proper accelerators or by use of retarders. Scorched or burnt stock is generally not processible. *Also known as Burning.*

Score \ˈskōr, ˈskȯr\ *v* (14c) To mark a line on a flat surface using a sharp-pointed instrument (*scriber*) and a straightedge. With a brittle material, the scoring may be deepened by repeated strokes, sometimes on both sides of the sheet, making it possible to cleanly break the sheet along the line of scoring. The same can be accomplished with brittle tubing with a triangular file, scoring only the outside surface.

Scotchcast Epoxide resin. Manufactured by Minnesota Mining & Manufacturing, US.

"Scotch-Tape" Test A method of evaluating the adhesion of a lacquer, paint, or printed label to a plastic substrate. Pressure-sensitive adhesive tape is applied to an area of the painted plastic article, which may first be cross-hatched with scored lines. Adhesion is considered to be adequate if no paint adheres to the tape when it is peeled off.

Scotopic Vision \skə-ˈtō-pik ˈvi-zhən\ The vision mediated by rods alone at very low levels of luminance; night vision.

Scouring (1) A wet process of cleaning by chemical and/or mechanical means. (2) An operation to remove the sizing and tint used on the warp yarn in weaving and, in general, to clean the fabric prior to dyeing.

SCR Abbreviation for either Saturable-Core Reactor or Silicon-Controlled Rectifier (see ▶ SCR Drive).

Scrap \ˈskrap\ *n* {*often attributive*} [ME, fr. ON *skrap* scraps; akin to ON *skrapa* to scrape] (14c) All products of a processing operation that are not present in the primary finished articles. This includes flash, runners, sprues, excess parison, and rejected articles. Scrap from thermosetting molding is generally not reusable. That from most thermoplastic operations can usually be reclaimed for reuse in the molder's own plant or can be sold to a commercial reclaimer.

Scraper Tool for scraping off paint or other adherent matter.

Scrap Grinder See ▶ Granulator.

Scrapless Thermoforming See ▶ Sheet Thermoforming.

Scratch Coat In three-coat plastering, the first coat of plaster, which is then scratched to provide a bond for the second (brown) coat.

Scratch Hardness The resistance of a material to scratching by another material. The test most often employed with plastics is the Bierbaum test, in which he specimen is moved laterally on the stage of a microscope under a loaded diamond point. The standard load is 0.0294 N. The width of the scratch is measured with a micrometer eyepiece and the hardness value is expressed as the quotient of the load divided by the scratch width. See also ▶ Mohs Hardness.

SCR Drive A variable-speed motor drive widely used today on new extruders in which a silicon-controlled rectifier (SCR) converts alternating current to run a direct-current motor. Unlike AC motors, DC motors have good torque characteristics over a wide speed range.

Screen \ˈskrēn\ *n* [ME *screne*, fr. MF *escren*, fr. MD *scherm*; akin to OHGr *skirm* shield; prob. akin to Sanskirt *carman* skin, *krṇāti* he injures] (14c) (1) A hollow, cylindrical, coarse-mesh wire device used in pickers and certain openers to form the loose staple stock into a sheet, or lap. The screen is mounted horizontally on a shaft on which it revolves freely. (2) A stencil used in screen printing. It is made of fine cloth, usually of silk or nylon, finely perforated in areas to form a design and mounted on a frame. The paste containing the dye is forced through the perforations onto the fabric, leaving the design. A series of screens, one for each color, is used for multicolored designs. Also see ▶ Printing.

Screen Analysis or Test See ▶ Sieve Analysis.

Screen Changer A device bolted to the head end of an extruder, between extruder and die adapter, that allows the operator to quickly replace a dirty screen pack with a clean one, usually without having to rethread the downstream line.

Screening See ▶ Sieving.

Screen Pack See ▶ Extruder Screen Pack.

Screen Painting See ▶ Silk-Screening.

Screen Print See ▶ Silk-Screening.

Screen Printing (Silk-screen printing) A printing process widely used on plastic bottles and other articles, employing as a stencil a taut woven fabric secured in a frame, the fabric being coated ins elected areas with a masking material that is not penetrated by the ink being used. The stencil fabric is commonly called a "silk screen" even though silk is rarely used today. Nylon is most often used, and screens of copper, stainless steel and many other materials are suitable. The screen is placed above the part to be decorated, and a flexible squeegee forces ink through the openings in the screen onto the surface of the plastic article. Multicolor work requires multiple screens and impressions. See ▶ Printing. (Printing ink handbook, 5th edn. National Association of Printing Ink Manufacturers. Kluwer Academic, London, 1999)

Screen Process Printing See ▶ Silk Screen Printing.

Screw In extrusion, the shaft provided with a helical channel that coveys the material from the feed hopper through the barrel, working it vigorously and causing it to melt, then developing pressure through pumping action to force the molten plastic through the die. See ▶ Extruder Screw for more details.

Screw Characteristic The relationship between volumetric flow rate in a melt-metering extrusion screw and the head pressure. For a melt screw operating isothermally, it is given by the equation:

$$Q = \alpha N - \beta \Delta P/(\eta L)$$

where

N = the screw rotational speed,

ΔP = the rise in pressure along the screw, equal to the head pressure,

L = the axial length of the screw,

η = the melt viscosity based on temperature and shear rate in the screw channel, and

α and β are constants related to the screw dimensions. (Strong AB (2000) Plastics materials and processing. Prentice Hall, Columbus, OH; Rosato's plastics encyclopedia and dictionary. Rosato DV (ed). Hanser-Gardner, New York, 1992) See also ▶ Die Characteristics.

Screw Conveyor A device for moving and metering the flow of solid particles at a fairly well controlled rate, consisting of a trough or sometimes a closed tube within which rotates a deep-flighted screw. Unlike an extruder screw, a conveyor screw normally operates only partly full and is a low-friction, low-torque device. A conveyor can be jacketed for heating or cooling. Screw conveyors are sometimes used as feeders for extruders and injection machines when it is desired to operate them in the starved mode.

Screw Dislocation A dislocation in which layers of particles in a crystal are warped around a screw axis.

Screw Extruder See ▶ Extruder.

Screw Feeds Feed systems in which the action of the screw generates pressure that causes flow. The system usually consists of a container with a closely fitting screw unit.

Screw Injection Molding See ▶ Injection Molding.

Screw Lead See second ▶ Lead.

Screw Melter (1) Screw extruder in which frictional forces between the screw and the heated barrel contribute to rapid melting of solid polymer. This configuration is capable of high throughput. (2) System in which a screw feed is used to feed polymer to a melt grid and to maintain a constant pressure at the grid.

Screw-Piston Injection Molding See ▶ Injection Molding.

Screw Pitch See ▶ Pitch (1).

Screw-Plasticating Injection Molding See ▶ Injection Molding.

Scrim \ˈskrim\ n [origin unknown] (1792) (1) A lightweight, open-weave, coarse fabric; the best qualities are made with two-ply yarns. Cotton scrim usually comes in white, cream, or ecru and is used for window curtains and as backing for carpets. (2) Fabric with open construction used as base fabric in the production of coated or laminated fabrics.

Scroop The sound of rustle or crunch that is characteristic of silk. Scroop is a natural property of silk, but may be induced in other fabrics to a degree by various treatments.

Scrub Board See ▶ Baseboard.

Scrub Resistance n The ability of a coating to resist being worn away or to maintain its original appearance when rubbed with a brush, sponge, or cloth, and an abrasive soap. For test method, refer to ASTM D 2486. Also known as Wet Abrasion Resistance.

Scrumbling n In painting, the operation of lightly rubbing a brush containing a small quantity of opaque or semi-opaque color over a surface to soften and blend tints that are too bright, or to produce a special effect; the coat may be so thin as to be semitransparent.

Sculptured Wallcovering Products molded of solid (usually synthetic) material in which the texture and shadows are real.

Scumble Glaze \ˈskəm-bəl ˈglāz\ n Transparent preparation used in the scumbling process.

Scumble Stain *n* Semitransparent stain for application over an opaque groundwork of paint. Brush, stipple or sponge may be used for manipulating the scrumble, or it may be combed, so that various effects, i.e., wood graining and other more formal patterns, are possible. In this, the nonflowing property of the scrumble greatly assists.

Scumming See ▶ Greasing.

Sealant \ ˈsē-lənt\ *n* (1944) A liquid, paste, or coating, or tape that fills small gaps between mating parts, e.g., pipe-thread sealant, or plugs small holds, stopping fluid leaks.

Seal Coat, Sealing Coat See ▶ Sealer.

Sealer *n* (1) A liquid coat which seals wood, plaster, etc., and prevents the surface from absorbing paint or varnish; may be transparent; may at as a primer for a following coat or as a finish for the surface. (2) A coat, applied in liquid form, which is laid over a tar-like substance to prevent its bleeding through an applied paint film. Also known as *Seal Coat or Sealing Coat.*
(Wicks ZN, Jones FN, Pappas SP (1999) Organic coatings science and technology, 2nd edn. Wiley-Interscience, New York) See ▶ Size, and ▶ Floor Sealer.

Sealing See ▶ Heat Sealing.

Sealing Coat See ▶ Sealer.

Sealing Wax *n* (14c) A resinous composition that is plastic when warm and is used for sealing.

Seaming *n* Joining the overlap of two pieces of fabric, usually near their edges.

Seamless *n* A term that describes a tubular knit fabric without seams, e.g., seamless hosiery.

Seamless Flooring *n* Fluid or trowel-applied flooring without aggregates. Syn: ▶ Monolithic Flooring.

Seam Mark *n* A particular type of pressure mark in the finished fabric. It is produced during finishing operations by the thickness of the seam used to join pieces for processing.

Seams *n* In hanging wallcoverings, there are three methods of joining seams. All three are satisfactory, but the "but" method produces the smoother finished job and is most often preferred. (1) In a "butt" the seam edges are fitted tightly together without any overlap. If the wallcovering comes untrimmed, selvages must be cut off at both sides of the roll. This leaves a flat, invisible seam, with no double thickness. (2) In the "lap" method, one strip is lapped over the selvage of another. Wall coverings are usually hung from left to right. In moving from left to right, the selvage is left intact on the right and trimmed on the left. The clean edge overlaps and covers the selvage edge. This procedure would be reversed in hanging from right to left. (3) The "wire-edge" method is a version of the "butt" technique, but will overlap about 1/16 of an inch (1.5 mm) into the pattern, so that no wall shows through in case the butting is not prefect. Here, also, both selvages must be removed. (Harris CM (2005) Dictionary of architecture and construction. McGraw-Hill, New York)

Seam Slippage *n* A defect consisting of separated yarns occurring when sewn fabrics pull apart at the seams. Seam slippage is more prone to occur in smooth-yarn fabrics produced from manufactured filament yarns.

Seam Welding *n* Any stitchless procedure for joining fabrics based on the use of thermoplastic resins or the direct welding of thermoplastic materials. Seam welding is an alternative to conventional needle-and-thread seaming operations that is extremely popular in the nonwoven field.

Sebacic Acid \si-ˈba-sik\ *n* [ISV, fr. L *sebaceus*] (1790) (1) (sebacylic acid, decanedioic acid) HOOC$(CH_2)_8$COOH. White leaflets derived from butadiene or castor oil, used as an intermediate in the production of plasticizers, alkyd resins, and certain nylons. (2) $(CH_2)_8(COOH)_2$. Used for the preparation of sebacate plasticizers and ass a stabilizer for alkyds. Mp, 133°C; bp, 295°C/100 mmHg.

SEBS *n* Abbreviations for block terpolymer Styrene-Ethylene/Butene-Styrene.

Secant Modulus \ ˈsē-ˌkant ˈmä-jə-ləs\ *n* The ratio of stress to corresponding strain at any specific point on a stress–strain curve. It is expressed in force per unit area, MPa or kpsi. This definition of modulus is useful for many plastics whose stress–strain relationships are nonlinear even at very low strains, exhibiting little or no Hooke's-law range. (Shah V (1998) Handbook of plastics testing technology. Wiley, New York)

sec-**Butyl Acetate** \ ˈsek ˈbyü-tᵊl ˈa-sə-ˌtāt] *n* (2-butyl acetate) $CH_3COOCH(CH_3)C_2H_5$. A solvent for nitrocellulose, ethyl cellulose, PVC, acrylics, polystyrene, phenolics and alkyd resins.

Second \ˈse-kənd *also* –kənt, *esp before a consonant* – kən, -kəŋ\ *n* [ME *secunde*, fr. ML *secunda*, fr. L, fem. of secundus second; from its being the second sexagesimal division of a unit, as a minute is the first] (14c) (1) (s) The SI basic unit of elapsed time, the duration of 9,192,631,770 periods of the radiation corresponding to the transition between the two hyperfine levels of the ground state of the cesium-133 atom. All other measures of time intervals are defined in terms of the second. (2) A finished product containing slight imperfections but without serious functional faults and therefore usable, but usually sold at a discount. (CRC handbook of chemistry and physics. Lide DR (ed) CRC, Boca Raton, FL, 2004 Version)

Secondary (1, *adj*) In organic chemistry, a functional group in which two of its hydrogen atoms have been replace by other groups as a secondary alcohol, > CHOH, or a secondary amine, > NH. Examples are isopropyl alcohol, $(CH_3)_2CHOH$, and dimethyl amine, $(CH_3)_2NH$. (Morrison RT, Boyd RN (1992) Organic chemistry, 6th edn. Prentice Hall, Englewood Cliffs, NJ)

Secondary Amine Value *n* The number of milligrams of potassium hydroxide (KOH) equivalent to the secondary amine basicity in 1 g of sample.

Secondary Backing *n* A layer of material, usually woven jute, polypropylene, vinyl, or latex foam, laminated to the underside of a carpet to improve body and dimensional stability.

Secondary Colors *n* (1831) Green, orange, and violet, each of which is obtained by mixing two primary colors. It is necessary to specify pigments or filters regarding primary and secondary colors. (Physics for industry, 2nd edn. McDonald, Roderick (eds) Society of Dyers and Colourists, West Yorkshire, England, 1997; Mayer R, Sheehan S (1991) Artist's handdbook of materials and techniques. Viking Adult, New York; Billmeyer FW, Saltzman M (1966) Principles of color technology. Wiley, New York)

Secondary Creek The nonrecoverable component of creep. Also see ▶ Delayed Deformation.

Secondary Emission *n* (1931) The emission of electrons from a surface that is bombarded by particles (as electrons or ions) from a primary source. (Giambattista A, Richardson R, Richardson RC, Richardson B (2003) College physics. McGraw Hill Science, New York)

Secondary Forces Forces generated by the interactions between atoms other than a covalent bond. Secondary forces include hydrogen bonding, ionic interaction, and dispersion forces.

Secondary Plasticizer *n* (extender plasticizer) A plasticizer that is less compatible with a given resin that is a ▶ Primary Plasticizer and thus would exude or cause surface tackiness if used in excess of a certain concentration. Secondary plasticizers are used in conjunction with primaries to reduce cost or to obtain improvement in electrical or low-temperature properties.

Secondary Reference Standard *n* A standard which has been calibrated relative to a primary standard and which, when used with the calibration values, essentially reproduces the primary standard.

Secondary Treatment *n* The second step in most waste-treatment systems, in which bacteria consume the organic parts of the wastes.

Second-Order Transition *n* A thermal transition that involves a change in heat capacity, but does not have a latent heat. The glass transition is a second order transition. (Groenewoud WM (2001) Characterization of polymers by thermal analysis. Elsevier Science and Technology Books, New York; Physical properties of polymers handbook. Mark JE (ed). Springer, New York, 1996) See ▶ Glass Transition.

Second-Order Transition Temperature *n* The temperature at which the non crystalline (amorphous) portions of polymer melt or become plastic. An inflection point or change is stress–strain properties occurs at this point; however, for most fibers, this change is small. (Groenewoud WM (2001) Characterization of polymers by thermal analysis. Elsevier Science and Technology Books, New York)

Seconds *n* (1) Imperfect fabrics (woven or knitted) containing flaws in the weave, finish, or dyeing, and sold as "seconds." (2) See ▶ Yarn Quality.

Second-Surface Decorating *n* A decorating process used with transparent plastics in which the decoration is applied to the back of the part so that it is visible from the front but is not exposed to possible damage.

Second Virial Coefficient *n* The second constant of the virial coefficient describes the interaction between two molecules. For polymer solutions, the second virial coefficient A_2 is related to the excluded volume by the expression

$$A_2 = Nu^2/2M_2^2$$

where N is Avogadro's number, and M_2 is the solute molecular weight. The name "virial coefficient" comes from the "virial theorem" that was much used toward the end of the nineteenth century. This theorem states:

$$\text{average of } (mv)^2/2 = -\text{average of } 0.5(Xx+Yy+Zz)$$

Here m is the mass of the particles, v is their velocity, x, y and z are their coordinates, and X, Y and Z are the components of the forces that act upon them. The expression on the right-hand side was call "virial" because forces were considered [*vis* (Latin) = force]. The virial could be expanded into a series whose coefficients were consequently the virial coefficients. (Elias HG (1977) Macromolecules vol 1–2. Plenum, New York; Kamide K, Dobashi T (2000) Physical chemistry of polymer solutions. Elsevier, New York)

Section Beam *n* (1) A large, flanged roll upon which warp yarn is wound at the beam warper in preparation for slashing. (2) Small flanged or unflanged beams assembled side-by-side on the shaft of a warp beam for further processing.

Section Mark *n* A fabric defect consisting of marks running warpwise in an evenly repeating pattern, caused by the improper setting of sections in silk system (or indirect) warping.

Section Modulus *n* (Z) In a beam under load, the quotient of the moment of inertia of the beam's cross section about its neutral axis divided by the distance from the neutral axis to the outermost surface of the beam (I/c). The bending moment divided by the section modulus gives the maximum stress in the beam at any point along it.

Sediment \se-də-mənt\ *n* [MF, fr. L *sedimentum* settling, fr. *sedēre* to sit, sink down] (1547) Any solid which can settle or be centrifuged from the main portion of the liquid, for example, foots, meal, grain, gum.

Seebeck Effect *n* The electrical phenomenon responsible for the action of a ▶ Thermocouple. If a circuit consists of two metals, one junction being hotter than the other, a current flows in the circuit. The magnitude and direction of the current depend on the metals chosen and on the difference in temperature between the junctions. If the cold junction is held at a known constant temperature, e.g., 0°C, the current becomes a measure of the temperature at the other junction. (CRC handbook of chemistry and physics. Lide DR (ed). CRC, Boca Raton, FL, 2004 Version; Giambattista A, Richardson R, Richardson RC, Richardson B (2003) College physics. McGraw Hill Science, New York)

Seediness \sēd-ē-nəs\ *n* (1) Particles in a coating caused by reactive pigments and acid components of the vehicle. (2) Defect in a clear varnish or lacquer caused by small particles sometimes visible when examined by transmitted light. On application, varnished or lacquered surfaces may present a bitty, specky or sandy appearance due to this defect.

Seeding *n* Formation of small, undesirable particles or granules in a paint, varnish or lacquer.

Seeds *n* Undesirable particles which develop in a liquid coating by partial gelation of the vehicle or by agglomeration of pigment particles. See ▶ Seediness.

Seedy \sēd-dē\ *adj* (1574) Descriptive of a paint finish that is not smooth owing to undispersed pigment particles or insoluble gel particles in the paint.

Seersucker \sir-sə-kər\ *n* [Hindi *śīrśaker*, fr. Per *shīro-shakar*, lit., milk and sugar] (1722) Lightweight fabric, made of cotton or manufactured fiber, having crinkled stripes made by weaving some of the warp threads slack and others tight. Woven seersucker is more expensive than imitations made by chemical treatment.

Segregated Oils *n* Drying oils which are obtained from the semidrying types by removal of nondrying constituents. This removal may be effected by selective crystallization, solvent extraction or vacuum distillation.

Segregation *n* (1) A close succession of parallel, rather narrow and sharply defined, wavy lines of color on the surface of a plastic, said color differing in shade from surrounding areas, and creating the impression that components of the plastic have separated. (2) In 1952, P. V. Danckwerts introduced the terms Scale of Segregation and Intensity of Segregation to define the state of a heterogeneous mixture and to quantify the effectiveness of mixing processes.

Selected Ordinate Method *n* A method for determining the CIE tristimulus values by adding the reflectances or transmittances at unequally spaced intervals of wavelength, so selected for each of the tristimulus values that the summation of the reflectances or transmittances give the same tristimulus values obtained by integration of weighted regularly spaced wavelength values. The accuracy of this technique is poor relative to that obtained with the weighted ordinate method unless an extremely large number of readings at different wavelengths is used (e.g., more than 100 for each tristimulus value). (Colour physics for industry, 2nd edn. McDonald, Roderick (eds) Society of Dyers and Colourists, West Yorkshire, England, 1997)

Selenium Red \sə-lē-nē-əm\ *n* See ▶ Cadmium Red.

Self-Adhesion *adj* (1958) See ▶ Autohesion.

Self-Crosslinking See ▶ Self-Curing and ▶ Ambient Cure.

Self-Curing *adj* Undergoing cure (crosslinking) without the application of heat. See ▶ Self-Vulcanizing and Ambient Cure.

Self-Dissociation *adj* The production of cations and anions by dissociation of solvent molecules without interaction with other species.

Self-Etch Primer See ▶ Wash Primer.

Self-Extinguishing *adj* A somewhat loosely used term denoting the ability of a material to cease burning after the source of flame has been removed. PRV, vinyl chloride-acetate copolymers, polyvinylidene chloride, some nylons, and casein plastics are examples of self-extinguishing materials. (Troitzsch J (2004) Plastics flammability handbook: Principle, regulations, testing and approval. Hanser-Gardner, New York; Tests for comparative flammability of liquids, UI 340. Laboratories Incorporated Underwriters, New York, 1997) See also ▶ Flammability and ▶ Flame Retardant.

Self-Ignition Temperature See ▶ Autoignition Temperature.

Self-Lifting Resistance *adj* The ability of dried films of organic coating materials to resist deformation upon application of an additional coat of the same material. For a method of test, see Federal Standard 141a, Method 6252.

Self-Luminous Paint See ▶ Luminous Paint.

Self-Polishing Wax See ▶ Liquid Emulsion Polymer Coating and ▶ Liquid Water Emulsion Wax.

Self-Priming *adj* Use of same coating for primer and for subsequent coats. It may be thinned differently for the various coats.

Self-Sealing Paint *n* A paint which, when applied over a surface of varying porosity, seals the surface and yet dries with a uniform color and sheen.

Self-Thermosetting See ▶ Self-Curing.

Self-Tone *n* Wallcovering in which shades of one color are featured. Habitually used for damasks and in flocking.

Self-Twist Yarn *n* An inherently twist-stable, two-ply structure having a ply twist that is alternately S- and Z-directed along the yarn.

Self-Vulcanizing *adj* Pertaining to an adhesive that undergoes vulcanization without the application of heat. See ▶ Self-Curing.

Sellmeier Equation *n* An empirical relationship between refractive index and wavelength for a particular transparent medium. The equation is used to determine the dispersion of light in the medium. It was first proposed in 1871 by W. Sellmeier, and was a development of the work of Augustin Cauchy on Cauchy's equation for modeling dispersion.

The usual form of the equation for glasses is $n^2(\lambda) = 1 + \frac{B_1\lambda^2}{\lambda^2 - C_1} + \frac{B_2\lambda^2}{\lambda^2 - C_2} + \frac{B_3\lambda^2}{\lambda^2 - C_3}$, where *n* is the refractive index, λ is the wavelength, and $B_{1,2,3}$ and $C_{1,2,3}$ are experimentally determined *Sellmeier coefficients* (Schott (2007) Refractive index and dispersion. Technical information document TIE-29). These coefficients are usually quoted for λ in micrometers. Note that this λ is the vacuum wavelength; not that in the material itself, which is $\lambda/n(\lambda)$. A different form of the equation is sometimes used for certain types of materials, e.g. crystals. The same relationships are applied to polymeric materials.

As an example, the coefficients for a common borosilicate crown glass known as *BK7* are shown below:

Coefficient	Value
B_1	1.03961212
B_2	.0231792344
B_3	1.01046945
C_1	$6.00069867 \times 10^{-3}$ μm^2
C_2	$2.00179144 \times 10^{-2}$ μm^2
C_3	1.03560653×10^2 μm^2

The Sellmeier coefficients for many common optical glasses can be found in the Schott Glass catalogue, or in the Ohara catalogue.

For common optical glasses, the refractive index calculated with the three-term Sellmeier equation deviates from the actual refractive index by less than 5×10^{-6} over the wavelengths range of 365 nm to 2.3 μm, which is of the order of the homogeneity of a glass sample. Additional terms are sometimes added to make the calculation even more precise. In its most general form, the Sellmeier equation is given as $n^2(\lambda) = 1 + \sum_i \frac{B_i\lambda^2}{\lambda^2 - C_i}$, with each term of the sum representing an absorption resonance of strength B1 at a wavelength $\sqrt{C_i}$. For example, the coefficients for BK7 above correspond to two absorption resonances in the ultraviolet, and one in the mid-infrared region. Close to each absorption peak, the equation gives non-physical values of n = ±∞, and in these wavelength regions a more precise model of dispersion such as Helmholtz's must be used.

If all terms are specified for a material, at long wavelengths far from the absorption peaks the value of *n* tends to $n \approx \sqrt{1 + \sum iBi} \approx \sqrt{\varepsilon_r}$, where ε_r is the relative dielectric constant of the medium.

The Sellmeier equation can also be given in another form:

$$n^2(\lambda) = A = \frac{B_1\lambda^2}{\lambda^2 - C_1} + \frac{b_1\lambda^2}{\lambda^2 - C_2}$$

Here, the coefficient A is an approximation of the short-wavelength (e.g., ultraviolet) absorption contributions to the refractive index at longer wavelengths. Other variants of the Sellmeier equation exist that can account for a material's refractive index change due to temperature, pressure, and other parameters. (Refractive index and dispersion. Schott technical information document TIE-29 (2007); Sellmeier W (1871) Zur Erkärung der abnormen Farbenfolge im spectrum einiger Substanzen. Annalen der Physik und Chemie 219:272–282)

Selvage \ |sel-vij\ *n* [ME *selvage*, prob. fr. MD *selvegge*, selvage, fr. *selv* self + *egge* edge; akin to OE *self* and to OE *ecg* edge] (15c) The narrow edge of woven fabric that runs parallel to the warp. It is made with stronger yarns in a tighter construction than the body of the fabric to prevent raveling. A fast selvage encloses all or part of the picks, and a selvage is not fast when the filling threads are cut at the fabric edge after every pick. Also spelled "*Selvedge*."

Selvedge See ▶ Selvage.

Semi-Automatic Molding Machine *n* A machine in which only part of the operation is controlled by the direct actions of a person. The automatic segments are controlled by the machine's instruments or by a computer according to a predetermined program.

Semibatch Reactors *n* In a semibatch reactor, some of the reactants are loaded into the reactor as a single charge and the remaining reactants are then fed gradually. Alternatively, a semibatch reactor may be loaded with reactants in a single charge, and then some of the products are drawn from the reactor.

Semiconductor \- |dək-tər\ *n* (1838) A substance whose electrical conductivity increases with increasing temperature.

Semicrystalline \- |kris-tə-lən\ *adj* (1816) Describing a polymer having both amorphous and crystalline regions in the range 20–80%. Most so-called "crystalline polymers" are, in most articles made from them, actually semicrystalline.

Semicure *adj* A preliminary, incomplete cure given to certain rubber articles to cause the rubber to set in some desired shape. Full vulcanization is subsequently completed by a final cure.

Semidrying Oil Oils which possesses the characteristics of a drying oil but to a less degree; for example, upon being exposed to oxygen it only partially hardens or is changed to a sticky mass. There is no definite line of demarcation between drying and semidrying oils.

Semigloss \ |se-mē- |gläs\ *adj* (1937) A gloss range between high gloss and eggshell, approximately 35–70 on the 60° gloss scale. See ▶ Gloss.

Semigloss Lacquer \ |la-kər\ Lacquers having the appearance of having been frictionally rubbed. Usually arrived at through the introduction of a pigment (inert). A type of satin finish.

Semipermeable Membrane \- |pər-mē-ə-bəl |men- |brān\ *n* A thin partition between two solutions that allows some molecules to pass through but not others. See ▶ Osmometer and Osmosis.

Semipositive Mold A mold with a plunger that fits loosely within the cavity as the mold begins to close, allowing excess material to escape as flash. As the mold nears complete closing, the plunger fits more closely to exert full molding pressure on the material. The semipositive mold combines the free flow of material inherent in a flash-type mold with the high density of moldings obtained with a positive mold.

Semirigid Plastic *n* For the purposes of general classification, a plastic that has a modulus of elasticity in either flexure or tension of between 70 and 700 MPa at 23°C and 50% relative humidity when tested in accordance with ASTM D 747, D 790, D 638, or D 882 (ASTM D 883). See also ▶ Rigid Plastic.

Semitransparent Stain *n* A coating which is formulated to change the color of a surface, but not conceal the surface. See also ▶ Stain and ▶ Opaque Stains.

Senegal Gum \ |se-ni- |gól\ *n* Form of gum Arabic obtained from Senegal. It is a natural exudation from *Acacia Senegal* and is water-soluble. (Industrial gums: Polysaccharides and their derivatives. Whistler JN, BeMiller JN (eds). Elsevier Science and Technology Books, 1992) See ▶ Acacia Gum.

Sensation \sen- |sā-shən\ *n* [ML *sensation-*, *sensatio*, fr. LL, understanding, idea fr. L *sensus*] (1615) Mode of mental functioning that is directly associated with stimulation of the organism.

Sensitizer \ |sen(t)-sə- |tīz-ər\ *n* [*sensiti*ve + -*ize*] *n* (ca. 1859) An agent which cleanses the ink-repellant, nonprinting areas of lithographic printing plates.

Separate Application Adhesive See ▶ Adhesive, Separate Application.

Sepia \ |sē-pē-ə\ *n* [L, cuttlefish, ink, fr. Gk *sēpia*] (1821) Dark brown artists' pigment, which occurs as a natural secretion in cuttlefish.

Sequential Arrangement *n* The arrangement of mer units in a polymer. A ▶ Head-To-Tail Polymer exemplifies one possible sequence.

Sequential Winding See ▶ Biaxial Winding.

Sequestering Agent *n* A chemical that prevents metallic ions from precipitating from solutions of anions that would normally, without the sequestering agent being present, precipitate those ions. See also ▶ Chelate and ▶ Chelating Agent.

Sequestrant *n* Any compound that will inactivate a metallic ion by forming a water-soluble complex in which the metal is held in a nonionizable form. This results in prevention of the usual precipitation reactions of the metal.

Sequestration See ▶ Chelation.

Serge \ˈsərj\ n [ME sarge, fr. MF, fr. (assumed) VL sarica, alter. of L serica, fem. of sericus silken] (14c) (1) Any smooth-faced cloth made with a two-up and two-down twill weave.

Serging n (1) Overcasting the cut edge of a fabric to prevent raveling. (2) Finishing the edge of a carpet by oversewing rather than binding. Generally, the sides of a carpet are serged and the ends bound.

Sericin \ˈser-ə-sən\ n [ISV, fr. L sericum silk] (ca. 1868) Silk gum. The gelatinous protein that cements the fibroin filaments in a silk fiber. It is removed in the process called degumming.

Sericite See ▶ Aluminum Potassium Silicate.

Serpentine \ˈsər-pən-ˌtēn\ n [ME, fr. ML serpentina, serpentinum, fr. LL, fem. & neuter of serpentinus resembling a serpent] (15c) A type of asbestos containing ▶ Chrysotile. A mineral or rock consisting essentially of a hydrous magnesium silicate usually having a dull green color and often a mottled appearance. See ▶ Asbestos.

Served Yarn n In aerospace textiles, a reinforcing yarn such as graphite or glass around which two different yarns are wound, i.e., one in the Z direction and one in the S direction, etc., for protection or compaction of the yarn bundle.

Service Factor n (1) See ▶ Factor of Safety. (2) The fraction (or percentage) of planned operating time during which a device or system satisfactorily accomplishes its anticipated mission.

Service Life n The time over which a part or system will continuously and satisfactorily perform its designed functions under stated service conditions. Service life may be determined by actual life testing or may be estimated by extrapolation from shorter-term testing, or from short-term testing under more severe conditions, as at higher temperature via ▶ Time-Temperature Equivalence.

Sesame Oil \ˈse-sə-mē also ˈse-zə-\ n (1870) Pale yellow oil, almost odorless, from the seeds of cultivated varieties of Sesamum indium. Sp gr, 0.916; Saponification value, 188–193; iodine value, 103–122. Syn: ▶ Benne, ▶ Teel or ▶ Gingelli Oil.

Set n (1) To convert an adhesive into a fixed or hardened state by chemical or physical action, such as condensation, polymerization, oxidation, vulcanization, gelation, hydration or evaporation of volatile constituents. (2) Condition of a paint or varnish film when it has dried to a point where, for all practical purposes, it ceases to flow. See ▶ Cure and ▶ Dry.

Set v To become at least partly fixed or hardened by chemical or physical action, such as condensation, polymerization, oxidation, vulcanization, gelation, hydration, or evaporation of volatile constituents. See also ▶ Cure. (2, n) See ▶ Permanent Set.

Setaflash Closed Tester n A closed cup flash point tester in which a small sample of the material is heated in a crucible or cup, and at a definite temperature, the vapor-air mixture which forms is ignited by a dipping flame. Reference ASTM D 3278.

Set Mark n A fabric defect consisting of narrow bars or bands across the full width of the fabric that may appear either as a tight, loose, or corduroy effect caused by loom stops improperly reset by the weaver. Set marks are sometimes caused by the weaver ripping out filling yarn and then not properly adjusting the pick wheel to obtain the proper relation between the fell of the cloth and the reed.

Set Off vt An undesirable transfer of ink from the printed sheet to the back of the sheet adjacent to it.

Set Off of Ink n The transferring or smearing of ink from freshly printed matter to another surface with which the undried print comes in contact, particularly from the face of one sheet onto the back of the sheet on top of it in the delivery pile or rewind roll.

Setoglaucine See ▶ Rhoduline Blue 66.

Set Point n The value set in a control instrument at which a control action will be taken when the instrument's sensor signals that the quantity controlled has just passed above (or below) the set value.

Setpoint Setting Agent Syn: ▶ Curing Agent.

Setting of Ink n The initial phase of the drying process by slight absorption of the vehicle by the paper, wherein printed sheets, though not fully dry, can be handled without smudging.

Setting Temperature See ▶ Temperature, Setting.

Setting Time n Of adhesives, castings, or hand layups and sprayups, the time from application until the material has become firm and handleable and will no longer flow by gravity, usually a time less than that required to dry completely or to reach a full cure or full strength. See ▶ Time, Setting.

Setting Up n (1) Conversion of liquid paint during storage to a gel-like or pseudosolid condition. The process is usually reversible by agitation and thinning but may be permanent when chemically reactive pigments or highly polymerized media are involved. (2) The thickening which occurs when paint stands in a open can. (3) The increasing viscosity of a paint film as it dries.

Settling n The sinking of pigments, extruders or other solid matter in a paint on standing in a container, with a consequent accumulation on the bottom of the can.

Set-to-Touch Time *n* The time required for the coating to reach a point where the adhesion to an external object is less than the internal cohesion of the film. See ▶ Drying Time.

Set Yarns False-twist yarns stabilized to produce bulk.

Sewing Thread See ▶ Thread.

Seydel Converter *n* Tow-to-top processing equipment. Seydel combines the prestretching and breaking process in one machine.

S-Finishing *n* A finishing process applied to acetate and triacetate fabrics using a sodium hydroxide solution to give surface saponification; i.e., the fiber "skin" is converted to cellulose. It improves the hand and reduces the tendency to acquire a static charge.

SFRP *n* Stable free-radical polymerization. (Odian G (2004) Principles of polymerization. Wiley-Intersciene, New York)

S Glass *n* A specialty glass composition containing 64% silica, 25% alumina, and 10% magnesia that provides high strength – 4.5 GPa in fibers – sometimes used as reinforcement for laminates where high specific strength is wanted. See also ▶ Glass-Fiber Reinforcement.

Sgraffito \zgra-ˈfē-(ˌ)tō\ *n* [It, fr. pp of *sgraffire* to scratch, produce sgraffito] (ca. 1730) Decoration by cutting away parts of a surface layer (as of plaster or clay) to expose a different colored ground.

Sgraffito Painting *n* Form of mural decoration in which a design is produced by scratching through a white or tinted topcoat of limewash to expose the darker colored mortar underneath.

Shade \ˈshād\ *n* [ME, fr. OE *sceadu*; akin to OHGr *scato* shadow, Gk *skotos* darkness] (before 12c) (1) In ink manufacture, commonly used as a Syn: ▶ Hue. See also ▶ Shading and Blending. (2) Term employed to describe a particular hue, or variation of a primary hue, such as a red shade blue. (3) Term used to describe the depth of color, such as a pale shade. (4) Term used to describe a mixture with black (or gray) as opposed to a tint which is a mixture with white. (5) Verb used to describe the process of making a small modification in a color by the addition of colorant(s).

Shade Bar See ▶ Mixed End and ▶ Filling.

Shade Cloth *n* A plain-weave cotton or linen fabric that is heavily sized and is often given oil treatment to make it opaque. The fabric is used for curtains and shades.

Shade Filling *n* A defect consisting of a bar running across the fabric caused by a difference in appearance of the filling yarn, and occurring at a quill change or knot.

Shading *n* In cut-pile fabrics, an apparent change in color when the pile is bent, caused by differences in the way light is reflected off the bent fibers. This phenomenon is a characteristic of pile fabrics, not a defect.

Shading and Blending *n* Altering the color of a paint slightly by the addition of black tinting color to create a decorative effect of graduated colors when applied to adjacent areas. The lap areas often are brushed and rolled to achieve a subtle blending. See also ▶ Shade.

Shadow Printing See ▶ Printing, Wrap Printing.

Shaft *n* (1) A term often used with reference to satins indicating the number of harnesses employed to produce the weave. (2) See ▶ Harness.

Shaft Mark A fabric defect characterized by a number of floating ends, usually caused by a broken harness strap on the loom.

Shag Carpet *n* A loosely tufted carpet construction with cut pile 1–5 in. in length and with greater than normal spacing between tufts.

Shale Oil \ˈshā(ə)l\ *n* (1857) An oil obtained by destructive distillation from shale tar, which, in turn, is obtained by distilling shale. Shale is a bituminous type of mineral, mined similarly to bituminous coal, and is found in many places throughout the world. Shale oil varies in specific gravity from 0.750–0.850. (Asphalt science and technology. Usmani AM (ed). Marcel Dekker, 1997)

Shank \ˈshaŋk\ *n* [ME *shanke*, fr. OE *scanca*; akin to ON *shakkr* crooked, Gk *skazein* to limp] (before 12c) The section of an extruder screw to the rear of the flighted sections. The forward part of the shank contains a milled keyway holding the key that engages the tubular drive shaft and may act as a shear safety protecting the screw from over torquing, while the rear part of the shank engages the radial and thrust bearings.

Shared-Electron Bond See ▶ Covalent Bond.

Sharkskin \ˈshärk-ˌskin\ *n* (1851) (1) An irregularity of the surface of an extrudate in the form of finely spaced sharp ridges perpendicular to the extrusion direction, believed to be caused by a relaxation effect on the melt at the die exit. (2) A hard-finished, twill fabric, woolen or worsted, made of simple weaves with a two-color arrangement of warp and filling yarns. (3) A plain-weave sportswear fabric made of dull-luster acetate or triacetate yarns.

Sharp Coat *n* A coat of white lead in oil, thinned liberally with turpentine or white spirit. A sharp coat used for the treating of new plaster following the trowel is frequently referred to as "sharp color".

Sharp Paint *n* Rapid-drying paint which yields a flat film for use as a seal coat. It is usually highly pigmented and contains a minimum amount of binder. Sharp paints dry off to a tack-free condition chiefly by evaporation of solvent. They are used as priming or sealing coats.

Shaw Pot *n* A name used in the early years of the industry for the original transfer-molding machine. It consisted of a conventional hydraulic press with a pot suspended above the mold. Material was charged into pot, then forced into the mold by the closing of the press.

Shear \shir\ *n* (before 12c) The movement, in a fluid or solid body, of a layer parallel to adjacent layers. See also ▶ Shear Strain and ▶ Shearing Stress.

Shear Degradation *n* (rheomalaxis) Chain scission of a polymer caused by subjecting it to an intense shear field, such as exists in the close clearances of extruders and internal mixers.

Shear Flow *n* The flow caused by the relative parallel or concentric motion of the surfaces confining a liquid, as in an extruder screw; or caused by a pressure drop, in the direction of flow, from the entrance of a flow passage to its exit, as in a die. Sometimes the two basic driving modes coexist, as in the metering sections of most extrusion screws and in wire-covering dies. In the direction crosswise to any laminar flow, successive layers slide past each other in shear.

Shear Heating Syn: ▶ Viscous Dissipation.

Sheariness *n* Variation in gloss or sheen on a paint surface which should have been uniform in this respect. This variation is due to differences in film thickness.

Shearing *n* A dry finishing operation in which projecting fibers are mechanically cut or trimmed from the face of the fabric. Woolen and worsted fabrics are almost always sheared. Shearing is also widely employed on other fabrics, especially on napped and pile fabrics where the amount varies according to the desired height of the nap or pile. For flat-finished fabrics such as gabardine, a very close shearing is given.

Shearing Strength *n* The maximum shear stress which a material is capable of sustaining. Shear strength is calculated from the maximum load during a shear or torsion test and is based on the original dimensions of the cross section of the specimen.

Shearing Stress *n* The tangential shearing force which is acting on a material to produce motion or flow in it.

Shear Modulus *n* (G, modulus of rigidity) The ratio of shear stress to shear strain within the proportional limit of a material. If the material exhibits no proportional region, a ▶ Secant Modulus is used.

Shear Rate *n* (shear-strain rate, velocity gradient) The rate of change of shear strain with time. In concentric-cylinder flow where the gap between the cylinders is much smaller than the cylinder radii, shear rate is almost uniform throughout the fluid and is given by $\pi(R_1 + R_2)N/(R_2 - R_1)$, where R_1 and R_2 are the radii of the cylinders, one rotating, the other stationary, and N is the rotational speed in revolutions per second. The universally used unit of shear rate is s^{-1}. In tube flow, the shear rate varies from zero at the center to its maximum at the tube wall where, for a Newtonian liquid, it takes on the value $4Q/\pi R3$. Q = the volumetric flow rate and R = the tube radius.

Shear Strain *n* The amount of movement of one layer relative to an adjacent layer divided by the layer thickness. This may be expressed as an angle of shear, in radians.

Shear Strength *n* The maximum Shear Stress required to shear the specimen in such a manner that the moving portion completely clears the stationary portion.

Shear Stress *n* Force per unit area acting in the plane of the area to which the force is applied. In an elastic body, shear stress is equal to shear modulus times shear strain. In an inelastic fluid, shear stress is equal to viscosity times the shear rate. In viscoelastic materials, shear stress will be a function of both shear strain and shear rate.

Shear Thickening *n* A rheological flow characteristic evidenced by an increase in viscosity with increasing rates of shear, an increase of flow resistance with agitation. Shear thickening is also known as *Dilatant Flow*. See also ▶ Dilatancy.

Shear Thinning See ▶ Thixotropy.

Sheary See ▶ Sheariness.

Sheath-Core Fibers *n* Bicomponent fibers of either two polymer types, or two variants of the same polymer. One polymer forms a core and the other surrounds it as a sheath.

Sheathing *n* Structural insulating board for use in housing and other building construction, which may be integrally treated, impregnated or coated to give it additional water resistance.

Shed *n* A path through and perpendicular to the warp in the loom. It is formed by raising some warp threads by means of their harnesses while others are left down. The shuttle passes through the shed to insert the filling.

Shedding *n* (1) The operation of forming a shed in weaving. (2) A loss of nominal length staple at any process in a staple yarn plant.

Sheen \shēn\ *n* (1602) An attribute of object mode of appearance which is similar to luster; gloss with poor distinctness-of-image reflectance. Frequently, in the paint industry, sheen is used synonomously with "low-angle sheen", a characteristic where a material appears to be matte when illuminated and viewed near to the perpendicular, but appears to be glossy when illuminated and viewed at an angle near to the grazing angle, such as 85° off the perpendicular. Sheen is therefore frequently evaluated in terms of gloss measurements made on a 75° or 85° gloss meter. See ▶ Gloss and ▶ Luster.

Sheers \ shirs\ *n* (ca. 1920) Transparent, lightweight fabrics of different constructions and yarns, especially those of silk and manufactured fibers. Examples are chiffons, some crepes, georgette, and voile.

Sheet \ shēt\ *n* [ME *shete*, fr. OE *scyte*; akin to OE *scēat* edge, OHGr *scōz* flap, skirt] (before 12c) (sheeting) Sheet is distinguished from film in the plastics and packaging industries according to thickness: a web under 0.25 mm thick is usually called film, whereas thicker webs are called sheet. Sheeting is most commonly made by extrusion, casting, and calendering, but some, such as decorative laminates, are compression molded.

Sheet Die *n* A heavy-walled, extremely rigid steel structure, bolted to an extruder head, whose inner passages form the molten plastic leaving an extruder screw into the shape of a flat sheet. Most modern sheet dies are of the coat hanger type with multi-zone temperature control, and contain adjustable choker bars and die lips for close control of lateral variation in the sheet thickness. See ▶ Coat Hanger Die, ▶ Choker Bar, and ▶ Flexible-Lip Die.

Sheeter Lines *n* Parallel scratches or projecting ridges distributed over considerable area of a plastic sheet such as might be produced during a slicing operation.

Sheet Line *n* The entire assembly necessary to produce plastic sheet, including the extruder, die, polishing-roll stand, cooling conveyor, pull rolls, and winder, or cutter and stacker, and all associated controls.

Sheet-Molding Compound *n* (SMC) A fiberglass-reinforced thermosetting compound in sheet form, usually rolled into coils interleaved with plastic film to prevent Autoadhesion. This term was chosen to replace the term ▶ Prepreg which was deemed to be confusing and insufficiently definitive. SMC can be molded into complex shapes with little scrap, and is low in cost.

Sheet Thermoforming *n* \-thər-mə-fórm-\ (thermoforming) The process of forming a thermoplastic sheet into a three-dimensional shape by clamping the sheet in a frame, heating it to render it soft and flowable, then applying differential pressure to make the sheet conform to the shape of a mold or die positioned below the frame. (Whittington's dictionary of plastics. Carley, James F (ed). Technomic, 1993) When the pressure is applied entirely by vacuum drawn through tiny holes in the mold surface, the process is called *vacuum forming* (Strong AB (2000) Plastics materials and processing. Prentice Hall, Columbus, OH) When above-atmospheric pressure is used to partially perform the sheet, the process becomes *air-assist vacuum forming*. In another variation, mechanical pressure is applied by a plug to perform the sheet prior to applying vacuum (*plug-assist forming*). In the *drape-forming* modification, the softened sheet is lowered to drape over the high points of a male mold prior to applying vacuum. Still other modifications are *plug-and-ring forming* (using a plug as the male mold and a ring matching the outside contour of the finished article); *ridge forming* (the plug is replaced with a skeleton frame); *slip forming* or *air-slip forming* (the sheet is held in pressure pads that permit it to slip inward as forming progresses); and *bubble forming* (the sheet is blown by air into a blister, then pushed into a mold by means of a plug). In *free forming* the sheet is formed entirely by gentle inflation with air and touches no mold, a method used to make cockpit canopies.

Shelf Life *n* (1927) (storage life) The length of time over which a product will remain suitable for its intended use during storage under specific conditions. (See ASTM, www.astm.org) The term is applied to some finished products as well as to raw materials. Premixes and prepregs, being reactive materials, enjoy considerably longer shelf lives if refrigerated. See also ▶ Storage Stability.

Shelf Life *n* The resistance to deterioration by oxygen and ozone in the air, by heat and light, or by internal chemical action. See ▶ Storage Life.

Shell *n* \ shel\ *n* [ME, fr. OE *sciell*; akin to OE *scealu* shell, ON *skel*, Gk *skallein* to hoe] (before 12c) A principal set of electron energies in an atom; designated by K, L, M, N etc, or *n* (principal quantum number) = 1, 2, 3, 4, etc.

Shellac *n* \shə-lak\ *n* [¹*shell* + *lac*] (1713) Alcohol-soluble, orange-colored resin which is the secretion of the female of the insect (*Laccifer lacca*, *Coccus lacca*) found in great quantities in India and Indochina. The Shellac appears as an incrustation on the twigs of certain trees. After feeding, the insect produces through its pores a gummy substance that hardens into a protective covering called lac. The lac is washed and dried, and when dissolved in ethanol will form a liquid suitable for application to furniture, etc. An esterification begins when dissolved in ethanol, but this process prevents from shellac from completely drying and provides a sticky-gum type of coating.. It forms the basis of "French Polish" and several other types of spirit varnish and is used for adhesive purposes. An often heard complaint about shellac is that it will not completely dry. Shellac has been largely replaced by better performing synthetic varnishes, etc. (Wicks ZN, Jones FN,

Pappas SP (1999) Organic coatings science and technology, 2nd edn. Wiley-Interscience, New York; Langenheim JH (2003) Plant resins: Chemistry, evolution ecology and ethnobotany. Timber, Portland, OR)

Shell Molding *n* In metal foundries, a process of casting metal objects in thin molds made from sand or a ceramic powder mixed with a thermosetting-resin binder. Some authors have misused the term by equating it to plastics processes such as dipping and slush casting.

Shellolic Acid *n* Very complex constituent of shellac believed to consist of a tricyclic nucleus, with two carboxyl and two hydroxyl groups.

Sherardizing *n* Method of coating steel or cast iron articles in intimate contact with zinc. The articles are heated with zinc dust for several hours. The zinc forms an alloy, at the interface of the ferrous surface, thus producing a thin, tightly adherent coating.

Shier *n* A short length of a single pick that appears to be cut out of the plane of the fabric.

Shift Factor *n* The amount by which the logarithm of the modulus (or compliance) of a plastic, measured at temperature T (K) must be shifted along the time axis to bring it onto a single curve with the modulus measured at T_g, the glass-transition temperature; the shift factor relationships is:

$$aT = \frac{-17.4(T - T_g)}{51.6 + (T - T_g)}.$$

(Physical properties of polymers handbook. Mark JE (ed). Springer, New York, 1996) See ▶ Time-Temperature Equivalence and ▶ Williams-Landell-Ferry Equation.

Shiner *n* A relatively short streak caused by a lustrous section of a filament yarn. The principal cause is excessive tension applied to a yarn during processing.

Shingle Stain *n* A low-viscosity, pigmented, penetrating paint for use on wood shingles to provide color and protection against moisture penetration.

Shire See ▶ Shier.

Shish-Kebab Structure \ˈshish-kə-ˌbäb\ *n* A term borrowed from the broiling kitchen to designate a polymeric microstructure in which the random-coil chain of an amorphous polymer (shish) has been interlaced with crystalline cross segments (kebab) produced by shearing the polymer in the molten condition or in solution. The string-of-lumps microstructure so resembles the edible delicacy that its borrowed name seems quite apt. (Elias HG (1977) Macromolecules vol 1–2. Plenum, New York)

Shoe See ▶ Chase.

Shoe Fold *n* A manner of folding fabric. The piece is folded from both ends into 12- or 16-folds. The length of the fold depends upon the length of the piece.

Shogged Stitch See ▶ Racked Stitch.

Shop Coat *n* One or more coats applied in a shop or plant prior to shipment to the site of erection or fabrication, where the field or finishing coat is applied.

Shop Painting See ▶ Shop Coat.

Shop Primer See ▶ Prefabrication Primer.

Shop Priming *n* Act of priming new wood or metal on the maker's premises prior to delivery to the purchaser.

Shore A See ▶ Shore Hardness.

Shore D See ▶ Shore Hardness.

Shore Hardness See ▶ Indentation Hardness.

Short \ˈshórt\ *adj* [ME, fr. OE *sceort*; akin to OHGr *scurz* short, ON *skortr* lack] (before 12c) (1) In reinforced plastics, an imperfection caused by an absence of surface film in some areas or by lighter, unfused particles of material showing through a covering surface film, accompanied possibly by thin-skinned blisters. (2) Lacking toughness or elasticity. A dried film is referred to as being short when it is quite brittle. The undried material is referred to as being short when it is crumbly and has no flowing properties; opposite of length.

Short-Beam Shear Strength *n* Shear strength of reinforced-plastics materials as measured in three-point bending of a specimen whose length between end supports is about five times the specimen depth. ASTM D 2344 describes one such test for flat specimens and short arcs cut from relatively large rings, while ASTM D 4475 describes a similar test for pultruded round rod.

Short-Chain Branching *n* The presence of 2- to 4-carbon side chains along the backbone of a polymer molecule. In an average low-density polyethylene molecule, both ethyl and butyl side chains, mostly the latter, are believed to be present, with a total of 50 in a typical molecule, in addition to one much longer branch.

Short-Cut Staple *n* Staple fiber less than 0.75-in. long. Typically used in wet-laid nonwoven processes to make fabrics, or as reinforcement in plastics, concrete, asphalt, and other materials.

Short Fibers *n* A somewhat relative term: inorganic fibers from 12 to 38 mm long are referred to as "short staple," those longer than 76 mm, "long staple." In fiber-reinforced thermoplastics sold in pellet form, the fibers are less than 12 mm long, yet provided substantial reinforcement. Studies have shown that most of the possible strength gain is achieved with an aspect ratio of about 200 or more. Since glass-fiber diameters are about 10 μm, a length of 2 mm should be sufficient. For *modulus* improvement, greater lengths are beneficial.

Short Ink *n* A highly pigmented ink which is stiff and cannot be drawn out to a thread.

Short-Liquor Dyeing See ▶ Dyeing.

Shortness *n* A qualitative term that describes an adhesive that does not string cotton, or otherwise form filaments or threads during application. That property of a printing ink which is characterized by a lack of flow. Opposite of length.

Short Oil Alkyd *n* An alkyd resin containing les than 40% oil in solids.

Short Oil Varnish *n* A varnish containing little oil in comparison with the amount of resin present, less than 15 gal oil/100 lb (1.25 L oil/kg) resin.

Short Shot *n* In injection molding, failure to fill all cavities of the mold completely, caused by too low melt temperature, too low injection pressure, insufficient plastication time, too constricted gates, too viscous resin, inadequate venting of cavities, etc. Short shots are often made deliberately when testing a new multicavity mold to reveal the pattern of runner flow and the sequence of cavity filling.

Shortstopper *n* A term used for an agent added to a polymerization-reaction mixture to inhibit or terminate polymerization.

Shot \ˈshät\ *n* (1) One complete cycle of a molding machine. (2) In woven pile floor coverings, the number of filling yarns per row of tufts. (3) Imprecise Syn: ▶ Shot Weight.

Shot Blasting See ▶ Shot.

Shot Capacity *n* The maximum weight of plastic that can be delivered to an injection mold by one stroke of the ram or screw. In the case of screw-injection molding machines that are not equipped with backflow-preventing valves at the end of the screw, slippage of material may occur in the screw flights and may not be reckoned with in calculations of shot capacity that are based on cubic displacement.

Shot Peening *n* Blasting with round iron shot, round steel shot, or any material which retains its spherical shape for peening purposes. *Also known as Shot Blasting*.

Shot Weight *n* (shot) In injection and transfer molding, the entire mass of plastic delivered in one complete filling of the mold, including the molded parts, sprue, runners, cull, and flash.

Show Card Colors See ▶ Poster Color.

Show Through *n* (1) See ▶ Photographing. (2) The transparency of printed sheets which permits printing to be seen from the other side of the sheet.

Shredding \ˈshred-, *esp Southern* ˈsred-\ *v* {shredded; shredding} (before 12c) The separation of compressed fibers in pulp sheets prior to acetylation in acetate manufacture.

Shrinkage \ˈshriŋ-kij, *esp Southern* ˈsriŋ-\ *n* (1800) Disruption of the level plane of the finished surface, resulting in a loss of gloss or wrinkling. Contraction of the wooden substrate, frequently resulting in cracking and/or checking of the coated surface.

Shrinkage Allowance *n* The dimensional allowance that must be made in molds to compensate for shrinkage of the plastic compound on cooling. The ASTM method for determining shrinkage from molded bars or disks and mold dimensions is D 955. This method does not provide for additional shrinkage that may occur as molded materials age beyond the first 48 h after removal from the mold.

Shrinkage Force *n* The force generated by thermoplastic materials when they are subjected to elevated temperatures.

Shrinkage Pool *n* An irregular, slightly depressed area on the surface of a molding caused by uneven shrinkage before hardening is complete.

Shrink Film *n* (1) Films which shrink when heated and are useful for packaging articles. (2) The prestretched or oriented film used in ▶ Shrink Packaging. See also ▶ Heat-Shrinkable Film.

Shrink Fit *n* A method of joining circular and annular parts in which the outer member, having a slightly smaller inside diameter than the inner member's outside diameter, is heated, causing it to expand, then slipped into place over the inner member and allowed to cool. Alternatively, one can chill the inner member in liquid nitrogen, slip it into the outer member, and let is warm. Care must be taken in designing the joint to have the final stresses in both members well below yield values so as not to lose the joint to creep over time. Compare ▶ Snap Fit.

Shrink Fixture Syn: Cooling Fixtures.

Shrink Mark See ▶ Sink Mark.

Shrink Packaging *n* A method of wrapping articles utilizing prestretched (oriented) films that are warmed to cause them to shrink tightly around the enclosed articles. First, the article is placed in a loose envelope of two layers of film, usually in the form of a V-folded strip. This envelope is heated sealed around the edges and detached from the strip, both of which operations can be done with an L-shaped, thermal-impulse sealer and cutter. The package is then conveyed through a hot-air oven or other heating device to shrink the film.

Shrink Tunnel *n* An oven in the form of a tunnel mounted over, or containing a continuous conveyor belt, used to shrink oriented films in the shrink-packaging process.

Shrink Wrap *vt* (1966) (stretch wrap) The use of plastic films for unitizing several boxes or items loaded on a pallet. Film may simply be stretched over the materials to be protected, pulled tight and secured; or shrunk by application of heat. ASTM D 4649 (Section 15.09) is a guide to the selection of such materials. See also ▶ Shrink Packaging.

Shrival Finish See ▶ Ripple Finish.

Shutter \shə-tər\ *n* (1542) A movable screen or cover used to cover an opening, especially a window.

Shuttle \shə-tᵊl\ *n* [ME *shittle*, prob. fr. OE *scytel* bar, bolt, akin to ON *skutill* bolt, OE *scēotan* to shoot] (14c) A boat-shaped device, usually made of wood with a metal tip that carries filling yarns through the shed in the weaving process. It is the most common weft-insertion device. The shuttle holds a quill, or pin, on which the filling yarn is wound. It is equipped with an eyelet at one end to control rate. The filling yarn is furnished during the weaving operation.

Shuttle Chafe Mark *n* A fabric defect that is usually seen as groups of short, fine lines across the fabric, often running for some distance in the piece and usually in the same area. Although these marks run in the direction of the filling, they are actually caused by the shuttle rubbing across and damaging the warp ends, producing a dull, chalky appearance.

Shuttleless Loom *n* A loom in which some device other then a shuttle is used for weft insertion. Also see ▶ Loom and ▶ Weft Insertion.

Si *n* Chemical symbol for the element silicon.

SI *n* (1) Abbreviation for ▶ Silicon or ▶ Polydimethylsiloxane. (2) Abbreviation for "International Systems of Units," derived from the official French name, *Le Systèm International d'Unitès*. An internationally agreed coherent system of units, derived from the MKS system, now in use for all scientific purposes and thereby replacing the cgs system and the fps system. The seven basic units are: the *metric* (symbol m), *kilogram* (kg), *second* (s), *ampere* (A), *Kelvin* (K), *mole* (mol), and *candela* (cd). The *radian* (rad) and *steradian* (sr) are supplementary units. Derived units include the *hertz* (Hz), *newton* (N), *joule* (J), *watt* (W), *coulomb* (C), *volt* (V), *farad* (F), *ohm* (Ω), *weber* (wb), *tesla* (T), *henry* (H), *lumen* (lm), and *lux* (lx).

Siamese Blow \sī-ə-mēz, -mēs\ *n* A colloquial term denoting the process of blow molding two or more objects or parts of objects in a single blowing mold, then cutting them apart.

SiB *n* Abbreviation for ▶ Polycarboranesiloxane.

Siccatives *n* Any reagent which catalyzes or promotes oxidation of oils; a drier. See ▶ Driers.

Side Bar *n* A loose piece used to carry one or more molding pins, and operated from outside the mold (seldom seen today).

Side-Draw Pin *n* A projecting mold element used to core a hole in a direction other than the direction of mold closing, and which must be retracted before the mold is opened and the part ejected.

Siding *n* The finish covering an exterior wall of a frame building; the siding may be a cladding material such as wood, aluminum, or cement asbestos (but not masonry); applied vertically or horizontally. Also see ▶ Clapboard, ▶ Drop Siding. Syn: ▶ Weatherboarding.

Siding Shingle *n* A shingle of any of a number of materials such as wood, cement asbestos, etc., used as a protective exterior wall covering over sheathing.

Siemens \sē-mənz\ *n* {plural siemens} [Werner von Siemens † 1892 German electrical engineer] (ca. 1933) (S) The SI unit of electrical conductance, equal to and replacing the *mho*, and the reciprocal of the OHM. 1 Siemens = 1 ampere/volt (A/V).

Sienna \sē-e-nə\ *n* [It *terra di Siena*, lit., Siena earth, fr. *Siena*, Italy] (1787) A red variety of iron oxide pigment.

Sierra Leone Copal \sē-er-ə-lē-ōn kō-pəl\ *n* Fossil copal of African origin, It becomes soluble in drying oils after running, and because of its pale color, which is maintained after running, is used to some extent in grinding media for whites and tinted whites, silver baking varnishes, etc.

Sieve \siv\ *n* [ME *sive*, fr. OE *sife*, akin to OH Gr *sib* sieve] (before 12c) In laboratory work, an apparatus in which the apertures are square, for separating sizes of material.

Sieve Analysis *n* (screen analysis) The separation of particulate solids into sequentially finer size fractions by placing a weighed sample into the topmost of a stack of graded standard sieves, mechanically shaking and tapping the stack for 10 min, then weighing the material collected on each sieve and the pan beneath the lowest, finest sieve. The procedure is described in ASTM D 1921. *Also known as Screen Analysis*.

Sieve Fraction *n* The mass fraction of a sieve-analyzed powder found between two successive screens in a ▶ Sieve Analysis. For example, one might say, "The −0.420-, +0.250-mm fraction was 15.27 percent." See also ▶ Mesh Number.

Sieve Number *n* A number used to designate the size of a sieve, usually the approximate number of sieve cross wires per linear inch.

Sieving *n* Separation of a mixture of various-sized particles, either dry or suspended in a liquid, into two or more portions, by passing through screens of specified mesh sizes. See ▶ Sieve Analysis. *Also known as Screening.*

Sigma-Blade Mixer *n* A type of ▶ Internal Mixer having blades that are (roughly) S-shaped. See ▶ Internal Mixer.

Particle size conversion

Sieve designation		Nominal sieve opening		
Standard	*Mesh*	*inches*	*mm*	*Microns*
25.4 mm	1 in.	1.00	25.4	25,400
22.6 mm	7/8 in.	0.875	22.6	22,600
19.0 mm	3/4 in.	0.750	19.0	19,000
16.0 mm	5/8 in.	0.625	16.0	16,000
13.5 mm	0.530 in.	0.530	13.5	13,500
12.7 mm	1/2 in.	0.500	12.7	12,700
11.2 mm	7/16 in.	0.438	11.2	11,200
9.51 mm	3/8 in.	0.375	9.51	9,510
8.00 mm	5/16 in.	0.312	8.00	8,000
6.73 mm	0.265 in.	0.265	6.73	6,730
6.35 mm	1/4 in.	0.250	6.35	6,350
5.66 mm	No. 3 1/2	0.223	5.66	5,660
4.76 mm	No. 4	0.187	4.76	4,760
4.00 mm	No. 5	0.157	4.00	4,000
3.36 mm	No. 6	0.132	3.36	3,360
2.83 mm	No. 7	0.111	2.83	2,830
2.38 mm	No. 8	0.0937	2.38	2,380
2.00 mm	No. 10	0.0787	2.00	2,000
1.68 mm	No. 12	0.0661	1.68	1,680
1.41 mm	No. 14	0.0555	1.41	1,410
1.19 mm	No. 16	0.0469	1.19	1,190
1.00 mm	No. 18	0.0394	1.00	1,000
841 µm	No. 20	0.0331	0.841	841
707 µm	No. 25	0.0278	0.707	707
595 µm	No. 30	0.0234	0.595	595
500 µm	No. 35	0.0197	0.500	500
420 µm	No. 40	0.0165	0.420	420
354 µm	No. 45	0.0139	0.354	354
297 µm	No. 50	0.0117	0.297	297
250 µm	No. 60	0.0098	0.250	250
210 µm	No. 70	0.0083	0.210	210
177 µm	No. 80	0.0070	0.177	177
149 µm	No. 100	0.0059	0.149	149
125 µm	No. 120	0.0049	0.125	125
105 µm	No. 140	0.0041	0.105	105

Particle size conversion (Continued)

Sieve designation		Nominal sieve opening		
Standard	Mesh	inches	mm	Microns
88 μm	No. 170	0.0035	0.088	88
74 μm	No. 200	0.0029	0.074	74
63 μm	No. 230	0.0025	0.063	63
53 μm	No. 270	0.0021	0.053	53
44 μm	No. 325	0.0017	0.044	44
37 μm	No. 400	0.0015	0.037	37

Larger sieve openings (1–1/4 in.) have been designated by a sieve "mesh" size that corresponds to the size of the opening in inches. Smaller sieve "mesh" sizes of 3 1/2–400 are designated by the number of openings per linear inch in the sieve.
The following convention is used to characterize particle size by mesh designation:
- A "+" before the sieve mesh indicates the particles are retained by the sieve.
- A "−" before the sieve mesh indicates the particles pass through the sieve.
- Typically 90% or more of the particles will lie within the indicated range.

For example, if the particle size of a material is described as −4 + 40 mesh, then 90% or more of the material will pass through a 4-mesh sieve (particles smaller than 4.76 mm) and be retained by a 40-mesh sieve (particles larger than 0.420 mm). If a material is described as −40 mesh, then 90% or more of the material will pass through a 40-mesh sieve (particles smaller than 0.420 mm).
This information is also provided on page T848 of the Aldrich 2003–2004 Catalog/Handbook of Fine Chemicals.
Copyright © 2007 Sigma-Aldrich Co. Reproduction forbidden without permission.
Sigma-Aldrich brand products are sold exclusively through Sigma-Aldrich, Inc. Best viewed in IE6 or higher.

Sigma (σ) Bond A covalent bond in which the electron charge cloud of a shared pair is centered on and symmetrical around the bond axis.

Signature \ˈsig-nə-ˌchúr\ *n* [MF or ML; MF, fr ML *signatura*, fr. L *signatus*, pp of *signare* to sign, seal] (1536) In web printing and binding the name given to a printed sheet after it has been folded.

Sign of Double Refraction *n* An empirical classification of crystals; it is positive for uniaxial crystals when $\varepsilon > \omega$, for biaxial crystals when $\gamma - \beta > \beta - \alpha$. It is negative for uniaxial crystals when $\varepsilon < \omega$, for biaxial crystals when $\gamma - \beta < \beta - \alpha$.

Sign of Elongation *n* This refers to the elongation of a substance in relation to refractive indices. If it is elongated in the direction of the high refractive index, it is said to be positive; if it is elongated in the direction of the low refractive index, it is negative.

Silane \ˈsi-ˌlān\ *n* [ISV *sil*icon + meth*ane*] (1916) Common name for silicon tetrahydride. It is used as a doping agent for solid state devices, production of amorphous silicon.

$$SiH_4$$

Silane Coupling Agent *n* Any silane or oxysilane that has the ability to bond inorganic materials such as glass, mineral fillers, metals, and metallic oxides to organic resins. The adhesion mechanism is due to two groups in the oxysilane structure. (Silanes and other coupling agents, vol 3. Mittal KL (ed). VSP International Science, New York, 2004) The $Si(OR_3)$ portion reacts with the inorganic reinforcement, while the organofunctional (vinyl–, amino–, epoxy–, etc.) group reacts with the resin. The coupling agent may be applied to the inorganic materials (e.g., glass fibers) as a pretreatment and/or added to the resin. Examples of silane coupling agents are:

N-β(Aminoethyl)-γ-aminopropyltrimethoxy silane
γ-Aminopropyltrimethoxy silane
Bis(B-hydroxyethyl)- γ-aminopropyltriethoxy silane
β-(3,4-Epoxycyclohexyl) ethyltrimethoxy silane
γ-glycidoxypropyltrimethoxysilane
γ-methacryloxypropyltrimethoxy silane
Sulfonylazidosilane
Vinyltrichlorosilane
Vinyltriethoxysilane
Vinyl-*tris*-(β-methoxyethyl)silane.

(Whittington's dictionary of plastics. Carley, James F z (ed). Technomic, 1993)

A newer class of silane coupling agents is known as *silyl peroxides*, represented by the general formula: $R'_m R''_{4-n-m} SI(OOR)_n$. A typical member of this family is vinyl-tris-(*t*-butylperoxy) silane. The coupling mechanism of the silyl peroxides, effected by heat only, is free-radical in nature. The conventional silanes require an eternal free-radical source and couple via an ionic mechanism initiate by hydrolysis.

Silanes *n* Group of compounds resembling the saturated hydrocarbons, in which the carbon is replaced by silicon. For example, monosilane SiH_4, cf. methane CH_4; disilane Si_2H_6; cf. ethane C_2H_6.

Silica \ˈsi-li-kə\ *n* [NL, fr. L *silic-*, *silex* hard stone, flint] (ca. 1801) (silicon dioxide) SiO_2. A substance occurring widely in minerals such as quartz, sand, flint, chalcedony, opal, agate, and many more. In powdered form it is used as a filler, especially in phenolic compounds for ablative nose cones of rockets. Synthetic silicas, made from sodium silicate or by heating silicon compounds, are useful in preventing ▶ Plate-Out. See also ▶ Fumed Silica.

$$O=Si=O$$

Silica, Amorphous \-ə-ˈmór-fəs\ *n* See ▶ Amorphous Silica.

Silica *n*, **Crystalline** SiO_2. Pigment White 27 (77811). Pigment grade crystalline silica is quartz sand that has been crushed, ground, and/or air classified. A reinforcing filler for paints, elastomers, sealants. Chemically inert, heat resistant, nonconductive; imparts burnish resistance, sheen uniformity, good flatting in latex paints. Density, 2.56 g/cm^3 (22.1 lb/gal); O.A., 24–36; particle size, 1.5–9.0 μm. Syn: ▶ Powdered Quartz, ▶ Silica Flour, ▶ Ground Silica.

Silica *n*, **Diatomaceous** \-ˌdī-ə-tə-ˈmā-shəs\ See ▶ Diatomaceous Silica.

Silica Flour See ▶ Silica, Crystalline.

Silica Gel *n* A form of colloidal silica consisting of grains having many fine pores and capable of adsorbing, and firmly retaining at room temperature, substantial quantities of water and some other compounds. It is used to dry gas streams and organic liquids to very low moisture levels. It can be reactivated by heating to temperatures above 100°C. Compare ▶ Molecular Sieve.

Silica, Microcrystalline \-ˈmī-krō-ˌkris-tə-lin\ *n* SiO_2 Pigment White 27 (77811). Extender pigment obtained from extensive geologic formations of hard, compact, homogeneous, microcrystalline silica located in the vicinity of Hot Springs, Arkansas. Can be considered as a very finely divided or "micro" form of quartz in a bound state of subdivision. Excellent dielectric and low abrasive properties. Density, 2.65 g/cm^3 (22.1 lb/gal); O.A., 20.0; mean particle size, 4.0 μm. "Novaculite" is a generic name. Syn: ▶ Novacite.

Silica, Soft See ▶ Amorphous Silica.

Silica, Synthetic *n* (**Aerogel and Hydrogel**) SiO_2. Extremely porous and light weight synthetic silicas which may contain up to 94% of "dead air" space that is enclosed by a tenuous webbing of microscopic silica filaments. See also ▶ Sodium Silicoaluminate.

Silica, Synthetic (Pyrogenic) SiO_2. A colloidal silica used in the rubber, plastic and paint industry. High purity, nonreactive pigment which imparts thixotropy, flatting and pigment suspension. Density, 2.2 g/cm^3; O.A., 230; particle size, 0.012 μm.

Silicate \ˈsi-lə-ˌkāt\ *n* [*silicic (acid)*] (1811) Any member of the very widely occurring compounds characterized by the presence of the elements, silicon, oxygen and one or more metals with or without hydrogen (e.g., talc).

Silicate Paints Water-paints based on sodium, potassium, or lithium silicate. Used in zinc-rich paints. They are characterized by their nonflammability. Care must be exercised in the selection of pigments used with the silicate because of its alkalinity.

Silicic Acid *n* (1817) Any of various weakly acid substances obtained as gelatinous masses by treating silicates with acids.

Silicon \ˈsi-li-kən\ *n* [NL *silica* + E *–on* (as in *carbon*)] (1817) Generic name for polymers with (—SiR_2—O—) links. Manufactured by Bayer, Germany; Dow, US; and General Electric, US.

Silicon Carbide *n* (1893) (SiC) In the form of crystals, produced in an electric furnace by reaction of carbon with sand, silicon carbide is a dense, extremely hard filler, used in some plastics to increase abrasion resistance, elastic modulus, and thermal conductivity.

Silicon Carbide Abrasive *n* Of the many crystalline forms of silicon carbide produced in electric furnaces, the hexagonal, or alpha crystals made up of masses of interlocking crystals, in the major type used for coated abrasives. It is greenish black iridescent, has high thermoconductivity, low thermoexpansion and great chemical stability, being unaffected by nitric, sulfuric, hydrochloric acids. (CRC handbook of chemistry and physics. Lide DR (ed). CRC, Boca Raton, FL, 2004 Version; Harrington BJ (2001) Industrial cleaning technology. Kluwer Academic, New York)

Silicon Carbide Waterproof Paper *n* Hard, sharp, coated abrasive particularly suited for lacquer, plastics and nonferrous metals, also for wet sanding primes, undercoats and between coats on wood and metal using water, oil or other lubricants. See ▶ Coated Abrasive.

Silicon Carbide Whiskers *n* These high-modulus SiC fibers are made by pyrolysis of organosilanes in hydrogen at 1,500–2,000°C and by other methods. The whiskers are very fine, with diameters of about 2.5 μm, density of 3.2 g/cm^3, ultimate strength of 20 GPa (3 Mpsi), and modulus of 480 GPa (more than twice that of steel).

Silicone \ˈsi-lə-ˌkōn\ *n* [*silic*on + -*one*] (1943) One of a large family of semi-organic polymers (*polyorgano siloxanes*) comprising chains of alternating silicon and oxygen atoms, modified with various organic groups attached to the silicon atoms. Depending on the nature of the attached organic groups, molecular weight, and the extent of crosslinking between chains, the polymers may be oily fluids ranging in viscosity from 0.001 to over 1,000 Pa s, or elastomers, or solid resins. (Handbook of adhesives. Skeist I (ed). Van Nostrand Reinhold, New York, 1990) The earliest silicones were dimethyl polysiloxanes, made by treating silicon derived from sand with methyl chloride in the presence of a catalyst to form a chlorosilane, hydrolyzing the chlorosilane to form a cyclic trimer of siloxane, then polymerizing the siloxane to form a dimethyl polysiloxanes. (Merck index, 13th edn. Merck, Whitehouse Station, NJ, 2001) Many modifications have been made including the incorporation of phenyl groups, halogen atoms, alkyds, epoxides, polyesters, and other organic compounds containing –OH groups. The silicone fluids are used as lubricants, mold-release agents, heat-transfer fluids, and water-repellent coatings. (Strong AB (2000) Plastics materials and processing. Prentice Hall, Columbus, OH) The elastomers, often called silicone rubbers and reinforced with inorganic fillers or fibers, are vulcanizable (crosslinkable) and offer superior resistance to high temperature and weathering. The silicone resins, possessing good electrical properties and strength at high temperature, are widely used for encapsulating and potting electrical components and in reinforced laminates. Silicone coatings offer superior weathering resistance in direct sunlight. (Wicks ZN, Jones FN, Pappas SP (1999) Organic coatings science and technology, 2nd edn. Wiley-Interscience, New York)

Silicone Foam *n* Foam based on fluid silicone resins is made by mixing the resins with a catalyst and blowing agent, pouring the mixture into molds, and curing at room temperature for about 10 h or at elevated temperatures for shorter periods. Silicone-foam sponge is made by mixing unvulcanized silicone rubber with a blowing agent and heating to the vulcanizing temperature.

Silicone Paint *n* Paint, based on silicone resins, that is resistant to very high temperatures and therefore useful on smokestacks, heaters, stoves, and electrical insulation; requires heat to cure or set; has a high resistance to chemical attack.

Silicone Plastics *n* Plastics based on silicone resins.

Silicone-Polycarbonate Copolymer *n* Introduced in 1969 by the General Electric Co, these thermoplastic copolymers vary from strong elastomers to rigid engineering plastics, depending on composition. They can be extruded, cast, or molded into optically clear films.

Silicone Resins *n* Group of resins containing a substantial amount of silicon, distinguished by their outstanding heat resistance, high water repellency, and chemical resistance. They are made by preparing dialkyl dichlorsilanes from silicon tetrachloride and the corresponding alkyl magnesium bromide. The silanes are converted to silanediols, which are polymerized into resinous products. Mixtures of silanediols and triols are copolymerized to yield thermosetting resins.

Silicone Rubber *n* (1944) A synthetic rubber made by vulcanizing an elastomeric silicone gum such as dimethyl silicone. A free-radical-generating catalyst such as benzoÿl peroxide is usually used as the vulcanizing agent. The tensile strength of unreinforced silicone rubber is low, about 350 kPa. Higher strengths are obtained by adding reinforcing fillers such as finely divided or fumed silica, or by putting crystallizing segments such as silphenylene into the polymer. (Allcock HR, Mark J, Lampe F (2003) Contemporary polymer chemistry. Prentice Hall, New York) See also ▶ Silicone.

Silicon Nitride *n* (1903) Any of several compounds of silicon and nitrogen; specifically a compound Si_3N_4 that is a hard ceramic used in high-temperature applications and in composites. (CRC handbook of chemistry and physics. Lide DR (ed). CRC, Boca Raton, FL, 2004 Version)

$$Si \equiv N$$

Silicon Nitride Whiskers *n* Very fine fibers of Si_3N_4 prepared by vapor-phase reaction of silicon and a silicate in nitrogen and hydrogen at 1,400°C. Density is 3.2 g/cm^3, ultimate strength is 4.8 GPa (700 kpsi), and modulus is 276 GPa. They are used as reinforcements in specialty laminates.

Silicosis \ˌsi-lə-ˈkō-səs\ *n* [NL, fr. silica + -osis] (1881) A form of pneumoconiosis resulting from occupational exposure to and inhalation of silica dust over a period of years; characterized by a slowly progressive fibrosis of the lungs, which may result in impairment of lung function; silicosis predisposes to pulmonary tuberculosis. Syn: ▶ Silicatosis. (Stedman's medical dictionary, 27th edn. Lippincott Williams and Wilkins, New York, 2000) A chronic disease of the lungs caused by the continued inhalation of silica dust. A disease due to breathing air containing silica characterized anatomically by generalized fibrous changes and the development of military modulation in both lungs, and clinically by shortness of breath, decreased chest expansion, lessened capacity for work, presence of fever, increased susceptibility to tuberculosis and by characteristic X ray findings.

Silk *adj {often attributive}* [ME, fr. OE *seolc*, prob. ultimately fr. Gk *sērikos* silken] (before 12c) British for *Satin Finish*. (Kadolph SJJ, Langford AL (2001) Textiles. Pearson Education, New York)

Silk Fiber *n* A fine, strong, continuous filament produced by the larva of certain insects, especially the silkworm, when constructing its cocoons. The silkworm secretes the silk as a viscous fluid from two large glands in the lateral part of the body. The fluid is extruded through a common spinneret to form a double filament cemented together. This double silk filament, which is composed of the protein fibroin, ranges in size from 1.75 to 4.0 denier, depending upon the species of worm and the country of origin. The filament of the cocoon is softened and loosened by immersion in warm water and is then reeled off. Although raw silk contains 20–30% of sericin, or silk glue, and is harsh and stiff, silk is soft and white when all of the glue has been removed by steeping and boiling in soap baths. Ecru is harsher, as it has only about 5% of the sericin removed. Silk is noted for its strength, resiliency, and elasticity. The major sources of commercial silk are Japan and China.

Silking *n* Surface defect characterized by parallel irregularities left on or in the dried surface of a glossy paint or varnish film, producing the appearance of silk. In dipping or flow coating, the irregularities appear in the direction of the flow in brushing, in the direction in which the film as finally brushed.

Silk-Screen *n* (1930) A printing process by which repeats of a motif or pattern can be made; the process involve the use of silk (or similar material of fine mesh) tightly stretched on frames; a separate frame is used for each color. Areas of the design to be printed in a given color are left open while the rest of he silk is treated with an insoluble coating. The ground-coated paper, fabric or other material on which the design is to be reproduced is laid out on long tables. The pattern is reproduced by drawing a suitable ink or paint across the screen with a rubber squeegee, which forces the color through the parts where the mesh is exposed, and the operator moves from one preset point to the next. By careful registering, a number of screens can be used in succession over the same design for multiple-color work.

Silk Screen Ink *n* Quick drying, full bodied, volatile inks used in the silk screen printing process.

Silk Screen (Screen Process) Printing *n* A process by which an image is transferred to a substrate by squeezing the ink through the unblocked areas of a metal or fiber screen. See ▶ Screen Printing.

Sill \sil\ *n* [ME *sille*, fr. OE *syll*; akin to OHGr *swelli* beam, threshold] (before 12c) The horizontal member that bears the upright portion of a frame; especially the base of a window.

Silopren *n* Polysiloxane rubber. Manufactured by Bayer, Germany.

Siloxane \sə-ˈläk-ˌsān\ *n* [*sili*con + *o*xygen + meth*ane*] (1917) (1) A chemical group with the structure shown: in which the Rs are usually alkyl and can be alkyl radicals or even just hydrogen. (2) A compound containing siloxane links. A simple representative is hexamethyldisiloxane, $(CH_3)_3SiOSi(CH_3)_3$. Siloxane links are common in silicone resins. See ▶ Silicone.

Silver Filler \ˈsil-vər ˈfil-lər\ *n* Particles or flakes of silver in the 1–15-μm size range, compounded with plastics to provide electrical conductivity. Flakes are mechanically flattened to provide layers of contiguous flakes and high conductivity. Silver oxide, sulfate, and carbonate are also good conductors and are used for the same purpose.

Silver Ink *n* A printing ink whose principal pigment consists of aluminum powder.

Silver Point *n* Method of drawing which consists of working with a silver tipped instrument on a specially prepared paper. The result is a drawing of great delicacy. Generally used by late medieval and renaissance artists, e.g., Leonardo da Vinci.

Silver-Spray Process *n* A metallizing process based on glass-mirror art. The plastic article is first cleaned and lacquered as in vacuum metallizing, then the lacquer is sensitized in an oxidizing aqueous solution of sulfuric acid and potassium dichromate. A silver-forming solution, e.g., silver nitrate and an aldehyde, is sprayed on the article with a two-nozzle spray gun so that the components mix at the surface. After rinsing and drying a final topcoat of lacquer is applied over the silver.

Silver White *n* (1) Any of several white pigments used in paints. (2) A very pure variety of white lead; French white, China white.

Silyl Peroxide See ▶ Silane Coupling Agent.

Sime, Setting *n* The period of time during which an assembly is subjected to heat or pressure, or both, to set the adhesive. See also ▶ Time, Curing; ▶ Time, Joint Conditioning; and ▶ Time, Drying.

Simon-Goodwin Charts n A series of charts for graphically calculating color differences using the modified Mac Adam Color Difference Equation. See ▶ Macadam Color Difference Equation and ▶ Foster Charts.

Simple Harmonic Motion n Periodic oscillatory motion in a straight line in which the restoring force is proportional to the displacement. If a point moves uniformly in a circle, the motion of its projection on the diameter (or any straight line in the same plane) is simple harmonic motion. If r is the radius of the reference circle, ω the angular velocity of the point in the circle, θ the angular displacement at the time t after the particle passes the mid-point of its path, the linear displacement,

$$x = r \sin \theta = r \sin \omega t$$

The velocity at the same instant,

$$v = r\omega \cos \theta = \omega \sqrt{r^2 - x^2}$$

The acceleration,

$$a = -\omega^2 x$$

The force for a mass m,

$$F = m\omega^2 x = -\frac{4\pi^2 mx}{T^2}$$

The period,

$$T = 2\pi \sqrt{\frac{x}{a}}$$

In the above equations the cgs system calls for x and r in centimeter, v in centimeter per second, a in centimeter per square second, T in second, m in grams, F in dynes, θ in radians and ω in the radians per second. (Handbook of chemistry and physics, 52nd edn. Weast RC (ed). The Chemical Rubber, Boca Raton, FL)

Simple Liquid See ▶ Newtonian Liquid.

Simple Machine n A contrivance for the transfer of energy and for increased convenience in the performance of work. *Mechanical advantage* is the ratio of the resistance overcome to the force applied. Velocity ratio is the ratio of the distance through which force is applied to the distance through which resistance is overcome. *Efficiency* is the ratio of the work done by a machine to the work done upon it. If a force f applied to a machine through a distance S results in a force F exerted by the machine through a distance s, neglecting friction,

$$fS - Fs$$

The theoretical mechanical advantage or velocity ratio in the above case is,

$$\frac{S}{s}$$

Actually, the force obtained from the machine will have a smaller value than will satisfy the equation above, If F' be the actual force obtained, the practical mechanical advantage will be

$$\frac{F'}{f}$$

The efficiency of the machine,

$$E = \frac{F's}{fS}$$

Simplified Flow Equation n An equation giving the delivery rate of a single-screw extruder as a function of the screw diameter, screw speed, channel depth in the metering section, and room-temperature density of the plastic. Underlying assumptions include: the screw is single-flighted in the metering section, with typical dimensional proportions and shape factors, and a lead angle of 17.7°; the feeding and plasticating capabilities of the screw (and drive) are sufficient to keep the metering section filled with melt; the feedstock is not preheated; and the total resistance to melt flow of screens and die does not entail excessive head pressure at the desired rate. See ▶ Net Flow.

SIMS Scanning Ion Mass Spectroscopy. Analysis of polymeric is a good application of SIMS, where metallic elements are often not observed. It is a microanalytical surface method of material examination. A SIMS spectrogram is shown. (Gooch JW (1997) Analysis and deformulation of polymeric materials. Plenum, New York)

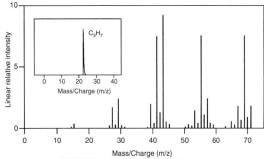

TOF-SIMS spectrogram of polypropylene specimen

Simulated Spun Yarns *n* Filament yarns that have been modified to have aesthetics similar to those of spun yarns. Simulated spun yarn have looped or hairy surfaces.

Singeling *n* The process of burning off protruding fibers from yarn or fabric by passing it over a flame or heated copper plates. Singeing gives the fabric a smooth surface and is necessary for fabrics that are to be printed and for fabrics where smooth finishes are desired. (Complete textile glossary. Celanese Corporation, Three Park Avenue, New York)

Single-Circuit Winding *n* A winding in which the filament path makes a complete traverse of the chamber, after which the following traverse lies immediately adjacent to the previous one.

Single-Knit Fabric *n* Also called plain knit, a fabric constructed with one needle bed and one set of needles.

Single Roll See ▶ Roll of Wallcovering.

Single-Screw Extruder See ▶ Extruder.

Single Spread See ▶ spread.

Single-Stage Resin (single-step resin) See ▶ Resol.

Singles Yarn *n* The simplest strand of textile material suitable for operations such as weaving and knitting. A singles yarn may be formed from fibers with more or less twist; from filaments with or without twist; from narrow strips of material such as paper, cellophane, or metal foil; or from monofilaments. When twist is present, it is all in the same direction. Also see ▶ Yarn.

Singling *n* A yarn defect caused by the breaking of one or more strands in a plying operation with resulting unevenness in the finished product.

Sinkage \ˈsiŋ-kij\ *n* (1883) The blotchy effect or loss of gloss due to absorption of the medium of a finishing coat by the undercoat. *Also called Sinking In.*

Sinker *n* In weave design, a blank square indicating a filling thread over a warp thread at the point of intersection.

Sinking *n* (hobbing) In mold making, pressing a hardened hob into annealed, soft mold steel or beryllium copper. See ▶ Hob.

Sinking In See ▶ Sinkage.

Sink Mark *n* (shrink mark) A shallow depression or dimple on the surface of an injection-molded article, usually in a thick section, caused by local internal shrinkage after the gate seals, or by a slightly short shot. Sinks can be diminished or prevented by reducing melt and mold temperatures, opening gates, filling faster, increasing pressure-holding time, and/or raising injection pressure.

Sinter \ˈsin-tər\ *v* [Gr *Sinter* slag, cinder, fr. OHGr *sintar*] (1871) {*transitive senses*} To cause to become a coherent mass by heating without melting.

Sinter Coating *n* A coating process in which the article to be coated is preheated to sintering temperature and immersed in a plastic powder, then is withdrawn and heated to a higher temperature to fuse the adhering sintered powder into a continuous skin on the article.

Sintering *n* The welding together of powdered plastic particles at temperatures just below the melting or fusion range. The particles are fused together to form a relatively strong mass, but the mass as a whole does not melt and may retain some porosity.

Sinter Molding *n* The process of compacting fine thermoplastic particles by applying pressure at temperatures a little below the melting range and holding pressure until the particles fuse together, often followed by further heating and post forming. Porous nylon bearings capable of absorbing lubricants are made by this method, and sinter molding is a main molding method for powders of polytetrafluoroethylene or ultra-high-molecular-weight polyethylene, which do not form true melts.

Sisal \ˈsī-səl\ *n* [*Sisal*, port in Yucatán, Mexico] (1843) A strong, durable, white fiber obtained from the *agave* plant grown in India, Indonesia, Mexico, and the West Indies. Chopped sisal enjoys some use as reinforcement in thermosetting resins.

Sissing See ▶ Cissing.

Size \ˈsīz\ *n* [ME *sise*] (15c) (1) Usually a liquid composition to prevent excessive absorption of paints into plaster, old wall paint and similar porous surfaces; also a liquid composition used as a first coat on metal to improve adhesion of succeeding coats (latter usage is limited to the metal decorating industry). The terms "sealer" or "size" are almost synonymous, but usage has established certain differences. A sealer is ordinarily a thin varnish or clear lacquer and is usually applied on wood and metal surfaces. Ordinary painter's size is a thin solution of glue, starch or other water-soluble substance and is usually applied on plaster surfaces, but size used in metal decorating is a thin varnish. Syn: Sizing. (2) An ink that dries with a sticky surface that will hold metallic or other powders. (3) Composition applied to paper to fill up the space between fibers and to produce a smooth surface. (4) A sealer used to prepare the wall before wallpaper is applied. (5) In antiquing, a thin pasty varnish liquid use for adhering gold leaf. It acts as a filler on porous surfaces and creates a more even effect with the gold leaf. (6) The formation of a dried layer or the film so formed on the surface of a mass. (7) In the plastics industry, a Syn: ▶ Coupling Agent. See ▶ Sizing.

Size-Exclusion Chromatography *n* (SEC, gel-permeation chromatography) A column-chromatography technique employing as the stationary phase a swollen gel made by polymerizing and crosslinking styrene in the presence of a diluent that is a nonsolvent for the styrene polymer. The polymer to be analyzed is introduced in dilute solution at the top of the column and then is eluted with pure solvent. The polymer molecules diffuse into the gel, and out of it, at rates depending inversely on their molecular size. As they emerge from the bottom of the column they are detected by the differential refractometer connected to a computer or recorder, where a plot of concentration versus retention time is developed. This is converted, through calibration information, to a molecular-weight distribution whose parameters are calculated and printed out by the computer. See also ▶ Chromatography.

Size Mark *n* A fabric defect that consists of a rough or frosted spin caused by uneven application or drying of the size.

Sizing *n* (1) The process of applying a material to a surface to fill pores to smooth it and reduce absorption of a subsequently applied adhesive or coating, or to otherwise modify the surface. (2) Determining dimensions during design or production. Also used as a Syn: ▶ Size.

Sizing *n* (1) A generic term for compounds that are applied to warp yarn to bind the fiber together and stiffen the yarn to provide abrasion resistance during weaving. Starch, gelatin, oil, wax, and manufactured polymers such as polyvinyl alcohol, polystyrene, polyacrylic acid, and polyacetates are employed. (2) The process of applying sizing compounds (also see ▶ Slashing) (3) The process of weighing sample lengths of yarn to determine the count.

Sizing Plate *n* In tubing and pipe extrusion, a plate with a central hole that may form the entrance to the water bath, and through which the still warm extrudate is passed to bring the outside diameter closer to its final dimension. At high line speeds, several plates having successively slightly smaller, smoothly finished openings may be used in sequence. Compare ▶ Sizing Ring.

Sizing Ring *n* (calibrator) In pipe extrusion, a hollow ring, slotted around its polished inside circumference, used to reduce the slightly oversize outside diameter of the warm pipe toward the desired final value. The core and slot of the ring are connected to a vacuum line. As the pipe passes through the ring, vacuum is applied to the core, sucking the pipe circumference against the slot and the ring's inside surface. Having made that seal, the operator raises the water level to submerge the pipe and cool it. Several rings of slightly decreasing inside diameters may be used in sequence for better control of pipe outside diameter and increased production rate. Compare ▶ Sizing Plate.

Skein \skān\ *n* [ME *skeyne*, fr. MF *escaigne*] (14c) A continuous strand of yarn or cord in the form of a collapsed coil. It may be of any specific length and is usually obtained by winding a definite number of turns on a reel under prescribed conditions. The circumference of the reel on which yarn is wound is usually 45–60 in. Also see ▶ Hank.

Skein Break Factor *n* The comparative breaking load of a skein of yarn adjusted for the linear density of the yarn expressed in an indirect system. It is the product of the breaking load of the skein and the yarn number expressed in an indirect system (e.g., pounds times cotton count). A statement of the skein break factor must indicate the number or wraps in the skein, if this is not otherwise apparent. Without specifying the number of wraps, a statement of the skein break factor is meaningless.

Skein Breaking Tenacity *n* The skein breaking load divided by the product of the yarn number in a direct numbering system and the number of strands placed under the tension (twice the number of wraps in the skein); preferably expressed in newtons per tex.

Skein Dyeing See ▶ Dyeing.

Skewness \skyü-nəs\ *n* (1894) The distance measured parallel to and along a selvage between the point at which a filling yarn meets this selvage and a perpendicular to the selvage from the point at which the same filling yarn meets the other selvage. Skewness may be expressed directly in inches or as a percentage of the width of the fabric at the point of measurement.

Skimmings *n* (**Tall Oil**) Curd, not acidified or otherwise processed, skimmed from the black liquor of the alkaline paper pump industry, from which tall oil is obtained.

Skin *n* (1) A relatively dense layer at the surface of a cellular polymeric material (ASTM D 883). (2) Film formed from a vehicle, liquid coating or ink composition or formed during storage. (3) An ungrounded, nonwashable, low-priced grade of wallpaper.

Skinning *n* Formation of a surface skin on paints or varnishes or printing inks in the container.

Skin Packaging *n* A variation of the thermoforming process in which the article to be packaged serves as the mold. The article is usually placed on a printed card prepared with an adhesive coating or mechanical surface treatment to seal the plastic film to the card. See also ▶ Blister Packaging.

Skipping *n* In coil coating, an irregular paint application usually occurring when improper contact is made between the applicator roll and the strip.

Skippy *n* Said of paint too heavy-bodied for uniform application, which causes the brush to skip on the surface, leaving some spots insufficiently coated and others with too heavy a coating.

Skips See ▶ Holidays.

Skirting Board See ▶ Baseboard.

Skiving \\ˈskīv-iŋ\\ *vt* {*skived; skiving*} [prob. of Scand origin; akin to ON *skīfa* to slice] (ca. 1825) Slicing off a thin layer. Skiving is the method by which veneers are cut from logs and by which polytetrafluoroethylene film is produced from cylindrical bars of ram-extruded or sinter-molded resin. PTFE film is made in this way because, unlike most thermoplastics, it cannot be directly extruded into film.

SKS *n* Copolymer from butadiene and styrene. Manufactured by the USSR.

Slab Stock *n* Large, thick sheets of plastics, usually formed by casting from syrup or melt and used for fabrication of larger structures such as tanks.

Slack End *n* An end woven under insufficient tension.

Slack Filling See ▶ Slack Pick.

Slack Melt Copal *n* Copal which has been run only to a limited extent, often at a comparatively low temperature.

Slack Mercerization \\ˌmərs-rə-ˈzā-shən\\ *n* A process for producing stretch in cellulosic fabrics.

Slack Pick *n* A single filling yarn woven under insufficient tension.

Slack Selvage *n* A self-descriptive fabric defect caused by incorrect balance of cloth structure between the ground and selvage or by the selvage ends being woven with insufficient tension.

Slack Thread See ▶ Slack End.

Slack Warp See ▶ Slack End.

Slaked Lime *n* Another name for calcium hydroxide, or quicklime which has been slaked by addition of water.

Slam-Off See ▶ Smash.

Slasher *n* A machine used to apply size to the warp ends, while transferring the warp yarns from section beams to the loom beam.

Slasher Sizing See ▶ Slashing.

Slashing *n* The process of sizing warp yarns on a slasher. Also see ▶ Sizing (1) and ▶ Slasher.

Slate \\ˈslāt\\ *n* [ME, fr. MF *esclat* splinter] (14c) A fine-grained metamorphic rock of varied composition, used in powdered form as a filler, especially in flooring compounds.

Slate Dust See ▶ Slate Powder.

Slate Flour See ▶ Slate Powder.

Slate Powder *n* Extender pigment obtained from slate. It is an aluminum silicate type of mineral, including a substantial amount of combined water. Used in wood and metal fillers. *Also called Slate Dust or Slate Flour.*

Sleazy \\ˈslē-zē *also* ˈslā-\\ *adj* [origin unknown] (ca. 1645) Thin, lacking firmness, open-meshed; usually describes poor-grade fabrics.

Sleepy *n* Description of a recently applied glossy coating which has lost its initial gloss other than by bloom and has become dulled or lacking in luster.

Sleeve Ejector *n* A bushing-type ▶ Knockout.

Sleeving *n* A braided, knit, or woven product or fabric in tubular or cylindrical form that is less than 4 in. in width (i.e., 8 in. in circumference).

Sley *n* The number of warp yarns per inch in a woven cloth on or off the loom.

Slide Waste *n* A yarn defect that is similar in appearance to a slub. It consists.

Slime \\ˈslīm\\ *n* [ME, fr. OE slīm; akin to MHGr slīm slime, L *limus* mud] (before 12c) A general name for any moist, sticky substance formed by fungi.

Sling Psychrometer \\ˈsliŋ sī-ˈkrä-mə-tər\\ *n* A psychrometer containing independently matched dry- and wet-bulb thermometers suitably mounted for swinging through the atmosphere, for simultaneously indicating dry- and wet-bulb temperatures.

Slip \\ˈslip\\ *v* [ME *slippen*, fr. MD or MLGr; akin to MHGr *slipfen* to slide, OHGr *slīfan* to smooth, and per. to Gk *olibros* slippery] (14c) (1) With reference to adhesives, slip is the ability to move or reposition the adherends after an adhesive has been applied to their surfaces. (2) Of plastic film, having low surface friction and sliding easily over another layer of film or over machine surfaces in film fabricating and packaging equipment.

Slip *n* (1) Film of coating is said to have slip when it has negligible surface tackiness, and appears to be lubricated. In actual fact, lubricants which are incompatible with the nonvolatile constituents of finishes are frequently added in order that they may be exuded to the surface and confer slip. Nongreasing gold and silver baking varnishes are typical products to which lubricants are added. (2) Denoting the ease with which two surfaces slide in contact with each other. NOTE – In a sense, it is the antithesis of friction, in that high coefficient of friction denotes poor slip and low coefficient of friction, good slip. (3) Vaporized coating material emitted from a spray gun.

Slip Agent *n* (**Compound**) (slip additive) An additive (e.g., oleamide) which exudes to the surface of a coating during and immediately after application and drying or baking. The additive coats the surface and provides the necessary lubricity to reduce the coefficient of friction and improve slip characteristics. Some experts feel that slip agents also act as ▶ Antiblocking Agents.

Slip Forming *n* (slip-ring forming) A variation of ▶ Sheet Thermoforming employing a sheet-clamping frame provided with tensioned pressure pads that permit the plastic sheet to slip inward as the part is being formed. This controlled slippage provides more uniform wall thickness in the formed article.

Slippage *n* The movement of adherents with respect to each other during the bonding process. Sliding or slipping of the filling threads over the warp ends (or vice versa), which leaves open spaces in the fabric. Slippage results from a loose weave or unevenly matched warp and filling.

Slip Sheet *n* [^{1}*slip*] (1903) A sheet of paper placed between two freshly printed sheets to prevent set-off or blocking.

Slit Die *n* (slot die, strip die) A ▶ Profile Die with nearly rectangular opening, used to produce an extrudate with a thin, moderately wide, rectangular cross section, but too narrow to be called sheeting. Sheet and film dies, which are more elaborate and have adjustable removable lips, are not considered to be slit dies. A die producing a *blocky* rectangular cross section is called a *bar-stock die*.

Slit-Film Yarn *n* Yarn of a flat, tape-like character produced by slitting an extruded film.

Slit Tape *n* A fabric, 12 in. or less in width, made by cutting wider fabric to the desired width. Slit tapes are made primarily of cotton, linen, jute, glass, or asbestos and are used principally for functional purposes.

Slitting *n* (1) The conversion of a given width of plastic film or sheeting to several smaller widths by means of knives. The operation can be performed as the material emerges from a production unit such as a calender, film-casting unit, or an (extrusion) roll stand (*in-line slitting*); by unwinding, slitting, and then rewinding of rolls; or by slitting of rolls without unwinding (*roll slicing*). Slitting knives may be actual straight-edge knives, or razor blades, or circular knives. (2) In coil coating, a process by which a wide strip is slit into narrower size prior to recoiling.

Sliver *1 is usually* ˈsli-vər, *2 is usually* ˈslī-\ *n* [ME *slivere*, fr. *sliven* to slice off, fr. OE *–slīfan*; akin to OE *-slīfan* to cut] (14c) A continuous strand of loosely assembled fibers without twist. Sliver is delivered by the card, the comber, or the drawing frame. The production of sliver is the first step in the textile operation that brings staple fiber into a form that can be drawn (or reduced in bulk) and eventually twisted into a spun yarn.

Sliver Knitting *n* Circular knitting coupled with the drawing-in of a sliver by the needles to produce a pile-like fabric, usually for high-pile coats or heavy linings.

Sloughed Filling See ▶ Looped Filling.

Slow Ray *n* The slow vibration of a crystal corresponding to the higher refractive index.

Slow Solvent *n* Solvent with a slow evaporation rate.

Slub \ˈsləb\ *vt* {*slubbed; slubbing*} [back-formation fr. *slubbing*] (1834) A yarn defect consisting of a lump or thick place on the yarn caused by lint or small lengths of yarn adhering to it. Generally, in filament yarn, a slub is the result of broken filaments that have stripped back from the end to which they are attached.

Slubber *n* A machine used in textile processes prior to spinning that reduces the sliver and inserts the first twist.

Slubbing *n* The product of the slubber, it is the intermediate stage between sliver and roving.

Slubbing Frame See ▶ Slubber.

Slub Catcher *n* A mechanical or electronic device designed to aid in the detection and removal of slubs or neps in yarns, usually during coning.

Slub Yarn *n* Any type of yarn that is irregular in diameter; the irregularity may be purposeful or the result of error. Also see ▶ Novelty Yarn, ▶ Nub Yarn, and ▶ Slub.

Slug \ˈsləg\ *n* [ME *slugge*, of Scand origin; akin to Norw dialect *slugga* to walk sluggishly] (15c) A thick place in a yarn or a piece of lint entangled in yarn, cord, or fabric.

Slug Molding *n* A process for making thin-walled containers of 200–300-mL capacity. Melt from an extruder is fed into a metering head that delivers a slug of precise mass into a cylindrical bushing. From the bushing the slug is propelled into a single-cavity mold by a high-speed ram that passes through the bushing. The finished part is removed by a mechanical arm and transferred to an air conveyor that takes it to an automatic stacking unit. Trimming is unnecessary.

Slur \ˈslər\ *n* [obs. E dialect *slur* thin mud, fr. ME *sloor*; akin to MHGr *slier* mud] ('609) A condition caused by slippage at the moment of impression between any two of the following: paper, printing plate, image carrier, or blanket.

Slurry *n* A mixture of water and any finely divided insoluble material such as clay or Portland cement.

Slurry Preforming *n* A method of preparing reinforcement performs by wet-processing techniques similar to those used in Pump Molding. For example, glass fibers suspended in water are directed onto a shaped screen that retains the fibers while allowing the water to pass through, thus forming a mat that has the shape of the object to be molded.

Slush Casting *n* A method of forming hollow objects, widely used for doll parts and squeeze toys, in which a hollow mold provided with a closable opening is filled with a fluid plastic mixture, usually a vinyl plastisol. Heat, applied to the mold before and/or after filling, causes a layer of material to gel against the inner mold surface. When the layer has attained the desired thickness, the excess fluid is poured out, and additional hat is applied to fuse the gelled layer. After cooling, the article is stripped from the mold. Molds for slush casting are thin-walled for rapid heat transfer and are made of electroformed copper or cast aluminum.

Slushing *n* Process by which a coating is liberally applied to surfaces which require protection but which are more or less hidden from view and not readily accessible for painting by ordinary methods. The paint or similar material is swilled on and the excess drained off.

Slushing Oil *n* Oil used on metals to form a protective coating against rust and tarnish. It should coat the surface completely, and yet be removable without undue labor.

Slush Molding *n* The preferred term for the process resembling Slush Casting but employing dry, sinterable powders.

SMA Abbreviation for ▶ Copolymers of Styrene and ▶ Maleic Anhydride.

Small-Angle X-Ray Scattering *n* (SAXS) A technique using high-intensity X-ray sources for determining a wide range of information on submicroscopic structures larger than 2 nm. Some applications with polymers are particle size and macromolecular structure of particles, ▶ Scale of Segregation in blends, void structures, and ▶ Specific Surface.

Smalt \smólt\ *n* [MF, fr. OIt *smalto*, of Gme origin; akin to OHGr *smelzan* to melt] (1558) A deep blue potassium-cobalt glass pigment made by fusing pure sand and potassium carbonate with cobalt oxide, grinding and powering. Smalt is sometimes applied to freshly coated surfaces to provide unusual decorating effects.

SMA PTB Alloy See ▶ Styrene Maleic Anhydride Copolymer PBT Alloy.

Smart Skin *n* (smart composite) A composite containing molded-in sensors and micro transmitters that enable aerospace engineers to detect in-flight changes in temperature, strain, ice thickness, and cracks.

Smash (1) A relatively large hole in fabric characterized by many broken warp ends and floating picks. One cause is the breaking of one or both harness straps, permitting the harness to drop and break out warp ends. (2) The breaking of many yarn ends in a beaming operation, usually as a result of mechanical failures.

SMC *n* Abbreviation for Sheet Molding Compound.

Smearing *n* The spreading of ink over areas of the plate and/or substrate where it is not wanted.

Smectic \smek-tik\ *adj* [L *smecticus* cleansing, having the properties of soap, fr. Gk *smēktikos*, fr. *smēchein* to clean] (1923) See ▶ Liquid-Crystal Polymer.

Smeraldino See ▶ Hydrated Chromium Oxide.

Smith-Ewart Kinetics *n* Emulsion polymerization kinetics describing the initiator that dissociates into free radicals, which can either travel into the micelle and start a polymerization directly (Smith-Ewart-Harkins theory) or react first with an emulsifier molecule, under transfer, forming an emulsifier radical, which then starts the polymerization (Medvedev theory).

Smoke \smōk\ *n* [ME, fr. OE *smoca*; akin to OE *smēocan* to emit smoke, MHGr *smouch* smoke, and prob. to Gk *smychein* to smolder] (before 12c) Carbon or soot particles less than 0.1 μm in size which result from the incomplete combustion of carbonaceous materials such as coal or oil. Air suspension (aerosol) particles, often originating from combustion or sublimation.

Smoke Chamber Test See ▶ Flammability Tests.

Smoked Sheet Smoked natural rubber.

Smoke Suppressants *n* An additive that, compounded into poorly burning polymers such as polyvinyl chloride, reduces smoke generation in fires. ASTM 2843 is a test for smoke generation, described under Flammability.

Smouldering *n* A slow, flameless, smoking burning of a fabric.

SMS Copolymer from styrene and α-methyl styrene. Abbreviation for ▶ Copolymers of Styrene and ▶ α-Methylstyrene.

Smudge *n* Mixture of residues of paints to which thinners are sometimes added. It is of unknown and variable quality and has no place in a normal paint system.

Sn *n* Chemical symbol for tin (Latin: *stannum*).

Snag *n* [of Scand origin; akin to ON *snagi* clothes peg] (ca. 1587) A pulled thread in knits. It is in the wale direction in warp knits and in the course direction in weft knits.

Snap-Back Forming See Vacuum Snap-Back Forming.

Snap Fit *n* A method of reversibly joining plastic parts in which one part has a concave element – usually a diameter – slightly smaller in an inside dimension than the outside dimension of the mating convex element of the second part. Behind the circle of interference the convex part will be relieved to a smaller diameter, the concave part to a larger one. To assemble, one forces (snaps) the two parts together. Once the zone of interference is passed, the momentarily high stress is partly or wholly relieved. Disassembly is easily accomplished by pulling the two members apart, but is unlikely to occur spontaneously. A ubiquitous application of the principle is seen in caps for polyolefin half- and whole-gallon milk bottles; another is the Tupperware® family of reusable food containers.

Snarl \ˈsnär(-ə)l\ v [²snarl] (14c) A short length of warp or filling yarn that has twisted on itself because of lively twist or insufficient tension. The snarling may occur during or prior to the weaving process.

S/N Curve Abbreviation for Stress at Failure versus number of cycles (curve). See ▶ Fatigue Curve.

Snell's Law n Snell's law of refraction states that if i is the angle of incidence through a transparent medium, r is the angle of refraction, the index of refraction n, then

$$n = \frac{\sin i}{\sin r}.$$

Snow Ball See ▶ Balling Up.

(SN)$_x$ See ▶ Sulfur Nitride Polymer.

Soaking n Treatment of rayon yarns in a lubricating and sizing solution preparatory to hard twisting.

Soap \ˈsōp\ n [ME sope, fr. OE sāpe; akin to OHGr seifa soap] (before 12c) The detergent obtained by the formation of a sodium or potassium salt of a fatty acid or mixture of fatty acids.

Soap, Metallic n Any product derived by reacting a fatty acid with a metal. Metallic soaps are widely used as stabilizers for plastics. The fatty acids commonly used are lauric, stearic, ricinoleic, naphthenic, octanoic (2-ethylhexanoic), rosin, and tall oil. Typical metals are aluminum, barium, calcium, cadmium, copper, iron, lead, magnesium, tin, and zinc.

Soap Resistance n The relative ability to withstand the action of soap. It is a property required of inks used in printing soap wrappers.

Soaps Metallic salts of fatty acids. There is also a tendency to apply the term to metallic salts of all organic acids. See ▶ Surfactants.

Soapstone \ˈsōp-ˌstōn\ n (ca. 1681) Another name for the mineral from which talc is obtained.

Society of Automotive Engineers n (SAE) Address: 400 Commonwealth Dr, Warrendale, PA 15906. Its 66,000 members are engineers, managers, and scientists in the field of self-propelled land, sea, air, and space vehicles. Among its publications is an annual handbook of standards, many of them shared with ANSI. The AMS 3000 series includes many standards on elastomers, casting resins, plastics extrusions, moldings, potting compounds, and reinforcing fibers.

Society of Plastics Engineers n (SPE) The leading international organization devoted to plastics engineering and technology, with 35,000 members, headquartered at 14 Fairfield Drive, Brookfield, CT 06804-3911. In 1993, SPE had 86 geographic sections, including 14 outside the US, and 20 special-interest divisions. SPE publishes several journals, holds many regional and national technical meetings each year, including its annual technical conference (ANTEC), has sponsored publication of scores of books on all aspects of plastics, supports plastics education at all levels, and founded the Plastics Institute of America, now a separate organization.

Society of the Plastics Industry n (SPI) An organization whose members include companies from all areas of plastics: materials producers, plastics sales people, processors and fabricators, designers and users of plastics products, and a few individual members. SPI's main address is 1275 K Street NW, Washington, DC 20005. SPI's Composites Institute is located at 355 Lexington Ave, New York, NY 10017. SPI sponsors a triennial international trade show. It has groups that interact with ▶ American National Standards Institute to promulgate standards for safe operation of plastics processing equipment and machinery and for certain plastics products such as pipe. The Composites Institute sponsors an annual technical conference on reinforced plastics. SPI lobbies for sensible legislation affecting plastics, and established and supports the Plastics Hall of Fame.

Sodium Aluminosilicate \ˈsō-dē-əm ə-ˌlü-mə-nō-ˈsi-lə-ˌkāt\ n See ▶ Sodium Silicoaluminate.

Sodium Aluminum Hydroxycarbonate n $NaAl(OH)_2CO_3$. Named Dawsonite (after the mineral of the same nominal composition) by Alcoa, this material is produced in the form of microfiber crystals useful for upgrading physical properties of thermoplastics. In PVC compounds it also acts as a smoke suppressant and HCl scavenger.

Sodium Aluminum Hydroxycarbonate

Sodium Azide \ˈā-ˌzīd\ *n* (ca. 1937) NaN₃. A poisonous crystalline salt used to make lead azide.

$$^-N=N^+=N^-$$
$$Na^+$$

Sodium Borate *n* Na₂B₄O₇. Powder of glass-like plates becoming opaque when exposed to air. Slowly soluble in water. See ▶ Borax.

Sodium Borohydride \-ˌbōr-ə-ˈhī-ˌdrīd\ *n* [*sodium + boron + hydride*] (1946) NaBH₄. A white crystalline powder, used as a blowing agent for foamed plastics such as rigid PVC and polystyrene, and for elastomers. The material decomposes at room temperature in the presence of water and an acidic medium, releasing hydrogen.

$$Na^+ \quad BH_4^-$$

Sodium Carbonate *n* (1868) Na₂CO₃. A sodium salt of carbonic acid used in making soaps and chemicals, in water softening, in cleaning and bleaching, and in photography.

Sodium Carboxymethyl Cellulose \-kär-ˌbäk-sē-ˌme-thəl ˈsel-yə-ˌlōs\ *n* See ▶ Carboxymethyl Cellulose.

Sodium Chlorate *n* (1885) NaClO₃. A colorless crystalline salt used as an oxidizing agent.

Sodium Chloride *n* (1868) An ionic crystalline chemical compound consisting of equal numbers of sodium and chlorine atoms.

$$Na^+ \quad Cl^-$$

Sodium Citrate *n* (1919) Na₃C₆H₅O₇. A crystalline salt used chiefly as a buffering agent, as an emulsifier, as an alkalizer and cathartic in pharmaceuticals, and as a blood anticoagulant.

Sodium Cyanide *n* (1885) NaCN. A white deliquescent poisonous salt used in electroplating, in fumigating, and in treating steel.

$$Na^+ \quad ^-C\equiv N$$

Sodium Dichromate \-(ˌ)dī-ˈkrō-ˌmāt\ *n* (1903) Na₂Cr₂O₇. A red crystalline salt used in tanning leather, in cleaning metals, and as an oxidizing agent.

Sodium Fluoride *n* (ca. 1903) NaF. A poisonous crystalline salt that is used in trace amounts in the fluoridation of water, in metallurgy, as a flux, and as a pesticide.

$$Na^+ \quad F^-$$

Sodium Fluoroacetate \-ˌflúr-ō-ˈa-sə-ˌtāt\ *n* (1945) A poisonous powdery compound.

Sodium Hydroxide *n* (1885) (lye, caustic soda) NaOH. A strong, cheap alkali, completely ionic, useful both as a reactant and a catalyst in polymer chemistry.

$$Na^+ \quad OH^-$$

Sodium Hypochlorite *n* (1885) NaOCl. An unstable salt produced usually in aqueous solution and used as a bleaching and disinfecting agent.

$$Cl—O^-$$
$$Na^+$$

Sodium Lithol Red See ▶ Lithol Red.

Sodium Metasilicate \-▪me-tə-▪si-lə-▪kāt\ *n* (ca. 1926) Na_2SiO_3. A toxic corrosive crystalline salt used mostly as a detergent or as a substitute for phosphates in detergent formulations.

Sodium Nitrate *n* (1885) $NaNO_3$. A deliquescent crystalline salt used as an oxidizing agent, a fertilizer, and in curing meat.

Sodium Nitrite *n* (ca. 1903) $NaNO_2$. A salt used in dye manufactuer.

Sodium Polyacrylate $n\,(-CH_2\overline{CH-})_n COONa$. A thickening agent, the sodium salt of polyacrylic acid.

Sodium Silicoaluminate *n* $9Na_2O \cdot 67SiO_2 \cdot 12Al_2O_3$. Ultrafine synthetic precipitate formed by reacting an aluminum salt with an alkali silicate. Density, 2.1 g/cm³ (17.5 lb/gal); O.A., 125; particle size, 0.1–0.4 μm. Syn: ▶ Sodium Aluminosilicate.

Sodium Stearate \-▪stē-ə-▪rāt\ *n* $NaOOCC_{17}H_{35}$. A water-soluble white powder, a soap, used as a nontoxic stabilizer.

Sodium Sulfate *n* (1885) Na_2SO_4. A bitter salt used in detergents, in the manufacture of wood pump and rayon, in dyeing and finishing textiles, and in its hydrated form as a cathartic.

Sodium Thiosulfate *n* (1885) $Na_2S_2O_3$. A hygroscopic crystalline salt used as a photographioc fixing agent and a reducing or bleaching agent.

Soffit \▪sä-fət\ *n* [F *soffite*, fr. It *soffitto*, fr. (assumed) VL *suffictus*, pp of L *suffigere* to fasten underneath] (1592) (1) The exposed undersurface of any overhead component of a building, such as a roof overhang, an arch, balcony, beam, cornice, lintel, or vault. See ▶ Eave. (2) In wallcoverings, border and soffit paper are the same. See ▶ Border.

Soft Distemper *n* Usually based on whiting, lightly bound with glue, size or other water-soluble binder so that the coating will withstand dry rubbing without disturbance but can easily be removed by water washing.

Softener *n* (1) A product designed to impart a soft mellowness to the fabric. Examples are glucose, glycerine, tallow, or any one of a number of quaternary ammonium compounds. (2) A substance that reduces the hardness of water by removing or sequestering the calcium and magnesium ions. (3) A substance used to reduce friction during mixing and processing when dry powders are added to polymers. See ▶ Blender.

Softening Point *n* The temperature at which a disc of a sample held within a horizontal ring is forced downward a distance of 25.4 mm (1 in.) under the weight of a steel ball as the sample is heated at a prescribed rate in a water or glycerol bath. See ▶ Ball and ▶ Ring Method.

Softening Range *n* A temperature interval over which a plastic changes from a rigid to a soft state or undergoes a rather sudden and substantial change in hardness (ISO). ASTM (www.astm.org) describes the Durrans' softening-point procedure for fully reactive epoxy resins. See also ▶ Ball-and-Ring Test, ▶ Melting Point, and ▶ Vicat Softening Point.

Softening Temperature See ▶ Softening Point.

Soft Ink *n* Printing composition exhibiting an almost complete absence of tack.

Soft Silica See ▶ Amorphous Silica.

Softwoods The botanical group of trees that have needles or scalelike leaves and are evergreen for the most part, cypress, larch, and tamarack being the exceptions. The term has no reference to the hardness of the wood. Softwoods are often referred to as conifers and, botanically, they are called gymnosperms.

Soil See ▶ Dirt.

Soil Burial Test *n* A test of resistance of textile material to certain microorganisms present in soil. The samples are buried in soil for an extended period, then removed and measured for strength loss.

Soiling The staining or smudging of textile materials resulting from the deposit of dirt, oil undesirable dye, etc.

Sol \ˈsäl, ˈsól\ *n* [-*sol* (as in *hydrosol*), 1 fr. *solution*] (1899) Short for hydrosol. A fluid colloidal system in which the continuous phase is a liquid. See ▶ Colloidal Solution.

Solid \ˈsä-ləd\ *n* (1) A printed area uniformly and completely covered with ink. (2) Nonvolatile matter in a coating composition, i.e., the ingredients of a coating composition which, after drying, are left behind and constitute the dry film. Also called *Nonvolatile Matter*.

Solid *n* (15c) A state of matter in which the relative motion of the molecules is restricted and they tend to retain a definite fixed position relative to each other, giving rise to crystal structure. A solid may be said to have a definite shape and volume.

Solid Angle *n* (ca. 1704) Measured by the ratio of the surface of the portion of a sphere enclosed by the conical surface forming the angle, to the square of the radius of the sphere. Unit of solid angle – the steradian, the solid angle which encloses a surface on the sphere equivalent to the square of the radius. Dimension, unity.

Solid Casting *n* The process of forming solid articles by pouring a fluid resin or dispersion into an open mold, causing the material to solidify by curing or by heating and cooling, them removing the formed article. Compare ▶ Casting.

Solid-Phase Forming *n* This term includes the shaping of plastic sheets or billets into three-dimensional articles either at room temperature or at higher temperatures up to the softening or melting range by processes resembling those used in the metals-working industry. Among plastics suitable for at least some of the solid-phase processes are acrylonitrile-butadiene-styrene resins, acetals, cellulosics, polyolefins, polycarbonates, polyphenylene oxides, and polysulfones. Brittle materials such as acrylics and polystyrene cannot be formed by solid-phase processes.

Solids *n* Nonvolatile matter in a coating composition, i.e., the ingredients of a coating composition which, after drying, are left behind and constitute the dry film. Also called *Nonvolatile Matter*.

Solids By Volume *n* The volume of the nonvolatile portion of a composition divided by the total volume, expressed as a per cent. Syn: ▶ Volume Solids.

Solids Content *n* The percentage by weight of the nonvolatile matter in an adhesive. NOTE – The actual percentage of the nonvolatile matter in an adhesive will vary considerably according to the analytical procedure that is used. A standard test method must be used to obtain consistent results.

Solids-Draining Screw *n* (barrier screw) Any of a half-dozen designs of screws for single-screw extruders whose intent is to separate generated melt from the bed of solids as the melt is formed, the two main goals being better contact of the remaining bed with the hot barrel and preventing unmelted solids from passing into the metering section and die. The first of these, and still in use, was the Maillefer Screw. Some others bear the names Barr, Dray and Lawrence, Ingen Housz, Kim, SPR (for Scientific Processing & Research), and Uniroyal.

Solid-State Controls *n* Control instruments or motor-drive controls whose circuitry employs transistors and kindred elements rather than mechanical or vacuum-tube devices. Practically all modern process instruments are of this type.

Solid-State Polymerization *n* A chain-growth polymerization initiated by exposing to ionizing radiation a crystalline monomeric substance. A large number of olefinic and cyclic solid monomers have been so polymerized, the crystalline monomer converting directly to the polymer with no obvious change in appearance of the solid. (Odian GC (2004) Principles of polymerization. Wiley, New York)

Soliton \ˈsä-lə-ˌtän\ *n* [*solitary* + ^{2}-*on*] (1965) A solitary wave (as in a gaseous plasma) that propagates with little loss of energy and retains its shape and speed after colliding with another such wave. Solitons have had many physical applications included order-disorder phase transitions, crystal dislocations, charge density waves, and Josephson junction transmission lines. (Ku CC, Liepins R (1987) Electrical properties of polymers. Hanser, New York)

Solprene See ▶ Plastomer.

Solubility *n* (1) The mass of a substance that will dissolve in a given volume or mass of a solvent to form a solution that is homogeneous to the molecular level, sometimes expressed as weight percent. (2) When no solvent and conditions are specified, the solubility in water at room temperature is usually meant.

Solubility *n* The concentration of a solute in a saturated solution; the maximum amount of solute which can be dissolved in a solution by simply adding it to a solvent at constant temperature.

Solubility Coefficient *n* The volume of a gas that can be dissolved by a unit volume of solvent at a fixed pressure and temperature.

Solubility Parameter *n* (SP, δ) The square root of the Cohesive Energy Density of a polymer, solvent, or adhesive. For solvents, SP equals the square root of the heat of vaporization per unit volume $(J/cm^3)^{0.5}$. Values for polymers (which do not vaporize) are found by indirect methods. If the solubility parameters of a polymer and solvent differ by 3 $(J/cm^3)^{0.5}$ or less, the polymer will probably dissolve in the solvent differ by $(J/cm^3)^{0.5}$ or less, the polymer will probably dissolve in the solvent. SPs for organic solvents range from 13 for neopentane to 30 methanol; for polymers, from 13 for polytetrafluoroethylene to 32 for Polyacrylonitrile. (Handbook of solvents. Wypych G (ed). Chemtec, New York, 2001; Barton AFM (1983) Handbook of solubility parameters and other cohesion parameters. CRC, Boca Raton, FL; Polymer handbook. Brandrup J, Immergut EH, Elias HG (eds). Interscience, New York, 1975)

Solubility Parameter Concept *n* Cohesive energy density of a plastic or solvent material. (Burrell H (1955) Solubility parameters. Interchem Rev 14(3):31; Hansen CM, Beerbower A (1974) Solubility parameters. Encylcopedia of chemial technology, Supplemental Volume. Hoy KL, Price BA, Martin (eds))

Solubility Product *n*, K_{sp} The equilibrium constant for an ionic solubility equilibrium. *Also called the Solubility Product Constant*. When a solid electrolyte (MA) dissolves at least two kinds (M and A) of particles (ions) are released to the solution (e.g., NaCl in water), then

$$MA \rightleftharpoons M^+ + A^-$$
$$\frac{[M^+][A^-]}{[MA]} = K'$$
$$[M^+][A^-] = K'[MA]$$
$$[M^+][A^-] = Ksp$$

K_{sp} is called the *solubility product*, $[M^+][A^-]$ is called the *ion product*, and at ionic equilibrium the solubility product is equal to the ion product. The solubility product is useful for expressing the solubility and ionic strength of a substance in aqueous solution (Russell JB (1980) General Chemistry, McGraw-Hill, New York).

Soluble *n* Capable of being dissolved, i.e., passing into solution.

Soluble Blue *n* A pigment dispersible in water that is made by treating an iron blue with sodium ferrocyanide or oxalic acid and used chiefly in permanent writing inks and laundry blues.

Soluble Drier See ▶ Liquid Driers; ▶ Drier.

Solute \\'säl-‚yüt\ *n* [L *solutus*, pp of *solvere*] (1893) That constituent, usually a solid substance, of a solution which is considered to be dissolved in the solvent. The solvent is usually present in larger percentage than the solute. A solution is saturated if it contains at given temperature as much of a solute as it can retain in the presence of an excess of that solute. A true solution is a mixture, liquid, solid, or gaseous, in which, the components are uniformly distributed throughout the mixture. The proportion of the constituents may be varied within certain limits.

Solution *n* A single-phase mixture of two or more component, homogeneous at the molecular level, such as a resin dissolved in a liquid, that forms more or less spontaneously, will not separate, and has no fixed proportions of the components. Polymer solutions are used in the plastics industry to apply coatings, for film casting, and for spinning fibers. The term also includes gas/gas, gas-in-liquid, liquid/liquid, and solid/solid solutions.

Solution Casting See ▶ Film Casting.

Solution Coating *n* Any coating process employing a solvent solution of a resin, as opposed to a dispersion, hot-melt or uncured thermosetting system. See also ▶ Spread Coating.

Solution Dyed See ▶ Dyeing, Mass-Colored.

Solution Polymerization *n* A polymerization process in which the monomer, or mixture of monomers, and the polymerization initiators are dissolved in a nonmonomeric solvent at the beginning of the polymerization reaction. The liquid is usually also a solvent for the resulting polymer or copolymer. Solution polymerization is most advantageous when the resulting polymeric solutions are to be used for coatings, lacquers, or adhesives. Vinyl acetate, olefins, styrene, and methyl methacrylate are the monomers most often employed. (Odian GC (2004) Principles of polymerization. Wiley, New York; Elias, Hans-Georg (2003) An introduction to plastics. Wiley, New York; Solomon DH (1969) Kinetics and mechanisms of polymerization series, vol 2 – Ring opening and volume 3 – Step growth. Marcel Dekker, New York)

Solution Polymers Solution polymers offer the absorbency of a granular polymer supplied in solution form. Solutions can be diluted with water prior to application, and can coat most substrates or used to saturated. After drying at a specific temperature of a specific time, the result is a coated substrate with superabsorbent functionality. For example, this chemistry can be applied directly onto wires & cables, though it is especially optimized for use on components such as rolled goods or sheeted substrates.

Solvation *vt* (1909) Adsorption of a microlayer or film of water or other solvent on individual dispersed particles of a solution or dispersion. It also applies to the action of plasticizers on resin dispersions in plastisols. The process of swelling, gelling, or solution of a resin by a solvent or plasticizer as a result of chemical compatibility. See ▶ Solubility Parameter.

Solvency *n* The degree to which a solvent holds a resin or other paint binder in solution. See ▶ Kauri-Butanol Value, ▶ Aniline Point, ▶ Solubility Parameter. Also called *Solvent Power*. (Handbook of solvents. Wypych G (ed) Chemtec, New York, 2001; Barton AFM (1983) Handbook of solubility parameters and other cohesion parameters. CRC, Boca Raton, FL)

Solvent *n* (1) Broadly defined, a liquid with the ability to dissolve other substances. Solvents are used by the plastics industry in three main ways. As intermediates, solvents are used in the production of many monomers and resins. In plastics processing, solvents are used in etching, welding, polishing, film casting, fiber spinning, and making laminates. Finally, solvents are widely used in adhesives, printing inks, and surface coatings for plastics, as well as those based on plastics and used on other materials. The major types of solvents listed in Table – Solvents, Viscosities, Densities and Vapor Densities used in all these applications are alcohols, esters, glycol ethers, ketones, aliphatic and aromatic hydrocarbons, chlorinated hydrocarbons, and nitroparaffins. (2) The constituent of a solution that is (usually) present in larger percentage; or, in the case of solutions of solids or gases in liquids, the constituent that is liquid in the pure state at room temperature. (Handbook of solvents. Wypych G (ed) Chemtec, New York, 2001)

Solvents, viscosities, densities and vapor densities

Field name	Temperature (°C)	Viscosity centistokes	Density (kg/L)	Vapor pressure (kPa)
Acetaldehyde	20	0.295	0.788	105
Acetaldehyde	30	0.275	0.748	148
Acetic acid	20	1.232	1.048	3.3
Acetic acid anhydride	20	0.88	1.084	1.3
Acetone	20	0.41	0.79	30
Allyl alcohol	20	1.603	0.852	2.4
Allyl alcohol	30	1.36	0.848	4.3
Allyl alcohol	40	1.067	0.844	7.4
Allyl chloride	20	0.354	0.94	30
Aluminium chloride [5% sol]	20	3.54	1.03	2.4
Aluminium nitrate [10% sol]	20	4.54	1.051	2.4
Aluminium sulfate [10% sol]	20	1.34	1.115	2.4
Amyl acetate	20	4.34	0.885	1.3
Aniline	10	6.4	1.03	0.5
Aniline	20	4.37	1.021	0.5
Beer	20	1.8	0.996	2.4
Benzene	20	0.744	0.879	14
Benzene	30	0.65	0.868	20.7
Benzene	40	0.58	0.858	30
Benzene	50	0.54	0.847	42.5
Benzene	60	0.51	0.836	60
Benzyl alcohol	20	5.52	1.045	0.5
Bromine	20	0.34	3.12	48
Butyl acetate	20	0.832	0.885	3.3
Butyl alcohol	20	3.64	0.81	5.4
Butyl alcohol	30	2.85	0.803	8.7
Butyric acid n	0	2.35	0.977	0.5

Solvents, viscosities, densities and vapor densities (Continued)

Field name	Temperature (°C)	Viscosity centistokes	Density (kg/L)	Vapor pressure (kPa)
Butyric acid n	10	1.93	0.967	0.5
Butyric acid n	20	1.61	0.957	0.5
Calcium chloride [25% sol]	20	3.9	1.227	2.4
Calcium chloride [5% sol]	20	1.161	1.037	2.4
Carbolic acid	20	11.3	1.078	0
Carbolic acid	30	9.7	1.069	0
Carbolic acid	40	7.95	1.059	0
Carbolic acid	50	6.15	1.05	0
Carbon disulfide	0	0.33	1.292	22
Carbon disulfide	10	0.316	1.277	33
Carbon disulfide	20	0.298	1.262	48
Carbon tetrachloride	20	0.612	1.595	20.7
Carbon tetrachloride	30	0.525	1.525	30
Castor oil	20	1017	0.96	0
Castor oil	30	580	0.955	0
Castor oil	40	315	0.95	0
Castor oil	50	200	0.945	0
Castor oil	60	115	0.94	0
China wood oil	20	308	0.933	0
China wood oil	30	200	0.926	0
China wood oil	40	120	0.918	0
Chloroform	20	0.38	1.489	30
Chloroform	30	0.38	1.471	43
Chloroform	40	0.37	1.452	62
Chloroform	50	0.36	1.434	87
Chloroform	60	0.35	1.415	120
Cotton seed oil	20	76	0.926	0
Cotton seed oil	30	50	0.921	0
Cotton seed oil	40	35	0.916	0
Chyclohexanol	20	71	0.952	0.5
Cyclohexanone	20	4.9	0.952	0.5
Cylinder oil	20	50000	0.94	0
Dioxan	20	2	1.03	0
Ethyl acetate	20	0.51	0.905	14
Ethyl alcohol	20	1.51	0.772	9
Ethyl alcohol	30	1.32	0.754	14
Ethyl alcohol	40	1.16	0.737	20.7
Ethyl glycol	20	2.3	0.93	0.5
Ethylene glycol	20	18	1.112	0.5
Ethylene glycol	30	16.5	1.104	0.5
Formic acid	20	1.5	1.22	5.4
Formic acid	30	1.38	1.208	8.7
Fuel oil (El) Extra light	20	6	0.85	0

Solvents, viscosities, densities and vapor densities (Continued)

Field name	Temperature (°C)	Viscosity centistokes	Density (kg/L)	Vapor pressure (kPa)
Fuel oil (l) light	20	16.5	0.91	0
Fuel oil (m) medium	20	520	0.99	0
Fuel oil (s) heavy	20	8000	0.99	0
Furfurol	20	1.45	1.16	0.5
Furfurol	30	1.25	1.149	1.5
Gear oil	20	3000	0.905	0
Glycerine	20	1183	1.261	0
Heptane	0	0.74	0.702	.02
Heptane	10	0.66	0.692	.03
Heptane	20	.6	0.682	.05
Heptane	30	.55	0.671	.08
Heptane	40	.51	0.661	0.1
Hexane	0	.62	0.678	.02
Hexane	10	.57	0.668	.03
Hexane	20	.51	0.658	.05
Hexane	30	.45	0.649	.08
Hexane	40	.4	0.639	0.1
Kerosine	20	2.4	0.804	0.5
Kerosine	30	1.85	0.78	0.5
Linseed oil	20	47	0.92	0
Machine oil - light	20	47	0.9	0
Machine oil - medium	20	850	0.94	0
Mercury	20	0.119	13.57	0
Methyl acetate	20	.44	0.959	48
Methyl acetate	30	.39	0.937	68
Methyl acetate	40	.35	0.916	95
Methyl alcohol	0	1.04	0.81	13.4
Methyl alcohol	10	0.855	0.801	20
Methyl alcohol	20	0.745	0.792	30
Methyl glycol	20	1.6	0.975	0
Methylene chloride	20	0.9	1.326	72
Milk	20	1.13	1.035	2.4
Nitro benzene	20	1.67	1.203	0.5
Nonane	0	1.35	0.733	0.5
Nonane	10	1.15	0.725	0.5
Nonane	20	1	0.717	0.5
Nonane	30	0.89	0.709	1.5
Nonane	40	0.79	0.701	2.4
Octane	0	1.05	0.719	0.5
Octane	10	0.9935	0.711	0.5
Octane	20	0.805	0.702	0.5
Octane	30	0.72	0.694	1.5
Octane	40	0.64	0.685	2.4

Solvents, viscosities, densities and vapor densities (Continued)

Field name	Temperature (°C)	Viscosity centistokes	Density (kg/L)	Vapor pressure (kPa)
Oil SAE 10W – 30	20	130	0.875	0
Oil SAE 10W	20	115	0.87	0
Oil SAE 20W-20	20	200	0.885	0
Oil SAE 30	20	350	0.89	0
Oil SAE 40	20	900	0.9	0
Oil SAE 50	20	950	0.902	0
Olive oil	20	91.5	0.91	0.5
Paraffin oil	20	2.4	0.804	0.5
Paraffin oil	30	1.85	0.78	0.5
Pentane	0	0.44	0.646	32
Pentane	10	0.39	0.636	50
Pentane	20	0.36	0.626	72
Pentane	30	0.34	0.616	101
Phenol	20	11.3	1.078	0.5
Phenol	30	9.7	1.069	0.5
Phenol	40	7.95	1.059	1
Phenol	50	6.15	1.05	1.6
Propanol	20	2.8	.804	2.4
Propanol	30	2.2	0.795	4.3
Propanol	40	1.7	0.786	7.4
Propanol	50	1.4	0.777	12.3
Propionic acid	20	1.13	0.99	0.5
Propylene glycol	20	54	1.038	0
Rapeseed	20	178	0.92	0
Sea water	0	1.774	1.028	0.6
Sea water	10	1.346	1.028	1.3
Sea water	100	0.229	0.984	101.3
Sea water	20	1.044	1.025	2.4
Sea water	30	0.822	1.023	4.3
Sea water	40	0.659	1.019	7.4
Sea water	50	0.536	1.015	12.3
Sea water	60	0.442	1.01	19.9
Sea water	70	0.369	1.004	31.2
Sea water	80	0.311	0.998	47.4
Sea water	90	0.265	0.991	70.1
Sodium chloride [25% sol]	20	2.4	1.19	2.4
Sodium hydroxide [20% sol]	20	4	1.226	2.4
Sodium hydroxide [30% sol]	20	10	1.33	2.4
Soya bean oil	20	75	0.926	0
Styrene	20	0.9	0.926	0.5
Suphuric acid	20	14.6	1.839	2.4
Terachloroethane	20	1.1	1.593	1.3
Tetrachloroethylene	20	0.95	1.621	3.3

Solvents, viscosities, densities and vapor densities (Continued)

Field name	Temperature (°C)	Viscosity centistokes	Density (kg/L)	Vapor pressure (kPa)
Toluene	20	0.68	0.867	5.4
Toluene	30	0.61	0.858	8.7
Toluene	40	0.55	0.849	13
Toluene	50	0.5	0.84	19.5
Toluene	60	0.46	0.831	28
Transformer oil	20	30	0.95	0
Trichloroethylene	20	0.96	1.463	14
Water	0	1.788	1	0.6
Water	10	1.307	1	1.3
Water	100	0.295	0.958	101.3
Water	20	1.002	0.998	2.4
Water	30	0.802	0.996	4.3
Water	40	0.662	0.992	7.4
Water	50	0.555	0.988	12.3
Water	60	0.475	0.983	19.9
Water	70	0.414	0.978	31.2
Water	80	0.365	0.972	47.4
Water	90	0.327	0.965	70.1
Xylene-o	20	0.93	0.864	0
Xylene-o	30	0.83	0.855	0
Xylene-o	40	0.74	0.847	0

Solvent Balance n The condition wherein a blend of solvents and/or diluents produce the desired properties of solvency and solvent evaporation. See ▶ Solvent Imbalance.

Solvent Bonding See ▶ Solvent Welding.

Solvent Cement See ▶ Adhesive.

Solvent Cementing n (solvent bonding, solvent welding) The process of joining articles made of thermoplastic resins by applying a solvent capable of softening the surfaces to be joined, and pressing the softened surfaces together until the bond has gained strength. Plastics joined by this method include acrylonitrile-butadiene-styrene resin, acrylics, cellulosics, polycarbonates, polystyrenes, and vinyls. (Handbook of adhesives. Skeist I (ed) Van Nostrand Reinhold, New York, 1990)

Solvent Diffusion n The migration of solvent molecules into or out of a polymer as driven by the concentration gradient. Diffusion is the rate-limiting process in drying plastics and is also important in Extraction Extrusion.

Solvent Dyeing See ▶ Dyeing.

Solvent Imbalance n The condition wherein the ratios of solvents and/or diluents are such that inadequate solvency or improper evaporation of volatiles results. See ▶ Solvent Balance.

Solvent Paste Wax n Similar to liquid solvent wax, except furnished in paste form. Occasionally colored.

Solvent Polishing n A method for improving the gloss of thermoplastic articles by immersion in, or spraying with a solvent that will dissolve surface irregularities, followed by evaporation of the solvent. The method is used primarily for cellulosics, for which dipping is suitable. Plastics that are subject to crazing, such as polystyrene, are usually sprayed rather than dipped.

Solvent Pop n Blistering caused by entrapped solvent.

Solvent Power n The strength of dissolving power of a solvent. See ▶ Solvency.

Solvent Release n The ability of a resin to influence the rate at which solvent evaporates from a coating.

Solvent Resistance n The ability of a plastic to withstand, unchanged, exposure to solvents. Plastics vary widely in their resistance to specific solvents.

Solvent Shock n The situation wherein some of the protective vehicle is washed off the fine pigment particles, allowing them to pull together into clusters or flocs or when flocs of resin form due to dilution with a solvent or diluent of insufficient strength.

Solvent Spinning See ▶ Spinning, Dry Spinning.

Solvent-Water Blends Solvation *n* The interaction of a solute with a solvent; the surrounding of solute particles by solvent particles.

Solvent Wax See ▶ Liquid Solvent Wax.

Solvent Welding See ▶ Solvent Cementing.

Sonic Modulus *n* The tensile/compressive modulus (E) estimated by measurement of sound-wave propagation in a material. ASTM Test C 769 (Section 15.01) describes such a method.

Sonic Velocity *n* The speed f sound in a material. In air at 20°C and 50% relative humidity, the velocity of a plane longitudinal sound wave is 344 m/s. In polymethyl methacrylate, polystyrene, and polyethylene it is 2,680, 2,350, and 1,950 m/s, respectively.

Soot-Chamber Test *n* (ASTM D 2741) A test evaluating the relative effectiveness of antistatic agents in blown polyethylene bottles. After conditioning at 23°C and 50% RH, each bottle is rubbed ten times with a paper towel to generate static charge, then placed in a test chamber at 15% RH for 2 h. Filter paper wetted with toluene is ignited and the smoke is blown into the test chamber. Fifteen minutes later the bottles are removed from the chamber and examined. Soot accumulation is judged to be clean (i.e., none), slight, moderate, or severe.

Sorbic Acid \ˈsór-bik-\ *n* [*sorb* fruit of the service or related trees, fr. F *sorbe*, fr. L *sorbum*] (1815) $CH_3CH=CHCH=CHCOOH$. Naturally occurring unsaturated monobasic acid. Mp, 134°C. Sorbic acid is used to improve characteristics of drying oils and to improve alkyds.

Sorbitol \ˈsór-bə-ˌtól\ *n* [*sorb* fruit of the service or related trees + *-itol*] (1895) $C_6H_8(OH)_6$. Polyhydric alcohol which has attracted some interest as a component of synthetic drying oils and alkyd-type resins and surfactants. Sp gr, 1.47; mp, 97.5°C.

Sorption *n* The process of one substance taking up and holding another by physical or chemical action. Usually the sorbed substance is mobile, a gas or vapor, and the sorbing phase is dense, a liquid or solid. However, components of liquids can also be sorbed by solids, as by a ▶ Molecular Sieve. Surface phenomenon which may be either absorption or adsorption or a combination of the two. Often used when the specific mechanism is not fully known. See also ▶ Absorption, ▶ Adsorption, ▶ Chemisorption, and ▶ Persorption. (Handbook of solvents. Wypych G (ed) Chemtec, New York, 2001)

Souring *n* Any treatment of textile materials in dilute acid. Its purpose is the neutralization of any alkali that is present.

Sov Pren *n* Poly(chloroprene), manufactured by the USSR.

Soya Bean Oil *n* (ca. 1916) A pale yellow semidrying oil extracted from the bean of the soybean plant, *Soja hispida*, a native of Asia, but also grown extensively throughout the world. Its drying properties are inferior to those of linseed but may be improved by suitable processing. When refined, it finds wide use as a component in both exterior and interior paints, but its widest use is in the preparation of alkyds. Coatings made with soybean oil are less prone to yellowing than those based on linseed oil. It is also called ▶ Soybean Oil.

Soybean Meal *n* The product of grinding soybean residue after extraction of oil, sometimes treated with formaldehyde to reduce moisture absorption. It is used as a filler, often in conjunction with wood flour, in thermosetting resins.

Soybean Oil *n* A pale yellow oil extracted from soybeans, used in epoxidized form as plasticizers and stabilizers for vinyl resins.

Sp *n* (0,1,2,3) Cleanliness of hand derusted steel according to SIS 0 55 900.

Space Dyeing See ▶ Dyeing.

Space Lattice A regular, repeating array of points in space.

Spachtling See ▶ Spackle.

Spackle \ˈspa-kəl\ *vt* {spackled; spackling} [*Spackle*] (1940) A paste, compound, or powder which can be mixed into a paste; used to fill holes, cracks, and defects in wood, plaster, wallboard, etc., to obtain a smooth surface. Also known as *Spachtling, Spackling, and Sparkling*.

Spandex® \ˈspan-ˌdeks\ {*trademark*} [anagram of *expands*] (1959) Generic name for a manufactured fiber in which the fiber-forming substance is a long-chain synthetic polymer comprised of at least 85% of a segmented polyurethane (Federal Trade Commission).

These fibers are used in garments to enhance stretch ability and resilience. Compared to its competitors for this purpose, spandex is stronger and stiffer, with better resistance to heat and oxidation.

Spandex® Fiber *n* A DuPont manufactured fiber in which the fiber-forming substance is a long chain synthetic polymer composed of at least 85% of a segmented polyurethane (FTC definition). Characteristics: Spandex is lighter in weight, more durable, and more supple than conventional elastic threads and has between two and three times their restraining power. Spandex is extruded in a multiplicity of fine filaments which immediately form a monofilament. It can be repeatedly stretched over 500% without breaking and still recover instantly to its original length. It does not suffer deterioration from oxidation as is the case with fine sizes of rubber thread, and it is not damaged by body oils, perspiration, lotions, or detergents. End Uses: Spandex is used in foundation garments, bathing suits, hose, and webbings.

Spandex® Polymers *n* Segmented polyurethane structure (Dupont, Inc.) with rubber-like elasticity.

Spangles See ▶ Glitter.

Spanishing *n* A printing process similar to ▶ Valley Printing. Ink is deposited on the bottoms and sides of depressed areas of an embossed plastic film.

Spanish Ocher \-ˈō-kər\ *n* See ▶ Iron Oxides, Natural.

Spanish Red Oxide Ferric oxide mineral pigment from the Malaga district of Spain. Of bright shade, with good hiding power. Ferric oxide, 82–86%.

Spanish White *n* Another name for Paris white. See ▶ Whiting and ▶ Calcium Carbonate, Natural.

Sparking See ▶ Spackle.

Spark Matching See ▶ Electrical-Discharge Machining.

Sparkproof Tools *n* Bronze beryllium tools or other nonmetallic tools to prevent spark formation in potentially explosive atmospheres, as with concentrations of hydrocarbon solvent vapors within the explosive limits.

Spark-Testing *n* Method of detecting holidays (flaws) on metallic substrates by means of spark-test tool.

Spar Varnish *n* A very durable waterproof varnish for exterior use. Obtained its name from use in coating ships' spars.

Spatter Finish *n* A decorative effect wherein spots, globules or spatters of contrasting color appear on, or within the surface of, a white or differently colored background. The effect may be achieved through the use of a multicolor, multiphase single coat or by spattering the contrasting colors onto a base coat.

Spattering *n* In antiquing, applying small specks of color to simulate textural qualities in wood. Usually done with a dry brush and little paint, or by flicking a toothbrush.

SPC *n* Abbreviation For ▶ Statistical Process Control.

SPE *n* Abbreviation for ▶ Society of Plastics Engineering.

Special Boiling Point Spirits See ▶ S.▶ B.▶ P. ▶ Spirits.

Special Industrial Solvents *n* A special class of proprietary solvents manufactured in accordance with authorized formulas. They can be sold only for industrial or manufacturing use in quantities of 50 gal or more, but otherwise no permit is required in the United States.

Specific Adhesion *n* Adhesion between two surfaces that are held together by valence forces of the same type as those that give rise to cohesion, as opposed to mechanical adhesion in which the adhesive holds the parts together by interlocking action. See ▶ Adhesion, ▶ Specific and ▶ Adhesion, Mechanical.

Specific Conductance See ▶ Conductivity.

Specific Gravity *n* (1666) (sp gr) The ratio of the density of a liquid or solid material to the density of water at 4°C, or other specified temperature. The notation $d^{t_1}_{t_2}$ is often used to specify the two temperatures, the subscript t_2 being that of water. For gases, specific gravity is ratioed on that of air at standard conditions, usually 101.325 kPa and 0°C. When buoyancy corrections are made to mass determinations of the densities, the term *absolute specific gravity* is used. Specific gravity is often misused as a Syn: density but the error isn't as serious as the parallel error with specific heat because, in the density unit still in most common use, g/cm^3, water's density is very close to 1.000 at room temperature. The ASTM test for specific gravity (and density) of plastics by displacement is D 792. See also ▶ Density and ▶ Bulk Density. In the SI, the term "relative density" is preferred.

Specific Heat *n* (1832) Strictly, specific heat is the ratio of the ▶ Heat Capacity of a substance to that of water at 15°C. In traditional cgs and English units, the heat units (calorie and British thermal unit) were defined by the heat capacity of water, making that of water at room temperature closely equal to 1.00 in either system. Thus, for other materials, specific heat and heat capacity were numerically equal. This fact led to the use, still ongoing, of "specific heat" when the property meant was heat capacity. In the SI system, the heat capacity of water is 4.18 J/g K, whereas specific heats are dimensionless and not affected by changes in units, so are no longer equal to heat capacities. Soon, it is hoped, this confusing and now useless term will fade out of the language of science and engineering.

Specific Heat of a Substance *n* The ratio of its thermal capacity to that of water at 15°C. Dimensions, unity. If the quantity of heat H calories is necessary to raise the temperature of m grams of a substance from t_1 to t_2°C, the specific heat, or more properly, thermal capacity of the substance,

$$s = \frac{H}{m(t_2 - t_1)}$$

Specific heat by the method of mixture – where a mass m_1 of the substance is heated to a temperature t_1, then placed in a mass of water m_2 at a temperature t_2 contained in a calorimeter with stirrer (or same material) of mass m_3, specific heat of the calorimeter $c_2 t_3$ the final temperature

$$m_1 s(t_1 - t_3) = (m_3 c + m_2)(t_3 - t_2).$$

Black's ioce calorimeter – If a body of mass m and temperature t melts a mass m' of ice, its temperature being reduced to 0°C, the specific heat of the substance is,

$$s = \frac{801 m'}{ml}$$

Bunsen's ice calorimeter – A body of mass m at temperature t causes a motion of the mercury column of l centimeters in a tube whose volume per unit length is v. The specific heat

$$s = \frac{884 l v}{ml}$$

Specific Humidity See ▶ Humidity Ratio.

Specific Inductive Capacity n The ratio of the capacitance of a condenser with a given substance as dielectric to the capacitance of the same condenser with air or a vacuum as dielectric is called the specific inductive capacity. The ratio of the dielectric constant of a substance to that of a vacuum. Syn: ▶ Permittivity.

Specific Insulation Resistance See ▶ Volume Resistivity.

Specific Intensity n Specific intensity of a retroreflector is the ratio of the luminous intensity of the reflector at a given observation angle to the illumination of the projected area (perpendicular to the direction of illumination) of the reflector. The units may be expressed as candles/ footcandle.

Specific Luminance n Specific luminance of a retroreflector is the ratio of the luminance of the reflector to the illumination of its actual area.

Specific Modulus n Elastic modulus (usually the tensile modulus) divided by density, the SI unit being Pa/(kg/m^3) (technically reducible to J/kg, which blurs its derivation and true nature). For example, the specific moduli of neat acetal resin (homopolymer) and glass-fiber-reinforced epoxy are, respectively, 2.3 and 11 MPa/(kg/m^3).

Specific Permeability n **(of a Film to Moisture)** The milligrams of water that permeate 1 cm^2 of film of 1 mm thickness each 24 h after a constant rate has been attained under the preferred conditions of 25°C and using 100% relative humidity inside the cup and a phosphorus pentoxide desiccated atmosphere outside the cup. The method requires that a moisture differential be established across the test specimen, and an apparatus provided which will permit weighing the amount of moisture which has permeated through the film under the test conditions. Reference ASTM D 1653.

Specific Rate Constant n The constant of proportionality in a rate law.

Specific Refractive Increment n The rate of change of refraction index (n) of a solution with concentration (c). Precise values of specific refractive increments are obtained using a differential refractometer.

Specific Rotation n If there are n grams of active substance in v cubic centimeters of solution and the light passes through l decimeters, r being the observed rotation in degrees, the specific rotation (for 1 cm),

$$[\alpha] = \frac{rv}{nl}$$

Specific Strength n Tensile strength (if some other is not specified) divided by density. The SI unit is Pa/(kg/m^3). For example, the specific strengths of neat acetal resin (homopolymer) and 30–33% glass-reinforced nylon 6/6 are, respectively, 48 and 118 kPa/(kg/m^3). Specific strength, like ▶ Specific Modulus is a more useful criterion than strength along for material selection when both strength and minimum weight are important in the design of a structure.

Specific Surface n Of porous solids, massed fibers, and particulate materials, the total surface area per unit of bulk volume or per unit mass. Specific surface is usually measured by gas adsorption or estimated from mercury-porosimetry measurements.

Specific Viscosity n Equal to the relative viscosity of the same solution minus one. It represents the increase in viscosity that may be contributed by the polymeric solute. The specific viscosity, η_{sp} is defined by the following equation:

$$\eta_{sp} = \eta_r - 1$$

where, η_r = relative viscosity. The fractional increase in the viscosity of a solvent resulting from dissolution of a polymer. The specific viscosity, η_{sp}, is found by subtracting one from the relative viscosity η_r. See ▶ Dilute-Solution Viscosity.

Specific Volume n Reciprocal of specific gravity. *Also known as Bulking Value.* Also referred to as the **molar**

volume, symbol V_m, is the volume occupied by 1 mole of a substance (chemical element or chemical compound) at a given temperature and pressure. It is equal to the molar mass (M) divided by the mass density (ρ). It has the SI unit cubic meters per mole (m³/mol) although it is more practical to use the units cubic decimeters per mole (dm³/mol) for gases and cubic centimeters per mole (cm³/mol) for liquids and solids. The molar volume of a substance can be found by measuring its mass density ρ and applying the relation,

$$V_m = \frac{M}{\rho}$$

For ideal gases, the molar volume is given by the ideal gas equation. This is a good approximation for many common gases at standard temperature and pressure. For crystalline solids, the molar volume can be determined by determining the unit cell volume (V_{cell}) calculated from unit cell parameters that are determined by x-ray crystallography.

The ideal gas equation can be rearranged to give an expression for the molar volume of an ideal gas:

$$V_m = \frac{M}{n} = \frac{R}{T}.$$

Hence, for a given temperature and pressure, the molar volume is the same for all ideal gases and is known to the same precision as the gas constant: $R = 8.314\,472(15)$ J/mol/K, that is a relative standard uncertainty of 1.7×10^{-6}, according to the 2006 CODATA recommended value. The molar volume of an ideal gas at 100 kPa (1 bar) is

$$22.710\,980(38) \text{ dm}^3/\text{mol at } 0°C$$
$$24.789\,598(42) \text{ dm}^3/\text{mol at } 25°C$$

▶ [edit] Crystalline solids

The unit cell volume (V_{cell}) for crystal may be calculated from the unit cell parameters, whose determination is the first step in an X-ray crystallography experiment (the calculation is performed automatically by the structure determination software). This is related to the molar volume by

$$V_m = \frac{N_A \, V_{cell}}{Z}$$

where N_A is the Avogadro constant and Z is the number of formula units in the unit cell. The result is normally reported as the "crystallographic density." (International Union of Pure and Applied Chemistry (1993). Quantities, units and symbols in physical chemistry, 2nd edn. Blackwell Science, Oxford. ISBN 0-632-03583-8, p 41. Electronic version)

Speck \spek\ *n* [ME *specke*, fr. OE *specca*] (before 12c) (1) A contaminant in polymer such as gels, metal, or dirt that shows up as a dark spot. (2) A small particle of foreign substance that has not been removed from the stock before spinning.

Specking *n* The removal of burrs, knots, and other objects that impair the finished appearance of woolens and worsteds.

Speckled Finish See ▶ Multicolor Finish.

Specks *n* Small particles of undispersed materials.

Specky *n* A term used to describe dyed woolen fabric with specks of undyed vegetable matter on the face. The specks can be removed by carbonizing or covered by speck dyeing.

Spectra \spek-trəm\ *n* [NL, fr. L, appearance] (1671) Plural of spectrum.

Spectral *n* Adjective referring to spectrum. See ▶ Spectrophotometric.

Spectral Match See ▶ Color Match.

Spectral Power Distribution Curve *n* Intensity of radiant energy as a function of wavelength, generally given in relative power terms.

Spectral Series *n* Spectral lines or groups of lines which occur in an orderly sequence.

Spectrochemical Series *n* A listing of ligands in order of their ability to cause *d*-orbital splitting in a complex.

Spectrogoniophotometer *n* Goniophotometer used to measure the geometric distribution of reflected or transmitted flux at individual wavelengths.

Spectrograph \spek-tra-graf\ *n* [ISV] (1884) A spectroscope equipped with a camera or some other device for recording the spectrum. Also see ▶ Spectroscope.

Spectrometer *n* [ISV] (1874) An instrument for identifying and comparing materials by the dispersing of light and the study of the spectra formed.

Spectrophotometer \fə-tä-mə-tər\ *n* [ISV] (1881) An instrument for measuring the brightness of various portions of spectra. One useful application of this instrument is the formulation of colorants to match a given sample under all types of illumination. The instrument produces a curve representing the amounts of light energy the specimen to be matched absorbs over a wide range of wavelengths. Matching this curve assures that the developed compound will appear to be the same color as the specimen under any lighting condition. Another important application of the principle in a different instrumental format is the field of quantitative chemical analysis. An example of an ultra-violet and visible spectrophotometer is the Lambda 950 Spectrophotometer (courtesy of PerkinElmer, Inc.) See also ▶ Colorimeter.

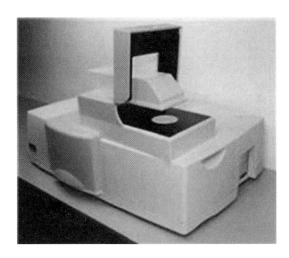

Spectrophotometer, Abridged *n* Photometric device for measuring the transmittance, reflectance, or relative emittance at discrete wavelengths. The particular discrete wavelengths may be selected by the choice of a series of narrow bandpass filters (such as interference filters), or masks, etc. No continuous spectral curve can be obtained from such an instrument.

Spectrophotometer, Filter See ▶ Filter Spectrophotometer and ▶ Spectrophotometer, Abridged.

Spectrophotometric *n* Adjective referring to spectrophotometry. The word "spectral" is frequently used in place of the longer "spectrophotometric," but they are not necessarily synonymous. Spectral is the adjective referring to spectrum and is the more general term, hence a spectrophotometric curve is also a spectral curve.

Spectrophotometric Curve *n* A curve measured on a spectrophotometer; hence a graph of relative reflectance or transmittance (or absorption) as the ordinate, plotted versus wavelength or frequency as the abscissa. The most common curves in the visible region use wavelength units of nanometers, with the short wavelength units at the left of the scale.

Spectrophotometric Hiding See ▶ Hiding, ▶ Spectrophotometric, ▶ Complete and ▶ Incomplete.

Spectrophotometry *n* [ISV] (1881) An analytical instrumental technique for measuring color values by the relative intensity of the component spectrum colors. Broadly, a technique which, by measuring the absorption (or reflection) of electromagnetic radiation from organic and inorganic substances, helps in their identification.

Spectroscope \-▮skōp\ *n* [ISV] (1861) An instrument for forming a spectrum for visual.

Spectroscopy *n* The study of electromagnetic waves that are absorbed or emitted by substances when excited by the arc, a spark, X rays, or a magnetic field. Each element emits light of characteristic wavelengths, by which minute quantities can be detected and estimated. An example of a spectrophotomer for the ultraviolet and visible spectrum is the PerkinElmer Lambda 25/35/45 UV/VIS (courtesy of PerkinElmer. Inc.). See also ▶ Nuclear Magnetic Resonance.

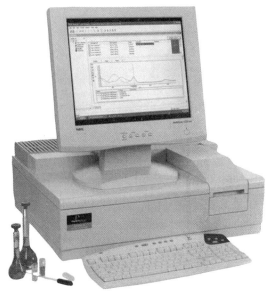

Spectrum *n* Spatial arrangement of components of radiant energy in order of their wavelengths, wave number or frequency.

Spectrum Locus *n* The curve, connecting points in a chromaticity diagram, that represents various wavelengths of the visible spectrum.

Specular Mirror-like.

Specular Gloss *n* The luminous fractional reflectance of a specimen in the specular direction. It is sometimes measured at 60° relative to a perfect mirror. The ASTM method for measuring specular gloss is D 523. See ▶ Reflectance, ▶ Specular and ▶ Gloss.

Specular Reflectance See ▶ Reflectance, Specular.

Specular Reflectance Excluded *n* (**SCE**) Measurement of reflectance made in such a way that the specular reflectance is excluded from the measurement, diffuse reflectance. The exclusion may be accomplished by using 0° (perpendicular) incidence on the samples, thereby reflecting the specular component of the reflectance back into the instrument, by use of black absorbers or light traps at the specular angle when the incident angle is not perpendicular, or in directional measurements by measuring at an angle different from the specular angle.

Specular Reflectance Included *n* (**SCI**) Measurement of the total reflectance from a surface, including the diffuse and specular reflectances.

Specular Reflection *n* Light striking a surface, and being reflected or turned back at an angle equal to the angle of incidence. See ▶ Gloss.

Specular Transmittance *n* (regular transmittance) Of a transparent plastic, the ratio of the light flux transmitted without diffusion to the flux incident (ASTM D 883). See also ▶ Light Transmittance and ▶ Transmittance, Specular.

Speed \spēd\ *n* [ME spede, fr. OE spēd; akin to HGr *spuot* prosperity, speed OE *spōwan* to succeed, L *spes* hope, L *spĕti* to be in time] (before 12c) Time rate of motion measured by the distance moved over in unit time. Cgs unit – 1 cm/s. Dimension $[l\ t^{-1}]$.

Speed Reducer *n* (gear reducer) The gearbox between motor and screw coupling of an extruder that reduces the high speed of the motor shaft to the much lower speed, with inversely higher torque, of the extruder screw. The reduction of about 12–1 normally requires two stages. The gears may be spur gears (cheapest and least desirable), helical gears, or double-helical (herringbone) gears. Where the extruder is expected to handle a wide variety of plastics and products using an assortment of screws, a change-gear reducer is recommended. It enables the operator to quickly accomplish large changes in the reduction ratio, while also permitting safe and efficient operation of the variable-speed motor-drive and gears. One may still find a worm-bear reducer in an antique extruder long past its rightful retirement age.

Spermaceti \spər-mə-ˈsē-tē\ *n* [ME *sperma cete*, fr. ML *sperma ceti* whale sperm] (15c) A waxy solid obtained from certain fish oils, e.g., sperm oil.

Spermaceti Wax \spər-mə-ˈsē-tē\ *n* Wax obtained from certain fish oils, e.g., sperm oil. It is characterized by a glistening white, crystalline appearance. Mp, 49°C; sp gr, 0.960; iodine value, 4; saponification value, 126.

Spew Groove Syn: ▶ Flash Groove.

Spewing *n* (1) Migration of components to the surface of a coating usually because of its compatibility. (2) The formation of a film, or the collection of particles, on a paint surface during the drying process; it results from the migration, to the surface, of the insoluble or incompatible portion(s) of the paint binder. Also called *Greasiness*.

Spew Line Syn: ▶ Parting Line.

Sphenoid \ˈsfē-ˌnóid\ *adj* [NL *sphenoides*, fr. Gk *sphēnoeidēs* wedge-shaped, fr. *sphēn* wedge] (1732) A crystal form made up of two nonparallel faces not symmetrical about a plane of symmetry.

Spherical Aberration *n* (1868) Occurs when light rays of one wavelength that travel near the center of the lens (or mirror) focus at a different point than light rays of the same wavelength near the periphery. The phenomenon is known as spherical aberration. For axial pencils the error is known as axial spherical aberration; for oblique pencils, coma.

Spherical Angle *n* (1678) The angle between two intersecting arcs of great circles of a sphere measured by the plane angle formed by the tangents to the arcs at the point of intersection.

Spherical Mirrors *n* If R is the radius of curvature, F principal focus, and f_1 and f_2 any two conjugate focal distances,

$$\frac{1}{f_1} + \frac{1}{f_2} = \frac{1}{F} = \frac{2}{R}$$

If the linear dimensions of the object and image be O and I respectively and a u and v their distances from the mirror,

$$\frac{O}{I} = \frac{u}{v}$$

Spherulite \ˈsfir-yə-ˌlīt\ *n* (1823) A rounded aggregate of radiating crystals with a fibrous appearance under the microscope. Spherulites have been observed in most crystalline plastics. They originate from a nucleus such as a particle of contaminant, residual catalyst particle, or change fluctuation in density. They may grow through stages: first needles, then bundles and sheaf-like aggregates, and finally the spherulites. Spherulites may range in diameter from a few tenths of a micrometer to several millimeters. They are birefringent, displaying a characteristic maltese-cross insignia when viewed through crossed polarizers.

SPI *n* Abbreviation for (The) ▶ Society Of The Plastics Industry.

Spider *n* (1) In a molding press, that part of an ejector mechanism that operates the ejector pins. (2) Within an end-fed extrusion die making a tubular section, the three or four legs extending from die to mandrel and supporting the mandrel at the center of the die. The spider legs may themselves be cored to pass a temperature-control fluid through the mandrel and into an extended calibrating mandrel attached to the die mandrel. also called *spider fins*. (2) In rotational casting, the grid-work of metallic members supporting cavities in a multicavity mold.

Spider Legs *n* Film defect where the coating material on an upright surface separates or breaks and the liquid runs down in long, crooked channels.

Spider Lines *n* In blow molding or pipe extrusion, visible marks parallel to the parison or pipe axis and corresponding to the positions of the spider legs. They are due to incomplete welding of the divided stream downstream of those legs. These lines are the exterior traces of weld surfaces that go through the annular wall, surfaces that are sometimes weaker than the material between them.

Spider Runners *n* (spider gating) A design for melt distribution to multiple cavities in an injection mold inn which the cavities are arranged in a circle and fed by runners radiating from a central sprue.

Spin *n* (1831) In nuclear physics, used to describe the angular momentum of elementary particles or of nuclei.

Spindle \spin-dᵊl\ *n* [ME *spindel*, fr. OE *spinel*; akin to OE *spinnan* to spin] (12c) A slender, upright, rotating rod on a spinning frame, roving frame, twister, winder, or similar machine. A bobbin is placed on the spindle to receive the yarn as the spindle is rotated at high speed.

Spin Drawing *n* (1) The reduction of roving during spinning by a roller drafting mechanism similar to that used on the roving frame. (2) Combined spinning and drawing in one operation in melt-spun fibers.

Spin Dyeing *n* (mass dyeing, dope dyeing) The process of coloring fibers or yarns by incorporating pigments or dyes in the material prior to spinning, either during or after polymerization of the material.

Spin Finish See ▶ Lubricant.

Spin Multiplier See ▶ Twist Multiplier.

Spinneret \spi-nə-ret\ *n* (1826) (1) An extrusion die consisting of a plate with many tiny holes, through which a plastic melt or solution is forced, to make fine fibers and filaments. Early spinneret holes were round and thus produced fibers of circular cross section. Today, spinneret holes have many different shapes, even annular ones, to produce fibers of corresponding cross sections. One purpose is to decrease the fiber-bundle density, giving added warmth, moisture permeability, and enhanced dye receptivity to the textile fabric. An important application of hollow fibers is in artificial kidneys for dialysis. Filaments emerging from the spinneret may be hardened by cooling in air or water, or by chemical action of solutions. (2) A spinneret hole.

Spinning *n* (1) The process of forming staple fibers by extruding polymers. There are three main variations of the process: *melt spinning*, *dry spinning*, and *wet spinning*. All employ extrusion dies with from one to thousands of tiny orifices, called *jets* or *spinnerets*. In melt spinning, the molten polymer is pumped first through sand-bed filters, then through the die orifices by small gear pumps operating at extremely high pressures. In both wet and dry spinning the polymer is dissolved in a solvent prior to extrusion. In dry spinning, the extrudate is subjected to a hot atmosphere that removes the solvent by evaporation. In wet spinning, the spinneret is immersed in a liquid that either diffuses the solvent or reacts with the fiber composition, precipitating it from solution. However produced, the fibers are then oriented to realize their optimal strength and modulus, four times or more that of the unoriented fibers. With larger fibers (see ▶ Monofilament) this is done by reheating the fiber in a carefully controlled oven, and while it is warm, stretching it with Godet stands operating at a large speed differential. Much staple fiber is also oriented in this way. With melt-spun staple, the trend is toward accomplishing most, if not all of the orientation by high-speed windup and drawing as the extruded filaments are leaving the spinnerets. (2) Method of application which distributes the paint over a flat surface by centrifugal action. Different types of processes and types of spinning are listed below.

1. **Yarn from Staple Fiber**: The formation of a yarn by a combination of drawing or drafting and twisting prepared strands of fibers, such as rovings.
2. **Filament Yarn**: In the spinning of manufactured filaments, fiber-forming substances in the plastic or molten state, or in solution, are forced through the fine orifices in a metallic plate called a spinneret, or jet, at a controlled rate. The solidified filaments are drawn-off by rotating rolls, or godets, and wound onto bobbins or pirns. There are several methods of spinning manufactured filaments:
   2a. **Dry Spinning** *n* The process in which a solution of the fiber-forming substance is extruded in a continuous stream into a heated chamber to remove the solvent, leaving the solid filament, as in the manufacture of acetate.
   2b. **Gel Spinning** *n* A spinning process in which the primary mechanism of solidification is the gelling of the polymer solution by cooling to

form a gel filament consisting of precipitated polymer and solvent. Solvent removal is accomplished following solidification by washing in a liquid bath. The resultant fibers can be drawn to give a product with high tensile strength and modulus.

2c. **Melt Spinning** *n* The process in which the fiber-forming substance is melted and extruded into air or other gas, or into a suitable liquid, where it is cooled and solidified, as in the manufacture of polyester or nylon.

2d. **Phase-Separation Spinning** *n* Extrusion of polymer and solvent at high temperature into a cooling zone. During the cooling process, a phase separation occurs, usually accompanied by crystallization of the solvent. Solvent can be removed before or after drawing.

2e. **Reaction Spinning** *n* Process in which an initial prepolymer is formed and then extruded into a reagent bath where polymerization and filament formation occur simultaneously. Spandex fibers can be made by this process.

2f. **Wet Spinning** *n* The process in which a solution of the fiber-forming substance is extruded into a liquid coagulating medium where the polymer is regenerated, as in the manufacture of viscose or Cuprammonium rayon.

3. **Yarn from Leaf and Bast Fiber** *n*: In the manufacture of leaf and bast fiber yarns, the terms "wet spinning" and "dry spinning" refer to the spinning of fibers in the wet state and in the air-dry state, respectively.

4. **Yarn from Filament Tow** *n*: The formation of a yarn from filament tow by a combination of cutting or breaking, drafting, and twisting in a single series of operations. *Also known as Converting.*

5. **Nonwoven Fabric** *n*: Fabrics can be produced directly from molten or dissolved fiber-forming substances by several continuous processes:

The process in which a fiber-forming substance in a volatile solvent is extruded from a high-temperature, high-pressure environment into lower temperature and pressure conditions, causing the solvent to rapidly evaporate, leaving a lacy, net-like fabric.

Spinning Frame *n* A machine used for spinning staple yarn. It drafts the roving to the desired size, inserts twist, and winds the yarn onto a bobbin. The term is generally used to indicate a ring spinning frame, although it does cover flyer spinning and cap spinning on the worsted system.

Spinning Solution *n* A solution of a fiber-forming polymer (e.g., cellulose acetate) in a suitable condition to be extruded by either dry spinning or wet spinning.

Spinning Twist *n* The twist added to yarn during spinning to give it strength and other desired characteristics.

Spin Quantum Number, m_s A quantum number which specifies the spin of an electron.

Spin Welding *n* (friction welding) A process for joining thermoplastic articles at circular mating surfaces by rotating one part in contact with the other until sufficient heat is generated by friction to produce a thin interfacial layer of melt. Rotation is stopped and pressure is maintained while the melt solidifies, completing the weld. The process can be performed manually in a drill press, or can be fully automated with devices for feeding, timing, controlling stroke and pressure of the press, and ejection. In a version of the process that is applicable to some noncircular joints, rotation is oscillatory through small, reversing arcs.

Spiral-Flow Test *n* A method of evaluating the molding flow of a resin to injection or transfer molding in which the melt is injected into a spiral runner of constant trapezoidal cross section with numbered and subdivided centimeters (or inches) marked along the runner. The mold is filled from a sprue at the center of the spiral and pressure is maintained until flow stops, the number just aft of the molded-spiral tip giving the flow distance. The spiral-flow test has been widely used since it was introduced in the early 1950s but has been standardized in the US only for thermosetting resins, in ASTM D 3123.

Spiral Grain *n* A form of cross grain in which the fibers extend spirally about instead of vertically along the bole of a tree or axis of a timber.

Spiral-Mandrel Die *n* The most popular die design for extrusion of blown film, in which the melt, fed into a manifold at the bottom of the die, flows up to the die lip via multiple spiral grooves. The design is capable of handling a range of resins, provides even distribution to the lip, thus helping to minimize circumferential variations in film thickness, and has relatively low pressure drop.

Spiral Mold Cooling *n* A method of cooling injection molds (or heating transfer molds) wherein the heat-transfer medium flows through a spiral passage in the body of the mold. In injection molds, the cooling medium is introduced at the center of the spiral, near the sprue section, because more heat is concentrated in this area.

Spiral Yarns *n* Specialty yarn made by winding heavier, slackly twisted yarn around a finer yarn with a hard twist to give a slubby appearance.

Spirit \\ˈspir-ət\\ *n* [ME, fr. OF or L; OF, fr. L *spiritus*, lit., breath, fr *spirare* to blow, breathe] (13c) (1) In the paint industry, this term is used somewhat loosely but generally refers to commercial ethyl alcohol normally sold as industrial methylated spirit. (2) The term "mineral spirits" (in the UK as "white spirit," mainly defines a mixture of aliphatic hydrocarbons with a proportion of aromatic hydrocarbons.

Spiriting Off *n* Final operation in a French polishing process by which the last traces of oil are removed by drawing a rag, dampened with methylated spirit, rapidly and repeatedly over the surface.

Spirit of Wine *n* (1753) See ▶ Ethyl Alcohol.

Spirits of Rosin See ▶ Rosin Spirit.

Spirits of Turpentine See ▶ Turpentine, Spirits of

Spirit-Soluble Dye *n* Dye soluble in alcohol and insoluble in water; usually azo colors lacking the group SO_2OH.

Spirit-Soluble Resin *n* A resin soluble in alcohol and insoluble in water.

Spirit Stains Solution s of suitable dyes in solvent which may also contain a proportion of resin as binder. See ▶ Stain.

Spirit Varnish *n* (1850) (1) Lacquer based on a solution of resin or resins in industrial methylated spirit. The more correct term would be "spirit lacquer". See ▶ Lacquer. (2) A varnish which is converted to a solid film by solvent evaporation. See ▶ Shellac.

Spitting See ▶ Flying; ▶ Misting.

splash See ▶ Splay.

Splay \\ˈsplā\\ *n* (splay marks, silver streaking) Scars or surface defects on injection molding arising from two main mechanisms. One is the injection of a high-velocity stream of molten material into the mold ahead of the general flow front. The prematurely injected melt, especially in the case of crystalline polymers with a sharp freezing point, cools and solidifies before the mold cavity is completely filled. These defects occur most frequently in the gate area, but may be washed into other areas of the cavity. Remedies are: increasing the mold temperature, local heating of the gate area, reduction of injection rate, and increasing the gate cross section. The second mechanism is release of moisture or residual monomer or solvent in the melt in the form of fine bubbles at the part surface. The remedy for this type of splay is predrying to remove volatiles from the feedstock before it enters the molder and/or careful handling of regrind to prevent moisture pickup during recycling. These precautions are especially important with hygroscopic resins sun as nylon 6/6.

Splicing *n* (1) The joining of two ends of yarn or cordage. There are several methods used, e.g., by interweaving the strands, by the use of knots, by tapering, lapping, and cementing the ends, etc. (2) A method of reinforcing knits, e.g., the heels and toes of hosiery, by introducing an additional yarn for strength.

Splinter *n* Two or more staple fibers adhering together, causing a stiff cluster that resists pulling apart in normal processing, and reacting in the yarn spinning process similarly to higher than nominal denier fiber.

Splinter Count *n* A measure of the number of coalesced fibers, mealy particles, or other such matter in staple fiber.

Split See ▶ Reed.

Split-Disk Method *n* (NOL ring test) A method of testing the tensile strength of tubular plastics and reinforced plastics embodied in ASTM D 2290.

Split End *n* (1) A defect in fabric caused by breakage of some of the singles yarns in a plied warp yarn. (2) A defect in manufactured filament yarn caused by breakage of some of the filaments.

Split-Film Fiber *n* A type of polypropylene fiber made by extruding a thin film, applying a high lengthwise draw to strongly orient it, at the same time weakening it in the transverse direction, pounding it to cause lateral fibrillation into narrow ribbons about 1 mm wide, and further orienting the ribbon fibers. These may be woven into a variety of products such as sacking, artificial turf, and indoor-outdoor carpet.

Split-Flow Metier See Split-Draft Metier.

Split Spray *n* Unsymmetrical spray pattern resulting in the application of bands of paint of uneven thickness, caused either by a defective spray or nozzle or by partial blockage of the nozzle or air passages of a spray gun.

Splitting *n* (1) A defect in a painted surface; results from the penetration of solvents, contained in a fresh coat of paint, into an older layer of paint over which it has been applied; likely to occur when the old layer has been sanded too much. (2) In the processing of tow, a defect in which the integrity of opened tow is disturbed by separation or division into two or more segments longitudinally. Splitting can be continuous or intermittent, long or short term. (3) In slashing, the separation of sized yarn ends before takeup on the slasher beam. (Paint film defects. Hess M, Morgans WM (1979) Wiley, New York)

Spontaneous Change \spän-ˈtā-nē-əs-\ *n* A change which is thermodynamically probable; a possible change.

Spontaneous Combustion *n* (1795) Self-ignition of combustible material through chemical action (as oxidation) of its constituents). *Known also as Spontaneous Ignition.*

Spool \ˈspül\ *n* [ME *spole*, fr. MF or MD; MF *espole*, fr. MD *spoele*; akin to OHGr *spuola* spool] (14c) A flanged wooden or metal cylinder upon which yarn, thread, or wire is wound. The spool has an axial hole for a pin or spindle used in winding. Also see ▶ Beam.

Spore \ˈspōr\ *n* [NL *spora* seed, spore, fr. Gk, act of sowing, seed, fr. *speirein* to sow] (1836) A simple reproductive body of lower plants. A spore is analogous to the seed of higher plants but lacks an embryo. It usually consists of one cell, but sometimes several spores are enclosed by a protective covering.

Spot Bonding See ▶ Bonding (2), ▶ Point Bonding.

Spot Finishing See ▶ Spotting In.

Spotting *n* Development of small areas on a painted surface which differ in color or gloss from the major portion of the work.

Spotting In *n* Rubbing down and refinishing small defective patches in coating. Syn: ▶ Retouching, ▶ Spot Finishing.

Spray-and-Wipe Painting *n* (fill-in marking) A decorating process for articles with depressed letters, figures, or designs. A lacquer or enamel is applied by spraying either the article's entire surface or a restricted area including the depression, then removing the excess wet paint by buffing or wiping the raised areas.

Spray Bonding See ▶ Bonding (2).

Spray Booth *n* An enclosed or semienclosed area used for the spray painting of fabricated items; may be equipped with a source of filtered air to keep the atmosphere dust-free, a waterfall backdrop to trap overspray, and an exhaust system to vent the fumes of the evaporating solvents.

Spray Coating *n* The application of a polymeric coating to a substrate by means of a spray gun. The process is used for coating any material with a plastic, and for coating plastics for decorative purposes. In the latter application, masks are usually employed to apply the coating only to selected areas. (Wicks WZ Jr, Jones FN, Pappas SP (1999) Organic coatings. Wiley-Interscience, New York) See also ▶ Airless Spray, ▶ Electrostatic Spray Coating, ▶ Flame Spray Coating, and ▶ Plasma-Spray Coating.

Spray Drier *n* A cylindrical or conical-bottomed chamber through which hot air rises and into which a resin emulsion is sprayed in small droplets. The solvent evaporates from the droplets as they fall and the dried-resin powder is removed from the chamber bottom by scraping or air blasting. Spray drying has been employed for emulsion-polymerized vinyl resin, and for amino and phenolic resins.

Sprayed-Metal Mold *n* A mold made by spraying molten metal onto a master form to obtain at shell of desired thickness, which may subsequently be backed with plaster, cement, casting resin, etc. Such molds are used most commonly in sheet-forming processes.

Spray Gun *n* (1920) A tool, operated with compressed air or fluid pressure, which expels paint, mortar, etc., through a small orifice, onto the surface being coated. Syn: ▶ Air Brush and ▶ Spraying Pistol. See ▶ Concrete Gun.

Spraying *n* (1) Method of application in which the coating material is broken up into a fine mist, which is directed onto the surface to be coated. This atomization process is usually, but not necessarily, effected by a compressed air jet. See ▶ Airless Spraying and ▶ Hot Spraying. (2) See ▶ Flying.

Spraying Pistol See ▶ Spray Gun.

Spray Molding See ▶ Sprayup.

Spray Mottle *n* Irregular surface of a sprayed film resembling the skin of an orange. The defect is due to the failure of the film to flow out to a level surface. See ▶ Orange Peel.

Spray Pattern *n* Configuration of spray with gun held in a steady manner.

Spray Spinning See ▶ Spun-Bonded Products.

Sprayup *n* A term coined in February, 1958, by the Engineering Editor of *Modern Plastics* to identify a new technique of reinforced-plastics molding that employed a gun that chops roving and impels the chopped strands through the focus of two resin sprays and onto the mold or mandrel. One resin stream contains catalyst, the other promoter, formulated to set quickly at room temperature. An advantage of the process is the ease with which section thickness may be varied in the molding to meet local strength and stiffness needs. The term sprayup was later extended to the foamed-plastics field, denoting the spraying of fast-reacting polyurethane or epoxy-resin systems onto surfaces where they react to foam and cure. In both areas the external mixing of streams by impingement avoids pot-life problems in spray equipment and tanks.

Spread *n* The quantity of adhesive per unit joint area applied to an adherend, usually expressed in points of adhesive per thousand square feet of joint area. (1) Single spread refers to application of adhesive to only one adherend of a joint, (2) double spread refers to application of an adhesive to both adherents of a joint.

Spreadable Life See ▶ Pot Life.

Spread Coating *n* A process for coating fabrics, sheet metals, etc., with fluid dispersions such as vinyl plastisols. The substrate is supported on a carrier, and the fluid material is applied to it just ahead of a blade or "doctor knife" that regulates the thickness of the coating. The deposit is then heated to fuse the coating to the substrate, often followed by embossing to impart texture.

Spreader *n* (1) See ▶ Torpedo. (2) Any device, such as a knife or roller, a part of spread-coating equipment that helps to produce an evenly thick coating.

Spreading *n* A thickening or enlarging of printed areas caused by bleeding or lateral penetration of ink.

Spreading Capacity *n* Average spreading rate to be expected from a paint when applied over a normal surface in a manner appropriate for that particular paint. Spreading rate will vary with the operator, the method of application, and the nature of the surface to be painted. See ▶ Spreading Rate.

Spreading Rate *n* The area of surface over which a unit volume of paint will spread, usually expressed in ft^2/gal or m^2/L. See ▶ Spreading Capacity.

Spread Stitch See Pinhole.

Spring Box Mold *n* A type of compression mold equipped with a spacing fork that prevents the shifting of bottom-loaded inserts or loss of fine details, and which is removed after partial compression.

Spring Needle *n* A knitting machine needle with a long, flexible hook, or beard, that allows the hook to be closed by an action known as pressing so that the loops can be cast off. The hook springs back to its original position when the presser bar is removed. Also see ▶ Latch Needle.

Springwood *n* The portion of the annual growth ring that is formed during the early part of the season's growth. It is less dense and weaker mechanically than summerwood.

Sprue \sprü\ *n* [origin unknown] (1880) (1) In injection and transfer molding, the main feed channel connecting the machine nozzle with the runners leading to the various cavities. The sprue is usually conical, widening slightly toward the mold, so that, as the mold opens, the plastic within the sprue remains attached to the runners and the sprue is cleared for the next shot. (2) The conical plastic stub that is formed within the sprue with ach shot and is removed with the runners.

Sprue Bushing *n* (British: sprue bush) A hardened steel insert in an injection mold that contains the tapered sprue passage and has a suitable seat, usually hemispherical, making a seal with the nozzle of the injection cylinder.

Sprue-Ejector Pin Syn: ▶ Sprue Puller.

Sprue Gating *n* (direct gating) In injection molding with single-cavity molds, particularly those with circular symmetry, filling from the center with the sprue connected directly to the gate and mold cavity.

Sprue Lock *n* In injection molding, a portion of the plastic composition that is held in the cold-slug well by an undercut, used to pull the sprue out of the sprue passage as the mold is opened. The sprue lock itself is pushed out of the mold by an ejector pin. When the undercut occurs on the cavity-block retainer plate, this pin is called the *sprue-ejector pin*.

Sprue Puller *n* A pin having a Z-shaped slot undercut in its end, by means of which it pulls the molded sprue out of the sprue passage.

Spun-Bonded Products *n* Nonwoven fabrics formed by filaments that have been extruded, drawn, then laid on a continuous belt. Bonding is accomplished by several methods such as by hot roll calendering or by passing the web through a saturated-steam chamber at an elevated pressure.

Spun-Bonded Sheet *n* A sheet structure resembling paper or felted fabric, made by heat sealing webs of randomly arranged, continuous thermoplastic fibers. Three such materials, of polyolefin and polyester fibers, were introduced by DuPont in 1968. Good qualities are all-directional tensile strength, high tear resistance, good flex life, and puncture resistance. Applications include wall coverings, book covers, tags, labels, mailing envelopes, industrial clothing, and filter media.

Spun-Dyed See ▶ Dyeing, Mass-Colored.

Spun Fabric *n* A fabric made from staple fibers that may contain one or a blend of two or more fiber types.

Spunlaced Fabric *n* A nonwoven fabric produced by entangling fibers in a repeating pattern to form a strong fabric free of binders.

Spunlike Filament Yarns See ▶ Simulated Spun Yarns.

Spun Roving *n* A glass-fiber strand, repeatedly doubled back on itself to form a roving, sometimes reinforced by one or more straight strands.

Spun Silk See ▶ Schappe.

Spun Yarn *n* (14c) (1) A yarn consisting of staple fibers usually bound together by twist. (2) A meltspun fiber before it is drawn.

Spur A term sometimes used to mean ▶ Sprue.

Sputtering See Vacuum Metallizing.

SQC *n* Abbreviation for *(Statistical)* Quality Control.

Square Construction See ▶ Balanced Cloth.

Squeegee A soft, flexible blade or roller used in wiping operations, particularly in ▶ Screen Printing and Layup Molding.

Squeegee Coat *n* Shellac, lacquer or similar materials applied in a heavy consistency or body by means of a rubber gasket or squeegee. Broom handles and lead pencils are coated in this manner.

Squeeze Molding *n* A process for making prototypes from sheet molding compounds (SMC) with inexpensive tooling and very low molding pressure, to develop designs for parts that will be produced by injection molding or from metal by die casting. An epoxy two-piece mold is prepared, details such as ribs, gussets, and bosses are positioned, then the mold is filled with reinforced SMC and pressed at 140–210 kPa until cured.

Squeeze Out *adj* Adhesive pressed out at the bond line due to pressure applied on the adherents.

Squeeze Rolls *n* Rolls used to apply pressure for removal of water or chemicals from fabric.

SRIM *n* Abbreviation for Structural Reaction Injection Molding. See ▶ Reaction Injection Molding.

SRP Abbreviation for Rubber-Reinforced Polystyrene. See ▶ Rubber Toughening.

SS *n* (1) Abbreviation for Single-Stage (▶ Resin). See ▶ Resol. (2) Abbreviation for ▶ Stainless Steel.

SSF Abbreviation for Seconds, Saybolt, Furol.

SSPC *n* Abbreviation for Steel Structures Painting Council.

SSU *n* Abbreviations for Seconds, Saybolt, Universal.

Stability \stə-ˈbi-lə-tē\ *n* {*plural* –ties} (14c) A term used to describe the tendency of a fiber or fabric to return to its original shape after being subjected to external influence, such as tension, heat, or chemicals.

Stabilized Fiber *n* Fiber that is heat or chemically treated to set the fiber properties and prevent deterioration, shrinkage, etc. Also see ▶ UV Absorber.

Stabilizer *n* A chemical added to some plastics to assist in maintaining the physical and chemical properties of the compound at suitable values throughout the processing and service life of the material and articles made therefrom. An *emulsion stabilizer* serves to keep emulsions and suspensions from separating. A *viscosity stabilizer* is used in vinyl dispersions to retard viscosity increase on aging. An agent used primarily to protect plastics and rubbers from deterioration by oxidation is called an ▶ Antioxidant. The remaining, and most important, types of stabilizers are those that protect plastics from the effects of heat and light. Such effects are evidenced by a change of color, ranging from slight yellowing to blackening; a progressive decrease in mechanical properties; a decrease in electrical properties; or undesirable surface conditions such as blisters, spew, or exudation of ingredients rendered incompatible by heat or light. Many resins are vulnerable to ultraviolet light because they are good absorbers of UV and because its photonic energy is high. Stabilizers that function primarily by absorbing UV are described under ▶ Ultraviolet Stabilizer. Thousands of compounds have been proposed as heat stabilizers and as combination heat and light stabilizes for various plastics. The principal classes of such compounds are: (1) *Group II metal salts of organic acids* (primarily the barium, cadmium, and zinc salts of fatty acids, and phenols, the most important group); (2) *Organotin Stabilizers*; (3) *Epoxy Stabilizers*; (4) *Salts of mineral acids*, e.g., carbonates, sulfates, silicates, phosphates, and phosphates; (5) other organic compounds of metals and metalloids, e.g., alcoholates and mercaptides. Heat stabilizers are used nearly exclusively with vinyl resins. See also ▶ Zinc Stabilizer.

Stabilizers *n* Materials added to a plastic to impede or retard degradation that is usually caused by heat or ultraviolet radiation.

Staging Life See ▶ Shelf Life.

Stain \ˈstān\ *n* [ME *steynen*, partly fr. MF *desteindre* to discolor & partly Scand origin; akin to ON *steina* to paint] (1583) (1) A transparent or semitransparent solution or suspension of coloring matter (dyes or pigments or both) in a vehicle, designed to color a surface by penetration without hiding it, or to color a material into which it is incorporated. True stains are classified as water stains, oil stains, and spirit stains, according to the nature of the vehicle. The so-called "varnish stains" are varnishes colored with a transparent material. These do not have the same power of penetration as the true stains and leave a colored coating on the surface. (2) An undesirable change of color resulting from foreign matter. (3) To change the color by use of a staining agent.

Stainer See ▶ Tinter and ▶ Colorist.

Staining *n* (1) The application of a liquid dye solution to a porous surface to impart color. (2) Discoloration produced by a rubber stock on organic finishes, lacquers, and fabrics, owing to the presence of discoloring type anti-oxidants in the compound. The condition is aggravated by exposure to heat, pressure, or sunlight. Also, discoloration of a surface due to migration of an ingredient from one rubber compound to an adjacent compound even through they may not be in actual contact with each other. (3) The undesired pickup of color by a fabric: (a) when immersed in water, dry cleaning solvent, or similar liquid medium that contains dyestuffs or coloring material not intended for coloring the fabric; or (b) by direct contact with other dyed material from which color is transferred by bleeding or sublimation.

Staining Power *n* Degree to which a colored pigment imparts color to a white pigment under defined conditions of test. The details of procedure for determining staining power, normally laid down in specifications for pigments, need to be carefully adhered to if consistent results are to be obtained. The corresponding property of a white pigment is "reducing power." *Also known as Tinting Strength.*

Stainless Steel *n* (1920) A generic term embracing over 70 standard ferrous alloys produced in either wrought or cast forms, containing from 12% to 30% chromium and from 0% to 22% nickel (18 Cr, 8 Ni is typical). Also present may be small percentages of other elements such as carbon, columbium, copper, molybdenum, tantalum, and titanium. Stainless and semi-stainless steels have been used in plastics processing equipment where corrosion is a problem. A stainless steel has sometimes been chosen instead of chrome plating in environments where the plating has to be frequently repeated. Stainless type 17–4 PH, containing 17% Cr, 4% Cu and 0.35% Co, which is hardenable, has served well in extrusion dies for PVC.

Stainless Steel Fiber *n* Textile fibers made of stainless steel. Steel fibers are used for antistatic purposes in carpets, for tire belt construction, and for high-temperature or heat-resistant end uses.

Stain Resistance *n* The ability of a plastic material to resist staining caused by traffic, coffee, tea, blood, waxing compounds, grease deposits, and other staining agents, In the case of plasticized PVC, the most severe staining is caused by shoe polish, tobacco smoke, lipstick, nail polish, ketchup, and mustard. The degree of staining can be reduced to some extent by use of certain plasticizers, e.g., butyl benzyl phthalate, diethylene glycol dibenzoate, dipropylene glycol dibenzoate, and 2,3,4-trimethyl-1, 3-pentanediol monoisobutyrate benzoate.

Staircase Method See ▶ Up-and-Down Method.

Staking *n* A term sometimes used for the process of forming a head on a protruding portion of a plastic article for the purpose of holding a surrounding part in place. Ultrasonic heating of the protrusion facilitates the staking operation.

Stalk *n* A European term for Spruce.

Stamping See ▶ Die Cutting.

Stamp Molding *n* A compression-molding variation in which the mold is closed so suddenly that the plastic is impacted rather than pressed, as in forging.

Standard \ˈstan-dərd\ *n* [ME, fr. OF *estandard* rallying point, standard, of Gme origin; akin to OE *standan* to stand and to OE *ord* point] (12c) A reference point or a practice established by general agreement.

Standard Atmosphere *n* Air maintained at 70°F (21°C) and 65% relative humidity. When international testing is involved, a standard temperature of 20°C or, by agreement, 27°C may be used. Special humidity and temperature conditions are sometimes prescribed for the testing of certain textiles for specific service predictions, resistance to water or biological action, etc.

Standard Condition *n* Standard condition is that reached by a specimen when it is in moisture equilibrium with a standard atmosphere. Standard condition is seldom realized in practice since laboratory atmospheres are continually fluctuating between narrow limits, and it is not practical to wait for the attainment of moisture equilibrium which would require several days or more for tightly wound samples of high regain material. Practically, specimens are brought to moisture equilibrium in the standard atmosphere for testing as defined in these definitions. The term "standard condition" should not be used as a Syn: the concept of "standard atmosphere."

Standard Conditions for Gases *n* Measured volumes of gases are quite generally recalculated to 0°C temperature and 760 mm pressure, which have been arbitrarily chosen as standard conditions.

Standard Conditions for Testing *n* Temperature: 23 ± 1.1°C (73.5 ± 2°F); relative humidity: 50 ± 4%. These conditions have been established for the use of all departments and establishments of the Federal Government, except where otherwise specified by an applicable specification (Fed. Stnd. No. 1). NOTE – In ISO, standard conditions are 23 ± 2°C and 50 ± 5% relative humidity.

Standard Hydrogen Electrode *n* (**s.h.e.**) Electrodes consisting of a piece of platinum (coated with platinum black) in contact with hydrogen gas at 1 atm and with a solution in which the concentration of hydrogen ions is 1 *M*. (Actually, the hydrogen gas and hydrogen ions are at unit activity.) The voltage of the s.h.e. is defined as zero.

Standard Light Source *n* A reference light source whose spectral power distribution is known. See ▶ Light Source and ▶ Illuminant.

Standard Moisture Regain *n* Accepted moisture allowance for textile materials expressed in percentages of their dry weight.

Standard Normal Deviate *n* (*z*) The difference between any member of a normal distribution of measurements and the distribution (true) mean, divided by the standard deviation, i.e., $(x_i - \mu)/\sigma$. Compare this with the identically structured standardize measurement for a sample.

Standard Observer, 10° (1964) See ▶ observer, Standard.

Standard Observer, 2° (1931) See ▶ Observer, ▶ Standard.

Standard of Performance *n* The measure of pollution control required by law, in the Clean Air Act of 1970.

Standard Perfect White Diffuser See ▶ Perfect Diffuser.

Standard Sand *n* Natural silica sand from Ottawa, Illinois, graded to pass a No. 20 (850 μm) sieve and be retained on a No. 30 (600 μm) sieve. This sand shall be considered standard when not more than 15 g are retained on the No. 20 sieve and not more than 5 g pass the No. 30 sieve after 5 min of continuous sieving of a 100 g sample. Used in sand grinding and the falling sand abrasion test, ASTM D 968.

Standard, Secondary Standard *n* An amount of a substance whose content of specified elements or compounds is known within error limits that are narrower than those of the analytical method or instrument to be calibrated with the standard. (2) An object having one or more measurable properties whose values have been certified by an appropriate standards-issuing authority to be within error limits that are usually narrower than those of the measuring method or instrument that the object will be used to calibrate. Examples are gauge blocks ("Jo blocks") for micrometers, standard resistors and capacitors, standard cells (DC emf sources) for potentiometers, NIST-calibrated thermocouples and thermistors, balance weights, wavelength (or frequency) standards, tensile- and compressive-force standards, and time standards. Secondary standards that are traceable to government-maintained primary standards are the ultimate basis of all quantitative scientific and engineering work, of manufacturing-quality control, and of commerce. Standards and calibrations services are available from the National Institute of Standards and Technology, and from many private companies. See also ▶ Calibration.

Standard State *n* A reference state for specifying thermodynamic quantities, usually defined as the most stable form of the substance at 1 atm pressure. For a solute the standard state is the ideal, 1 *M* solution.

Standard, White Reflectance *n* In general usage, may refer to a physical white standard which is nearly a perfect white diffuser, such as pressed barium sulfate of high purity, pressed magnesium oxide or freshly smoked magnesium oxide. Other less perfectly diffusing or reflective materials, such as white ceramics, may be calibrated in reference to the perfect diffuser or to one of the near-perfect materials, and used in place of the material relative to which it is calibrated. Such a white reference standard should properly be called a secondary reference standard.

Standing Wire *n* A broad term describing fixed rods or strips extending through the loom reed, that control the height of the pile in a woven pile fabric.

Stand Oil *n* A drying oil which has been partially refined by allowing certain impurities to settle out after heat treatment. Generally used in the United Kingdom to describe linseed oil. So-called because it was originally made by letting linseed oil stand in the sun for a time, often as much as seven to 10 years, during which time a natural polymerization took place. The Dutch "Standolie" and German "Standöl" are counterparts of the English term which have been in use for hundreds of years, antidating the introduction and use of wood oil. See ▶ Oil, Heat-Bodied.

Stannous 2-Ethylhexanoate \\ˈsta-nəs ˈe-thəl-ˌhek-sə-ˈnō-āt\ *n* (stannous octanoate) A polymerization catalyst for urethane foam.

Staple \\ˈstā-pəl\ *n* [ME, fr. MF *estaple*, fr. MD *stapel* emporium] (14c) Natural fibers or cut lengths from filaments. The staple length of natural fibers varies from less than 1 in. as with some cotton fibers to several feet for some hard fibers. Manufactured staple fibers are cut to a definite length, from 8 inches down to about 1–1/2 in. (occasionally down to 1 in.), so that they can be processed on cotton, woolen, or worsted yarn spinning systems. The term staple (fiber) is used in the textile industry to distinguish natural or cut length manufactured fibers from filament.

Staple Fabric See ▶ Spun Fabric.

Staple Fiber *n* Short, spinnable fibers between 1 and 13 cm in length. See ▶ Staple.

Staple Processing *n* The conversion of staple into spun yarns suitable in evenness, size, twist, and strength for use in the weaving or knitting of fabrics. Also see ▶ Textile Processing.

Staple Yarn See ▶ Spun Yarn.

Starch Coating *n* Surface coating for flat paints and/or wallpaper. Made with colors that do not smear.

Starch Lump See ▶ Hard Size.

Starch, Permanent *n* An aqueous emulsion of a synthetic resin for application to fabrics which when ironed become stiff as if starched.

Stardust Imperfections in high gloss finishes caused by minute blemishes which give sparkling appearance.

Stark Effect *n* The splitting of a single spectrum line into multiple lines which occurs when the emitting material is placed in a strong electric field. The observed effect depends on the angle between the direction of the field and the direction of observation. The effect is due to the shifting of the energy states of certain orbits which all have the same energy in zero field.

Starter Strip *n* In the coil coating industry, a length of metal threaded through a line after shutdown. This is used repeatedly to attach new strip to be coated.

Start Up Mark See ▶ Set Mark.

Starved Area See ▶ Resin-Starved Area.

Starved Joint *n* A poorly bonded glue joint resulting from an insufficient quantity of glue in the joint. See ▶ Joint.

Starved Surface See ▶ Hungry Surface.

Starve Feeding *n* (starved feeding) In extrusion, regulating, with an auxiliary feeding device such as a weigh feeder or screw conveyor, the rate at which feedstock enters the feed port of the extruder so that the screw flights are less than full.

Starving Out *n* In the coil coating industry, similar to skipping in that it does not apply an even coating to the strip.

Statcoulomb \▎stat-▎kü-▎läm\ *n* The unit of electric charge in the metric system. 3×10^9 statcoulombs = 1 C.

Static \▎sta-tik\ *adj* [NL staticusm fr. Gk *statikos* causing to stand, skilled in weighing, fr. *histanai* to cause to stand, weigh] (1638) An accumulation of negative or positive electricity on the surface of fibers or fabrics because of inadequate electrical dissipation during processing. Static results in an electrical attraction or repulsion of the fibers relative to themselves, to machine parts, or to other materials, preventing the fiber from traveling in a normal path in the process.

Static Adhesion *n* In tire cord, the measurement of the strength of a cord-to-rubber bond under static conditions or very low strain rate.

Static Coefficient of Friction *n* The ratio of the force that is required to start the friction motion of one surface against another to the force, usually gravitational, acting perpendicular to the two surfaces in contact. Also called coefficient of friction, static.

Static Crack See ▶ Shier.

Static Eliminator *n* A mechanical or electrical device for draining off static electrical charge from plastics articles by creasing an ionized atmosphere in close proximity to the surface. Types include static bars, ionizing blowers and air guns, and radioactive elements. All except the latter operate on the principle that a high-voltage discharge from the applicator to ground creates an ionized atmosphere. A newer type employs ceramic microspheres containing radioisotopes that emit alpha particles. A layer of the microspheres is bonded to a substrate with a resinous binder, and the laminate is installed in an aluminum housing. By ionizing air this device provides a conductive path through which charge is drained.

Static Mixer *n* (motionless mixer) Any of several types of devices, used widely in the process industries and in plastics extrusion, that contain no moving parts but instead accomplish mixing by repeatedly splitting the melt (or other) stream and braiding or intertwining the streamlets. The scale of segregation of the different constituents in the flow or of hotter and cooler regions, is exponentially reduced in proportion to the number of stages. They are mainly used between extruder and die to insure the thermal homogeneity of the melt reaching the die lip, with consequent close control of extrudate thickness across the lip.

Stationary or Standing Waves *n* Waves which are produced in a medium by the simultaneous transmission, in opposite directions of two similar wave motions. Fixed points of minimum amplitude are called *nodes*. A *segment* extends from one node to the next. An *antinode* or *loop* is the point of maximum amplitude between two nodes.

Stationary Platen *n* In an injection molder, the large front plate to which the front plate of the mold is bolted. This platen does not normally move.

Statistical Process Control *n* (SPC) The application of statistical methods, both simple and advanced, to identifying and reducing sources of product-quality variation and output-quality limiters in production processes, the goal being to build quality ("zero defects") into the process rather than to remove defective items by inspection, as in (statistical) quality control.

Statistics *n* \stə-ᴗtis-tiks\ (1) A branch of mathematics dealing with the collection, analysis, interpretation, and presentation of masses of numerical data (2) A collection of quantitative data.

Statistical Terms and Symbols

Terms	Symbol
A member of	$\in$
Addition	$+$
Alternative hypothesis	H_1
Amalgamation	$\amalg$
And, conjunction	$\wedge$
Angle	$\angle$
Approaches a limit, definition	$\doteq$
Arithmetic mean operator	$<\ >$
Assertion	$\vdash$
Assignment	$:=$
Autocorrelation	$R_{xx}()$
Autocorrelation coefficient	$\rho_{xx}()$
Autocovariance	$C_{xx}()$
Base of natural logarithms (≈ 2.71828183)	e
Because	$\because$
Binomial coefficient	C_k^n
Composition	$\circ$
Conditional operator	$|$
Contour integral	$\oint \Box$
Convolution	$*$
Correlation coefficient	ρ_{xy}
Covariance	$C_{ov}()$
Cross-correlation	$R_{xy}()$
Cross-correlation coefficient	$\rho_{xy}()$
Cross-covariance	$C_{xy}()$
Cube root	$\sqrt[3]{\ }$
d'Alembertian operator	$\Box^2$
Degree	$°$
Delta	Δ
Direct sum, various	$\oplus$
Divides, divisible by	$|$
Division	$\div$ or $/$

Statistical Terms and Symbols (Continued)

Terms	Symbol
Double factorial	$!!$
Double prime	$''$
Empty set, null set	$\emptyset$
Equal angles	$\stackrel{\vee}{=}$
Event	E
Equals approximately, isomorphic to	$\cong$
Equal to	$=$
Equivalent to, congruent to	$\equiv$
Existential quantifier	$\exists$
Expectation value operator	$E\{\ \}$
Factorial	$!$
Gamma function	$\Gamma()$
Greater than	$>$
Greater than or equal to	$\geq$
Hence, therefore	$\therefore$
If and only if	$\Leftrightarrow$ or $\rightarrow$
Implies	$\Rightarrow$ or $\rightarrow$
Infinity	∞
Integral	$\int$
Intersection	$\cap$
Kurtosis	κ or γ_2
Laplacian operator	∇^2, Δ
Less than	$<$
Less than or equal to	$\leq$
Maps from	$\leftarrow$
Maps into	$\hookrightarrow$ Or $\hookleftarrow$
Maps to	$\mapsto$
Mean	μ
Minute	$'$
Minus or plus	$\mp$
Moment operator	$M\{\ \}$
Much greater than	$\gg$
Much less than	$\ll$
Multiplication	$\times$ or $\cdot$
Nabla, del	∇
Nearly equal to	$\approx$
Negation	$\neg$
nth root	$\sqrt[n]{\ }$
Not a member of	$\notin$
Not equal to	$\neq$
Not equivalent to, not congruent to	$\not\equiv$
Not parallel	$\nparallel$
Null hypothesis	H_0

Statistical Terms and Symbols (Continued)

Terms	Symbol
Or, disjunction	∨
Parallel	∥
Partial differential	∂
Perpindicular	⊥
Pi = 3.14159265	π
Plus or minus	±
Prime	≠
Probability or probability distribution	P
Product	Π
Ratio	:
Sample space	S or Ω
Second	≡
Similar to, asymptotically to	~
Skewness	γ or γ_1
Spherical angle	∢
Square root, radical	√
Square root of −1	i
Standard deviation	σ
Statistical weights	W
Subset of	⊂
Subset of or equal to	⊆
Subtraction	−
Summation	Σ
Superset of	⊃
Superset of or equal to	⊇
Tends to, maps to	→
Triple prime	‴
Union	∪
Universal quantifier	∀
Variance	σ^2 or Var()
Varies as, proportional to	∝
Various	⊖, ⊗, ⊙

Staudinger Index \ˈstäd-in-jər ˈin-deks\ *n* Intrinsic viscosity [η] is identically equal to,

$$\lim_{c_2 \to 0} \frac{\eta_{sp}}{c_2} \equiv [\eta]$$

where c_2 is solute concentration (g/cm³), and η_{sp} is the specific viscosity derived from solute concentration and relative viscosity; [η] is dependent on molecular weight and hydrodynamic volume of the solute or polymer (see ▶ Mark-Houwink Equation). (Staudinger H, Heuer WA (1930) Relationship between the viscosity and the molecular weight of polystyrene (German). Ber 63B:222–234)

Steady Flow *n* (steady-state flow) Any flow in which velocities throughout the stream do not vary with time.

Steady State *n* A condition of processes or parts of processes in which the state variables describing the process, e.g., temperature, pressure, compositions and velocities of streams, and amounts of materials residing in various process equipment, do not change with time. Most extrusion operations closely approximate the steady state except during startup and shutdown, whereas injection molding and sheet thermoforming are unsteady, *intermittent* processes.

Steam Chest *n* A steam-heated cabinet used in manufactured fiber production. Usually refers to the heated cabinet in which spin-drawing is done or to the cabinet around a stuffer-box crimper.

Steam-Distilled Pine Oil See ▶ Pine Oil.

Steam-Distilled Wood Turpentine See ▶ Turpentine.

Steam Molding *n* A process for molding plastic-foam parts from pre-expanded beads of polystyrene that contain a volatile hydrocarbon, e.g., isopentane, as a blowing agent. The steam is usually in direct contact with the beads, but with thin-wall moldings such as coffee cups, may be used indirectly to heat mold surfaces that contain the beads. The process is widely used for molding packaging elements for the electronic-equipment industry, but it can also make huge "logs" that are subsequently sliced into foam "lumber." See ▶ Polystyrene Foam.

Steam Plate See ▶ Force Plate.

Steam-Set Inks *n* Inks that dry by the application of steam or moisture.

Stearic Acid \stē-ˈar-ik\ *n* (1831) (octadecanoic acid) $CH_3(CH_2)_{16}COOH$. A saturated organic fatty acid obtained by the hydrolysis with sodium hydroxide (*saponification*) of beef tallow. Mp, 69°C; bp, 29°C/100 mmHg. The salts, amides, and esters of this acid have long played important roles in the realm of plastics as lubricants, slip agents, and plasticizers, and stearic acid itself is present in numerous plastics compounds.

Stearin \ˈstē-ə-rən\ *n* [F stéarine, fr. Gk *stear*] (1817) $C_3H_5(C_{18}H_{35}O_2)_3$. Glycerol tristearate, tristearin. Mol wt, 890.86; mp, 71.6°C; sp gr, 0.943. Also spelled "*Stearine*."

Stearin (Stearine) Pitch *n* Pitch obtained from stearin (Stearine), from the residues of distillation of fatty acids, vegetable oils, etc. They differ from other pitches or bitumens by the possibility of possessing both acid and saponification values and having a tendency to oxidize on exposure. Used in some inks to impart flow and good wetting qualities. Also spelled "*Stearine Pitch*."

Stearyl Methacrylate \-ˈme-ˈtha-krə-ˈlāt\ *n* A group name for compounds of the general formula $CH_2 = C(CH_3)OOC(CH_2)_nCH_3$, in which n is from 13 to 17. It is a polymerizable monomer for acrylic plastics.

Steatite \ˈstē-ə-ˈtīt\ *n* [L *steatitis*, a precious stone, fr. Gk, fr. *steat-*, *stear*] (1758) High purity talc containing maximum allowable proportions of 1.5% CaO, 1.5% Fe_2O_3 and 4% Al_2O_3 as impurities. See ▶ Magnesium Silicate, Nonfibrous.

Steel Blue *n* (1817) Type of Prussian blue. See ▶ Iron Blue.

Steel Plate Printing *n* An intaglio type of printing using metal plates with the image etched or engraved below the surface. *Also known as Steel Die Printing.*

Steel Plate (Steel Die) Printing *n* An intaglio type of printing using metal plates with the image etched or engraved below the surface.

Steel Structures Painting Council *n* (**SSPC**) The Society for Protection Coatings, address: 40 24th Street, 6th Floor, Pittsburg, PA 15222-4656, www.sspc.org.

Steel-Rule Die *n* A sharp-edged knife fashioned from thin steel strip, flexible enough to be shaped to complex outlines. It is used as the cutting element in ▶ Die Cutting.

Steel Wool *n* A matted mass of long, fine, steel fibers available in a variety of grades of coarseness. Used for cleaning and polishing surfaces, burnishing, removing film blemishes between coats and the internal dulling of coated surfaces.

Stefan-Boltzmann Law of Radiation *n* The energy radiated in unit time by a black body is given by, $E = K(T_4 - T_o^4)$, where *T* is the absolute temperature of the body, To the absolute temperature of the surroundings, and *K* a constant. (Holst GC (2003) Electro-optical imaging system performance. JDC; Handbook of infrared optical materials. Klocek P (ed) Marcel Deker, New York, 1991)

Stellite® {*trademark*} Trade name of the Hayes Stellite Co for a family of hard and corrosion-resistant metal alloys. Stellites 1 and 6 are used for hard-facing extruder screws by first machining a groove in what will become the flights' outer surfaces, then welding the alloy into the groove and grinding to final diameter and finish after machining the main screw channel. Both alloys are mainly cobalt, Stellite 1 containing about 31% chromium and 13% tungsten (W), while Stellite 6, the choice for use in Xaloy-306-lined extrusion cylinders, contains bout 28% Cr and 3–6% W. Hardness of Stellite 1 is 52–55 on the Rockwell C scale; that of Stellite 6 is a little less.

Stencil \ˈsten(t)-səl\ *n* [ME stanselen to ornament with sparkling colors, fr. MF *estanceler*, fr. *estancele* spark, fr. (assume.) VL *stincilla*, alter. of L *scintilla*] (1707) Method of applying a design by brushing ink or paint through a cutout overlay placed on the surface.

Stenter See ▶ Tenter Frame.

Stepladder \ˈstep-ˈla-dər\ *n* (1751) A ladder having flat steps, or trends, in place of rungs; usually provided with a supporting frame to steady it.

Stereobase Unit \ˈster-ē-ˈō-ˈbās ˈyü-nət\ *n* The base unit of a polymer, taking steric isomerism into account.

Stereoblock \-ˈblāk\ *n* A regular block (in a polymer) that can be described by one species of stereo repeating unit in a single sequential arrangement (IUPAC).

Stereoblock Polymer *n* A polymer whose molecules consist of stereoblocks connected linearly (IUPAC). See also ▶ Stereospecific.

Stereochemistry The study of the spatial geometries of molecules and polyatomic ions.

Stereochromy A comparatively new method of mural painting in which water glass serves as the connecting medium between the color and its substratum and also as a protective coating.

Stereograft Polymer *n* A polymer consisting of chains of an Atactic polymer grafted to chains of an isotactic polymer. For example, Atactic polystyrene can be grafted to isotactic polystyrene under suitable conditions.

Stereoisomerism \ˈster-ē-ō-ˈī-sə-mər\ *n* [ISV] (1894) A kind of isomerism in organic compounds arising from the fact that a carbon atom linked to four different groups can exist in two spatially different forms that, though chemically identical, are not superimposable. A simple example is lactic acid, $CH_3CH(OH-COOH)$.

Stereoisomers Molecules or polyatomic ions with the same atoms and the same bonds but differing in the geometrical orientations of the atoms and bonds.

Stereolithography \ˈster-ē-ō-li-ˈthä-grə-fē\ *n* The term sometimes used to describe the polymerization of acrylic and other liquid monomers using focused laser beams to form a three-dimensional solid object.

Stereolithography is useful for low cost production of small models. Stereolithography is one of the more commonly used rapid manufacturing and rapid prototyping technologies. It is considered to provide high accuracy and good surface finish. It involves building plastic parts a layer at a time by tracing a laser beam on the surface of a vat of liquid photo-polymer. The photopolymer is solidified by the laser light. Once a layer is completely traced, it is lowered a small distance into the liquid and a subsequent layer is traced, adhering to the previous layer. After many such layers are traced, a complete 3D model is formed. Some specific technologies require further curing of the polymer in an oven. For models that have delicate structures that must be supported against gravity to prevent collapse, fine supports may be added during fabrication, either automatically in software or manually, and subsequently removed. The devices used to perform stereolithography are called either SLAs for stereolithography-apparatus(es), or simply stereolithographs. (Photoinitiated polymerization. Belfield KD, Crivello JV (eds). American Chemical Society Publications, 2003)

Stereoregularity Alternate name for ▶ Tacticity.

Stereoregular Polymer *n* A regular polymer whose molecules can be described by only one species of ▶ Stereorepeating Unit in a single sequential arrangement. A stereoregular polymer is always a tactic polymer, but a tactic polymer need not have every site of stereoisomerism defined (ISO).

Stereorepeating Unit *n* A configurational repeating unit having defined configuration at all sites of stereoisomerism in the main chain of a polymer molecule (IUPAC). In stereoregular polypropylene, the two simplest possible stereorepeating units are: which would be the mers of the corresponding stereoregular polymers, the first being isotactic, the second, syndiotactic.

Stereoselective Polymerization *n* Polymerization in which a polymer molecule is formed from a mixture of stereoisomeric monomer molecules by incorporation of only one stereospecific species.

Stereospecific *n* Of polymerization catalysts, implying a specific or definite order of spatial arrangement of molecules in the polymer resulting from the catalyzed polymerization. This ordered regularity of molecules (▶ Tacticity), in contrast to the branched or random structure found in other plastics, permits close packing of the molecular segments and leads to high crystallinity, as in polypropylene. The adjective is sometimes applied imprecisely to polymers to mean tactic.

Stereospecific Polymerization *n* A polymerization in which a tactic polymer is formed.

Stereotypes *n* Printing plates cast in one piece from a heat-dried papier-mâché matrix containing the impression of the assembled type.

Steric Factor *n* The fraction of the collisions in an elementary process in which the colliding particles have the proper geometrical orientation with respect to each other to produce the activated complex.

Steric Hindrance *n* A characteristic of molecular structure in which the molecules have a spatial arrangement of their atoms such that a given reaction with another molecule is prevented or retarded.

Steric Number *n* The sum of the number of bonds and lone pairs around a bonded atom (VSEPR theory).

Sticker *n* (1) A distortion in the weave characterized by tight and slack places in the same warp yarns. The principal causes are rolled ends on the beam, warp ends restricted by broken filament slubs, and knots catching at lease rods, drop wires, heddles, or reeds. (2) See ▶ Hard Size.

Sticklac *n* Raw material obtained directly from the tree, from which shellac is derived.

Stick Shellac *n* Shellac in a solid, sticklike form, which is manufactured in clear form and in numerous colors and shades to match all woods. *Also called Wood Cement.*

Sticky See ▶ Pulling and ▶ Drag.

Stiffness *n* (1) The load per unit area required to elongate the film 1% from the first point in the stress–strain curve where the slope becomes constant. (2) A term relating to the ability of a material to resist bending while under stress. Resistance to the bending is called flexural stiffness, and may be defined as the product of the modulus of elasticity and the moment of inertia of the section. Compare ▶ Rigid Plastic.

Stiffness *n* Refers to the deformation behavior in the elastic region. Elastic or Hookean behavior implies that deformation effects due to load are completely recoverable – i.e., no permanent dimensional change occurs. In a theoretical sense this is seldom observed. Practical approximations are usually adequate. When plotted, the horizontal axis is strain, the vertical axis is stress. *Also known as Modulus of Elasticity.*

Stillingia Oil *n* Pale yellow, limpid, drying oil; peculiar odor; mustard-like taste. Slightly soluble in alcohol. Sp gr, 0.943–0.945; iodine number, 160; saponification value, 210. See ▶ Chinese Vegetable Tallow.

Still Life Subject for painting composed of inanimate objects, e.g., fruit, flowers, dead fish, etc.

Stipple \ˈsti-pəl\ *vt* [D *stippelen* to spot, dot] (ca. 1762) (1) To even out a coat of paint and remove brush marks and other imperfections, immediately after application, by systematically dabbing the surface with a soft stippling brush. (2) To produce a broken color or textured effect, either by applying spots of a different color or by disturbing the surface of the paint coat, e.g., with a stippling brush or rubber stippler.

Stippler *n* (1) A broad flat-based brush having stiff bristles for producing a texture on a surface such as soft plaster or paint. (2) Any tool (as a rubber sponge or a textured or tufted roller) used to create a stippled surface.

Stippling *n* (1) Method of painting by means of small dots, touches of color or pebbly-textured finish. See ▶ Pointillism. (2) Engraving by means of dots; distinct from engraving in lines. (3) In antiquing, to obtain an irregular effect with a mottled appearance. It usually entails the use of a sponge, newspaper, or points of dry brush.

Stir-In Resin *n* (dispersion resin, paste resin) A vinyl resin that does not require grinding or extremely high-shear mixing to effect dispersion in a plasticizer or to form a plastisol or organosols.

Stitch Bonding See ▶ Bonding (2).

Stitched *n* Stapling device used in coil coating to connect the end of a coil to the beginning of another coil.

Stitching *n* (stitch welding) (1) The progressive joining of thermoplastic film and sheets by successive applications of two small, mechanically operated electrodes, connected to the terminals of a radio-frequency generator, using a mechanism somewhat like that of an ordinary sewing machine. (2) The process of passing a fiber or thread through the thickness of fabric layers to secure them. In composite manufacture, stitching is used to make preforms or to improve damage tolerance of complex-shaped parts.

Stock *n* Paper or other material to be printed. Substrate.

Stock Dyeing See ▶ Dyeing.

Stockholm Pine Tar \ˈstäk-ˌhō(l)m-\ *n* See ▶ Pine Tar *n* Stockholm.

Stockinette \ˌstä-kə-ˈnet\ *n* [alter. of earlier *stocking net*] (1784) A knit fabric in tubular or flat form made with a plain stitch from yarns of wool, cotton, manufactured fibers, or a combination of these fibers. Stockinette fabrics are used for underwear, industrial applications, and other purposes. In heavier constructions, dyed and napped stockinette finds apparel uses. Also spelled stockinet.

Stock Temperature *n* In plastics processing, the temperature of the plastic (as opposed to temperatures of metal parts of the equipment) at any point. Often taken to mean, if not otherwise qualified, the temperature of the melt within an extruder had or leaving the nozzle of an injection molder. See ▶ Melt Temperature. Stock temperatures may be measured by sturdy thermocouples or thermistors inserted into the plastic stock, or by infrared instruments pointed at emerging extrudates, sheets being thermoformed, etc.

Stoddard Solvent \ˈstä-dərd-\ *n* (1) A petroleum distillate comprising 44% naphthenic, 39% paraffin's, and 17% aromatics, used as a diluent in PVC organosols. (2) Petroleum spirits with a minimum flash of 37.8°C (100°F) and low odor level.

Stoichiometric \ˌstói-kē-ō-ˈme-trik\ *adj* (1892) Pertaining to a mixture of chemical reactants, each ion or compound of which is present in the exact amount necessary to complete a reaction with no excess of any reactant. For example, each ingredient of a urethane-foam formula should be present in its stoichiometric quantity in order to assure a high-quality product.

Stoichiometric Balance *n* Mass balance of materials (non-energy) involving reactions and conversions.

Stoke \ˈstōk\ *n* The deprecated cgs unit of kinematic viscosity, equal to 10^{-4} m^2/s.

Stokes *n* The cgs unit of kinematic viscosity. A liquid with a viscosity of 1 poise has a kinematic viscosity of one stokes if its density is one. A measure of kinematic viscosity equal to poises divided by density.

Stokes' Law *n* (1) The mathematical equation derived by G. G. Stokes which relates the velocity of fall v of a spherical body under gravity g through a fluid medium to the radius r of the body, to the viscosity η of the medium and to the difference between the density of the solid body (D_s) and that of the medium (D_1):

$$v = \frac{2(D_s - D_1)r^2 g}{9\eta}$$

(2) The empirical law stating that the wavelength of light emitted by a fluorescent material is longer than that of the radiation used to excite the fluorescence. In modern language the emitted photons carry off less energy than is brought in by the exciting photons; the details accord with the energy conservation principle.

Stoll-Quartermaster Universal Ware Tester *n* A versatile testing apparatus for measuring wear resistance of fabrics, yarns, thread, etc. It can be equipped with either of two testing heads, one for testing abrasion resistance of flat surfaces and the other for testing resistance to flexing and abrasion.

Stop Motion *n* Any device that automatically stops a textile machine's operation on the occurrence of a yarn break, a high defect count, etc.

Stopper *n* Pigmented composition used for filling fine cracks and indentations to obtain a smooth, even surface preparatory to painting. Syn: ▶ Filler.

Storage Life *n* The period of time during which a packaged adhesive can be stored under specified temperature conditions and remain suitable for use. *Sometimes called Shelf Life*. A Syn: ▶ Shelf Life. See also ▶ Working Life.

Storage Modulus *n* In dynamic mechanical measurements, the part of the ▶ Complex Modulus that is in phase with the strain, with the symbol G' if the testing mode is shear, E' if it is tension or compression.

Storage Stability *n* General composite property of resistance to any change, generally in a closed container, over a period of time; color and liquid separation, formation of lumps, hard pigment settling, substantial changes in viscosity, pH, development of odor, etc., are examples of undesirable changes. See also ▶ Shelf Life.

Stormer Viscometer \ˈstórm-ər vis-ˈkä-mə-tər\ *n* A rotational-type instrument in which the test liquid fills a stationary, baffled cup and the rotating element may be a concentric cylinder or one of several paddle designs. The rotor is powered by adjustable falling weights through step-up gearing. Because of the complex geometry with any of the Stormer rotors, viscosity cannot be directly calculated from the weight and the time required for the rotor to make 100 revolutions. For each configuration, the instrument must be calibrated with liquids of known viscosity to establish the instrument constant for use with unknown liquids or slurries. Operation of the instrument, used mostly in the paint industry, is described in ASTM D 562 (Section 06.01). See ▶ Krebs-Stormer Viscometer.

Stoving (British) See ▶ Baking.

STP *n* Abbreviation for Standard Conditions Of Temperature And Pressure. In scientific work these are 0°C and 101.325 kPa (1 atm). American gas industries and some others often choose 70°F (21.1°C) as their standard temperature.

Straight-Line Pad Sander *n* A portable sanding machine consisting of a backup pad that moves to and fro in a straight line, with a stroke of about 5/16 in. on an average of about 3,200 times per minute. The resulting scratch pattern is ideal when sanding with the grain in wood and far less noticeable than the swirls from an orbital pad sander when the finish coat is applied on either wood or metal.

Strain \ˈstrān\ *v* [ME, fr. MF *estraindre*, fr. L *stringere* to bind or draw tight, press together, akin to Gk *strang-*, *stranx* drop squeezed out, *strangalē* halter] (14c) In tensile and compression testing, the ratio of the elongation to the gauge length of the test specimen, that is, increase, (or decrease) in length per unit of original length. The term is also used in a broader sense to denote a dimensionless number that characterizes the change in dimensions of an object during a deformation or flow process. In shear deformation, strain is the shear angle in radians. Strain is a dimensionless quantity which may be measured conveniently in %, in in. per in., in mm per mm, etc. See also ▶ Shear Strain and ▶ True Strain.

Strain Birefringence \-ˌbi-ri-ˈfrin-jən(t)s\ *n* Double refraction in a transparent material subjected to stress and accompanying strain. One of the techniques of experimental stress analysis is based on this phenomenon.

Strain Energy *n* The recoverable, elastic energy stored in a strained body and recovered quickly upon release of stress. Strain energy in a perfectly elastic material is equal to the area beneath the stress–strain curve up to the strain being considered. For Hooke's-law material, it is equal to 0.5 modulus strain^2. This area, which appears to have the dimensions of stress (Pa), is actually the strain energy per unit volume (J/m^3).

Strain Gauge *n* A small electrical element consisting of a very fine wire of many short runs and reverse turns embedded in a tape-like matrix. The strain gage is adhered to an object whose deformation under stress it is desired to measure. With complex objects, several gages may be attached at different locations and oriented in different directions. The gage undergoes the same strain as the object when stress is applied and the strain stretches the wire, which, through the many doublings-back, magnifies the change in length. The increase in resistance, proportional to object strain, is measured with a bridge circuit of which the gage is one branch.

Strain Hardening *n* An increase in hardness and strength caused by plastic deformation at temperatures lower than the re-crystallization range.

Straining *n* The mechanical separation of relatively coarse particles from a liquid as distinguished from the process of filtration.

Strain Recovery Curve See ▶ Tensile Hysteresis Curve.

Strain Relaxation *n* A misnomer for Creep. What may be mistakenly meant by this term is Stress Relaxation. See ▶ Creep.

Strand \ˈstrand\ *n* [ME *strond*] (15c) (1) A single fiber, filament, or monofilament. (2) An ordered assemblage of textile fibers having a high ratio of length to diameter and normally used as a unit; includes slivers, roving, single yarns, plies yarns, cords, braids, ropes, etc.

The number of filaments in a strand is usually 52, 102, or 204.

Strand Chopper *n* (1, pelletizer) A type of cutter into which multiple extruded and chilled strands are fed for cutting into short lengths about equal to the strand diameter. The strands are fed perpendicular to the edge of a thick stationary knife and are sheared off by several rotating knives bolted to a massive cylindrical head. The pellets so formed are discharged into a shipping container or conveyed into mass storage bins. In this way, molding powders are produced for extrusion or molding into finished products. Widely used by many smaller compounders and by reclaimers, this method of producing pellets has largely given way, in the plants of the large resin producers, to ▶ Underwater Pelletizing. (2) A device for chopping strands of fibrous reinforcement into lengths suitable for blowing onto a perform screen, compounding into premix, or combining with spayed resin in ▶ Sprayup.

Strapping \ˈstra-piŋ\ *adj* (1657) (1) Thin, flat, continuous-strip material available in widths from 6 to 25 mm, usually in coils, and designed to be used with hand-operated machines that dispense, tighten, and clamp the strip around a package or bundle of packages. The material may be steel, unidirectionally fiber-reinforced plastic, or oriented nylon or polypropylene. (2) High-strength hoisting straps woven from aramid fiber, replacing chains in many industries for handling loads with hoists.

Strasbourg Turpentine \ˈsträs-ˌbúrg *or Gr* ˈshträs-ˌbúrk\ *n* Oleoresinous exudation obtained from the white fir.

Straw \ˈstró\ *n* [ME, fr. OE strēaw; akin to OHGr strō straw, OE *strewian* to strew] (before 12c) A general term for plant fibers obtained from stems, stalks, leaves, bark, grass, etc. They are made into hats, bags, shoes, mats, etc., by weaving, plaiting, or braiding.

Streak \ˈstrēk\ *n* [ME *streke*, fr. OE *strica*; akin to OHGr *strich* line, L *striga* row] (before 12c) A discoloration (rust, oil, dye, grease, soap, etc.) extended as an irregular stripe in the cloth.

Streamline Flow *n* A streamline is a line in a fluid such that the tangent to it at every point is in the direction of the velocity of the fluid particle at that point, at the instant under consideration. When the motion of the fluid is such that, at any instant, continuous streamlines can be drawn through the whole length of its course, the fluid is said to be in streamline flow.

Strength \ˈstreŋ(k)th\ *n* [ME *strengthe*, fr. OE *strengthu*; akin to OHGr *strengi* strong] (before 12c) The strength of a pigment is its opacity or tinting power.

Strength Count Product See ▶ Break Factor.

Strength, Dry *n* The strength of an adhesive joint determined immediately after drying under specified conditions or after a period. See ▶ Strength.

Strength (of an Electrolyte) *n* The extent or degree of dissociation of an electrolyte in solution.

Strength, Tensile *n* (ca. 1864) (1) General: The strength shown by a specimen subjected to tension as distinct from torsion, compression, or shear. (2) Specific: The maximum tensile stress expressed in force per cross-sectional area of the unstrained specimen, for example, kilograms per square millimeter, pounds per square inch. (3) Ability a material possesses of resisting deformation by the application of a force or load.

Strength, Wet The strength of an adhesive joint determined immediately after removal from a liquid in which it has been immersed under specified conditions of time, temperature, and pressure. NOTE – The term is commonly used alone to designate strength after immersion in water. In the latex adhesives the term is also used to describe the joint strength when the adherents are brought together with the adhesive still in the wet state. See ▶ Strength, Dry.

Stress \ˈstres\ *n* [ME *stresse* stress, distress, short for *destresse*] (14c) (σ or τ) The force producing or tending to produce deformation in a body, divided by the area over which the force is acting. If the stress is tensile or compressive, the area is perpendicular to the stress; in shear it is parallel to the stress. The SI unit of stress is the pascal (Pa) equal to 1 newton per square meter (N/m^2). Practical stresses are usually in the range from 1 kPa to 1 MPa. In theoretical mechanics, the stress tensor in a body has nine possible components, three of which are tensile/compressive, the others shear. In many cases, all but one or two of the components are zero or not relevant to the behavior question of interest. NOTE 1 – Typical examples are tensile stress, shear stress and compressive stress. NOTE 2 – Stress usually reaches a maximum at the time of rupture.

Stress Concentration *n* The magnification of applied stress in the vicinity of a notch, hole, inclusion, or inside corner. Minimizing stress concentrations is an important aspect of plastics product design.

Stress Corrosion *n* The preferential attack of areas under stress in a corrosive environment, when environment alone would not have caused corrosion.

Stress Cracking *n* (stress crazing) External or internal cracking in a plastic caused by tensile stresses less than the short-time tensile strength. See ▶ Environmental Stress Cracking.

Stress-Intensity Factor *n* (K) A fracture-mechanics parameter that describes the magnifying of stress caused by a flaw in a material. It is defined by the equation,

$$K = f \cdot S(\pi a)^{0.5}$$

where f = a factor dependent on the flaw geometry and the structure in which it is contained, S = the nominal or average stress caused by the load, and a = half the length of the flaw measured normal to the direction of stress.

Stress Pattern *n* Distribution of applied or residual stress in a specimen, usually throughout its bulk. Applied stress is a stress induced by an outside force, e.g., by loading. Residual stress or stress memory may be a result of processing or exposure. The stress pattern can be made visible in transparent materials by polarized light.

Stress Relaxation *n* The decay of stress with time at constant strain. If a plastic specimen is strained and the recovery of the strain prevented, the chain segments of the molecules will tend to realign so as to lower the free energy of the system, sometimes by breaking covalent bonds. The elastic and retarded strains which would usually be recovered upon release of the stress are instead converted into unrecoverable strains when chain segments have been rearranged. The decrease in stress is often plotted against time in order to estimate when it will have fully decayed. The strains induced during processing of molten thermoplastics, particularly injection molding, often do not completely recover before cooling and become frozen into the material. At ambient temperatures these strains may slowly recover, as evidenced by warping and excessive shrinkage of parts. For this reason, some moldings with high molded-in strain are annealed while being securely held in shape-retaining fixtures, to permit stress relaxation. ASTM D 2991 describes a standard practice for testing stress relaxation of plastics.

Stress Rupture *n* The sudden, complete failure of a plastic member held under load. In laboratory testing, the temperature and rate of loading or, in longer-term tests, the time for which the load was sustained, should be stated along with the load and corresponding type and mode of stress. The mode of loading may be tensile, flexural, torsional, biaxial, or hydrostatic.

Stress–Strain Curves *n* A graphical representation, showing the relationship between the change in dimension (in the direction of the applied stress) of the specimen from the application of an external stress, and the magnitude of that stress. In tension tests of textile materials, the stress can be expressed either in units of force per unit cross-sectional area, or in force per unit linear density of the original specimen, and the strain can be expressed either as a fraction or as a percentage of the original specimen length.

Stress–Strain Diagram *n* The plot of stress on a test specimen, usually in tension or compression, versus the corresponding strain, usually carried to the point of failure. The test is usually carried out at constant crosshead speed, i.e., at a constant rate of nominal elongation. The stress plotted is usually the *nominal stress*, i.e., the measured force at any time divided by the original cross-sectional areas, normal to the force, over which the force is distributed. The strain plotted is usually also the nominal quantity, i.e., the increase in specimen gage length divided by the original length. See also ▶ True Stress and ▶ True Strain.

Stress Whitening *n* Whiteness seen in some plastics and rubbers that are subjected to extreme stretching, thought to be due to the formation of light-scattering Microvoids (*microcavitation*) within the material.

Stress Wrinkles *n* Distortions in the face of a laminate caused by uneven web tensions, slowness of adhesive setting, selective absorption of the adhesive, or by reaction of the adherends with materials in the adhesive.

Stretch \ strech\ *n* (1541) An extension of the length of a material to the point of rupture to determine its tensile strength. Stretch is generally measured as a percentage of the original length.

Stretch-Blow Molding *n* A blow-molding variant for making bottles from polyethylene terephthalate in which shaped performs (as distinct from simple tubular parisons) are injection molded, reheated through several zones, then inflated and stretched lengthwise to essentially the perform shape magnified many times, cooled, and released from the mold. The process may be done in a single stage, or in two essentially separate operations, or in an integrated two-stage method in a single machine (Rosato DV, DiMattia DP, Rosato AV (eds) (2003) Blow Molding Handbook: Technology, Performance, Markets, Economics: The Complete Blow Molding Operation, Hanser-Gardner Publications, Cincinnati, Ohio). See also ▶ Blow Molding.

Stretch Breaking *n* In conversion of tow-to-top, fibers are hot stretched and broken rather than cut to prevent some of the damage done by cutting.

Stretched Tape *n* Strong tape made of a crystalline plastic such as polypropylene or nylon, which has been unidirectionally oriented by warm stretching followed by cooling while under tension. Used mainly for ▶ Strapping.

Stretch-Film Wrapping See ▶ Shrink Wrap.

Stretch Forming *n* A sheet-forming technique in which a heated thermoplastic sheet is stretched over a mold and subsequently cooled. See also ▶ Drape Forming and ▶ Sheet Thermoforming.

Stretch Growth See Secondary Creep.

Stretching See ▶ Orientation.

Stretch Ratio *n* In making uniaxially oriented films and filaments, the length of a sample of stretched film or filament divided by its length before stretching. Also see ▶ Blow-Up Ratio and ▶ Draw Ratio.

Stretch Spinning *n* A term used in the manufacture of rayon. Rayon filaments are stretched while moist and before final coagulation to decrease their diameter and increase their strength.

Stretch Wrap See ▶ Shrink Wrap.

Stretch Yarn See ▶ Textured Yarns.

Striae \ˈstrī-ə\ *n* [L, furrow, channel] (1563) Surface or internal thread-like inhomogeneities in a transparent plastic.

Striation \strī-ˈā-shən\ *n* (ca. 1847) (1) In blow molding a rippling of thick parisons caused by a local weld-line effect in the melt, imparted by a spider leg. See ▶ Spider Lines. (2) In the mixing of contrasting colors of resin by viscous shear in the melt, as in extrusion, a layer of one color adjacent to one of the contrasting color. The thickness of the layer is called the *striation thickness* and the thinning of the layer is a measure of the effectiveness of mixing, corresponding to Danckwerts' ▶ Scale of Segregation. (3) Streaks or bands or various natures in fibers or fabrics.

Strié *n* A term describing any cloth having irregular stripes or streaks of practically the same color as the background.

Strike Through *n* The penetration of the vehicle of a printing ink through the sheet, so that it is apparent on the opposite side.

String *n* (Varnish) Varnish being polymerized will become viscous and form strings which suspend from an object when dipped in it and then withdrawn. The length of the string is an indication of the degree of polymerization.

Stringiness (1) The property of an adhesive that results in the formation of filaments or threads when adhesive transfer surfaces are separated (see also ▶ Webbing). NOTE – Transfer surfaces may be rolls, picker plates, stencils, etc. (2) The property of an ink which causes it to draw into filaments or threads.

Stringup See ▶ Threadup.

Stringy Selvage See ▶ Slack Selvage.

Strip \ˈstrip\ *n* [ME, per. fr. MLGr *strippe* strap] (15c) (1) A length of wallpaper cut to fit height of wall. (2) A single section of the design, in a scenic wallpaper. (3) A board of lumber less than 4 in. (10 cm) wide.

Strip Coating *n* Strip coating or coil coating is the method of application of a coating to raw metal strip before its fabrication into finished products.

Striping *n* In antiquing, painting a series of fine lines – in gold or a contrasting color – to enhance some basic design characteristics on a piece of furniture. It may be accomplished by a resist technique (such as taping a fine line effect over a surface, and then blocking out the rest of the area and spraying into the opening left by the tape). It may be painted free hand or with the use of a ruler or straight edge. It may be done with sizing and gold leaf. *Also called Dutch Gold.*

Strippable Coating *n* A temporary coating applied to finished articles to protect them from abrasion or corrosion during shipment and storage, which can be removed when desired without damaging the articles. Vinyl plastisols, applied by dipping, spraying, or roll coating, are often used for this purpose. See also ▶ Release Paper.

Strippable Paper *n* A chemically treated stock tough enough to resist tearing, with a special formulation which permits a release of the wallcovering from the adhesive. This makes it possible to remove an entire strip from the wall without wetting. It is referred to as dry strippable.

Stripper See ▶ Paint and Varnish Remover.

Stripper Plate In molding, a plate that strips a molded piece from core pins or a force plug. The plate is actuated by opening the mold.

Stripping *n* (1) Removing old paint, wallpaper, distemper, etc., by the use of a blowtorch, paint remover steam stripping appliance, stripping knife, or other scraping tools. (2) Sealing the joint between a metal sheet and a built-up roofing membrane. (3) Taping the joints between insulation boards. (4) A condition where the ink fails to adhere to and distribute uniformly on the metal rollers of the press. (5) A chemical process for removing color from dyed cloth by the use of various chemicals. Stripping is done when the color is unsatisfactory and the fabric is to be redyed. (6) The physical process of removing fiber that in embedded in the clothing of a card. (7) See ▶ Degumming.

Stripping Agent See Mold Wash.

Stripping Fork *n* (comb) A tool, usually of brass or laminated sheet, used to remove articles from molds.

Stripping Torque *n* (1) Of a self-tapping screw, the twisting moment in newton meter required to strip the threads formed by the screw in a softer material. (2) Of a molded-in insert, the torque required to break the mechanical bond between the plastic and the insert's knurled surface.

Stroke \ˈstrōk\ *n* [ME; akin to OE *strīcan* to stroke] (13c) (1) The movement of a hydraulic piston, press ram, or other reciprocating machine member. (2) The distance between the extremes of movement of such a member in normal operation.

Strong (Electrolyte) *n* Completely, or almost completely, dissociated in solution.

Strontium Chromate \ˈstran(t)-sh(ē-)əm ˈkrō-ˌmāt\ *n* $SrCrO_4$. Pigment Yellow 32 (77839). A bright yellow pigment of type similar to lead chromate except that it is not blackened by hydrogen sulfide. It is used in corrosion-resistant primers. Density, 3.67–3.82 g/cm³ (30.6–31.9 lb/gal); O.A., 20–33; particle size, 10–30 μm. Syn: ▶ Strontium Yellow.

Strontium White *n* Another name for the ground mineral, celestite.

Strontium Yellow See ▶ Strontium Chromate.

Structural Adhesive *n* A bonding agent used for transferring required loads between adherents exposed to service environments typical for the structure involved.

Structural Foam *n* A term originally used for cellular thermoplastic articles with integral solid skins having high strength-to-weight ratios and also used for high-density cellular plastics that are strong enough for structural applications. Today the term is more apt to mean the products of SRIM (at ▶ Reaction Injection Molding). In the original Union Carbide process called *structural-foam molding*, pellets of resin containing a blowing agent are fed into an extruder provided with an accumulator, where the melt is maintained above the foaming temperature but at a pressure high enough to preclude foaming. A piston in the accumulator forces a measured charge of molten resin into the mold, the volume of the charge being only about half of the mold volume, which is quickly filled, however, by the expansion of the gas. As the foam contacts the mold surfaces, the cells at those surfaces collapse to form a solid skin. Parts produced by this process are from three to four times as rigid as solid injection moldings of the same weight (because the latter are so much thinner). Many variations of the process have been developed, most employing injection molding or extrusion, with emphasis on elimination of swirls in the products.

Structural Formula *n* A drawing showing which atoms are bonded to each other in a molecule or polyatomic ion.

Structured Paint *n* Paint having a gel-like consistency which breaks down under the kind of shear exerted during brushing or roller application and re-forms when the shearing force is removed. When there is a time lag between the removal of the shearing force and the start of gel re-formation, the paint is said to be thixotropic. The thixotropic structure can vary from a very light gel to the fully gelled thixotropic paints of the nondrip type. See ▶ Viscosity.

Stucco \ˈstə-(ˌ)kō\ *n* [It, of Gr origin; akin to OHGr *stucki* piece, crust, OE *stocc* stock] (1598) (1) An exterior finish, usually textured; composed of Portland cement, lime, and sand, which are mixed with water. (2) A fine plaster used for decorative work or moldings. (3) Simulated stucco containing other materials, such as epoxy as a binder. (4) A partially or fully Calcined gypsum that has not yet been processed into a finished product.

Stud \ˈstəd\ *n* [ME *stode*, fr. OE *studu*; akin to MHGr *stud* prop, OE *stōw* place] (before 12c) Upright post in the framework of a wall for supporting lath, sheets of wallboard or the like.

Student's t *n* (t) The difference between the mean of a sample of measurements and the presumed true mean of the parent population from which the measurements were drawn, divided by the standard error of the sample mean, i.e.,

$$t = \frac{\bar{x} - \mu}{s/\sqrt{n}}$$

in which *s* is the sample standard deviation and *n* is the sample size. If the individual x_i are normally distributed, then *t* will have the t-distribution first formulated by W. S. Gosset in 1908 under the pseudonym, "Student." The *t*-statistic and its distribution are very useful in testing hypotheses bout population means and in setting up confidence internals for same. Tables of the percentage points of the t-distribution are widely available in handbooks and computer software.

Stuffer Box *n* A mechanism for crimping in which a fiber bundle (e.g., tow or filament yarn) is jammed against itself, causing it to crimp. By the suitable application of heat (usually wet steam) and pressure to the stuffed tow, a high and permanent crimp can be forced into the bundle. Also see ▶ Texturing.

Stuffers Extra yarns running in the warp direction through a woven fabric to increase the fabric's strength and weight.

S Twist See ▶ Twist, Direction of

Styrenated Alkyd See ▶ Alkyd Molding Compound.

Styrene \ˈstī-ˌrēn\ *n* [ISV] (1885) (vinyl benzene, phenylethylene, cinnamene) $C_6H_5CH = CH_2$. A colorless to yellowish oily liquid with a strong, sharp odor, produced by the catalytic dehydrogenation of ethylbenzene. Styrene monomer is easily polymerized by exposure to light, heat, or a peroxide catalyst, and even spontaneously, so a little inhibitor is added if it is to be stored. Styrene is a versatile comonomer, polymerizing readily with many other monomers, and is the active crosslinking monomer in most polyester laminating resins. Mol wt, 104.14; sp gr, 0.909; mp, $-33°C$; bp, 145–146°C. *Also known as Vinylbenzene, Styrol, Styrolene, Cinamene, and Phenylethylene.*

Styrene-Acrylonitrile Copolymer *n* (SAN) Any of a group of copolymers containing 70–80% styrene and 30–20% acrylonitrile and having higher strength, rigidity, and chemical resistance than straight polystyrene. These monomers may also be blended with butadiene to make a terpolymer or the butadiene may be grafted onto the SAN, either method producing ABS resin.

Styrene Butadiene *n* A group of thermo plastic elastomers. They are linear copolymers of styrene and butadiene, produced by lithium catalyzed solution polymerization, with a sandwich molecular structure containing a long Polybutadiene center surrounded by shorter polystyrene ends. A copolymer of styrene and butadiene made by emulsion polymerization for use in latex paints.

Styrene Butadiene Block Copolymer *n* Thermoplastic amorphous block polymer of butadiene and styrene having good impact strength, rigidity, gloss, compatibility with other styrenic resins, water resistance, and processibility. Used in food and display containers, toys, and shrink wrap.

Styrene Butadiene Copolymer *n* Thermoplastic polymers of butadiene and >50% styrene having good transparency, toughness, and processibility. Processed by extrusion, injection and blow molding, and thermoforming. Used in film wraps, disposable packaging, medical devices, toys, display racks, and office supplies.

Styrene Butadiene Latex See ▶ Butadiene Styrene Latex.

Styrene-Butadiene Rubber *n* (SBR, Buna-S, GR-S) A group of widely used synthetic rubbers comprising about three parts butadiene copolymerized with one part of styrene, with many modifications yielding a large variety of properties. The copolymers are first prepared as lattices, in which form they are sometimes used. The lattices can be coagulated to produce crumblike particles resembling natural crepe rubber. SBR has better abrasion resistance than natural rubber and has largely supplanted it in tire treads, though not in sidewalls.

Styrene Butadiene Thermoplastic *n* (S/B, SB) A group of thermoplastic elastomers introduced in 1965 (Shell chemical Co, Thermolastic®). They are linear block copolymers of styrene and butadiene, produced by lithium-catalyzed solution polymerization, with a sandwich molecular structure containing a long Polybutadiene center surrounded by shorter polystyrene ends. The materials are available in pellet form for extrusion, injection molding, and blow molding, and S/B sheets are thermoformable.

Styrene-Maleic *n* A anhydride copolymer is an alternating copolymer useful as a textile size an emulsifier.

Styrene-Maleic Anhydride Copolymer *n* An alternating copolymer useful as a textile size and emulsifier.

Styrene Maleic Anhydride Copolymer PBT Alloy *n* Thermoplastic alloy of styrene maleic anhydride copolymer and polybutylene terephthalate having improved dimensional stability and tensile strength. Processed by injection molding. Also called SMA PTB alloy.

Styrene Polymers See Strenic Resins.

Styrene Resin *n* Synthetic resin made from vinylbenzene.

Styrene-Rubber Plastic See ▶ High-Impact Polystyrene.

Styrenic-Acrylonitrile Copolymer See Styrene Acrylonitrile Copolymer.

Styrenic Resins *n* Sytrenic resins are thermoplastics prepared by free-radical polymerization of styrene alone or with other unsaturated monomers. The properties of styrenic resins vary widely with molecular structure, attaining the high performance level of engineering plastics. Processed by blow and injection molding, extrusion, thermoforming, film techniques and structural foam molding. Used heavily for the manufacture of automotive parts, household goods, packaging, films, tools, containers and pipes. Also called styrene resins, styrene polymers, styrene plastics.

Styrenic Plastic *n* (styrenic) A term that encompasses all the wide variety of thermoplastics in which styrene is the monomer or comonomer or in which polystyrene is a member of a polymer blend. See ▶ Polystyrene.

Styrenic Thermoplastic Elastomers *n* Linear or branched copolymers containing polystyrene end blocks and elastomer (e.g., isoprene rubber) middle blocks. Have a wide range of hardnesses, tensile strength, and elongation, and good low-temperature flexibility, dieclectric properties, and hydrolytic stability. Processed by injection and blow molding and extrusion. Used in coatings, sealants, impact modifiers, shoe soles, medical devices, tubing, electrical insulation, and auto parts. Also called TES.

Styroflex {*trademark*} Poly(styrene) (fibers), manufactured by Ndd. Seekabelwerke, Germany.

Styrofoam \ˈstī-rə-ˌfōm\ {*trademark*} Poly(styrene) (foam), manufactured by Dow.

Styrol *n* The name given to styrene by the chemist who first observed the monomer in 1839. The name was changed to styrene by German researchers around 1925.

Styron \ˈstī-rən\ *n* Poly(styrene), also copolymers, manufactured by Dow, US.

Styropor P *n* Poly(styrene) (foam), manufactured by BASF, Germany.

Sublimation \ˌsə-blə-ˈmā-shən\ *n* [ME, fr. ML *sublimatus*, pp of *sublimare*] (15c) A phase change in which a substance, such as a dye, passes directly from the solid to the vapor phase without passing through a liquid phase. This process is the basis for transfer printing.

Sublimed White Lead *n* Another name for basic lead sulfate (Gooch JW (1993) Lead-Based Paint Handbook, Plenum, New York).

Sublistic® Process {*trademark*} *n* A method of applying print designs to fabrics containing manufactured fibers by paper-transfer techniques. Developed by Sublistatic Corp. Also see ▶ Printing, ▶ Heat Transfer Printing.

Submarine Gate \ˈsəb-mə-ˌrēn\ *n* (tunnel gate) In injection molds, a type of edge gate where the opening from the runner into the mold is located below the parting line or mold surface, as opposed to conventional edge gating where the opening is machined into the surface of the mold or mold cavity. With submarine gates, the item is broken from the runner system on ejection from the mold. See also ▶ Gate.

Submicron Reinforcement *n* Very fine whiskers that, when used as reinforcements for thermoplastic resins, are short enough to flow through restricted gates without breaking, yet have aspect ratios well over 200, adequate for good isotropic strength development in the composite. However, they are costly, so are used only for special applications.

Subshell *n* The electrons within the same shell (energy level) of the atom are characterized by the same principal quantum number (n), and are further divided into groups according to the value of their azimuthal quantum numbers (l); the electrons which possess the same azimuthal quantum number for the same principal quantum number are considered to occupy the same subshell (or sublevel). The individual subshells are designated with the letters s, p, d, f, g, and hm as follows:

l value	Designation of subshell
0	s
1	p
2	d
3	f
4	g
5	h

An electron assigned to the *s*-subshell is called an *s*-electron, one assigned to the *p*-subshell is referred to as a *p*-electron, etc. In formulae of electron structure, the value of the principal quantum number (n) is prefixed to the letter indicating the azimuthal quantum number (l) of the electron; thus; e.g., a 4*f*-electron is an electron which has the principal quantum number 4 (i.e., assigned to the *N*-shell) and the orbital angular momentum 3 (*f*-subshell).

Sub Spirits *n* Mineral spirits. See ▶ Turpentine Substitute.

Substance *n* The weight in pounds of a ream (500 sheets) of paper cut to the standard size (17 × 22) for business papers (bond, ledger, mimeograph, duplicator, and manifold). For example, 20 lb. Similar to basis weight of other grades of paper.

Substitution Reaction *n* A reaction in which one atom or group of atoms is substituted for another in a molecule.

Substrate \ˈsəb-ˌstrāt\ *n* [ML *substratum*] (1807) (1) Any surface to which a coating or printing ink is applied. (2) The nonpainted material to which the first coat was applied; sometimes referred to as the original substrate. See ▶ Ground. (3) Base on which organic coloring agents are precipitated to form lakes. See ▶ Base (3). (4) Fabric to which coatings or other fabrics are applied. It can be of woven, knit, nonwoven, or weft-insertion construction. Generally, substrate properties are dependent both on fiber type and fabric construction. Usually the fabric is scoured, heat-set, and otherwise finished prior to coating or bonding. Many smooth-surfaced manufactured fiber fabrics require impregnation with a latex prior to coating to ensure adequate adhesion.

Subtractive Color Mixture See ▶ Colorant Mixture.

Subtractive Colorant Mixture *n*, **Complex** Colorant mixture which must take into account both the absorption and scattering of two or more of the individual pigments used in the mixture. See ▶ Kubelka-Munk Theory.

Subtractive Colorant Mixture, Simple *n* Colorant mixture which can be described by the single variable of absorption of radiant energy. The term can be used to describe mixture of liquids which have no scattering or of pigment mixtures where almost all of the scattering comes from one major component, frequently white pigment or a textile fiber, for example. See ▶ Beer's Law (Liquids) and ▶ Kubelka-Munk Theory.

Sub Turps See ▶ Turpentine Substitute.

Succinic Acid (ˌ)sək-ˈsi-nik-\ *n* [F *succinique*, fr. L *succinum* amber] (ca. 1790) $(CH_2)_2(COOH)_2$. A saturated dibasic acid. Mp, 184°C; bp, 235°C. Used in the preparation of alkyd type resins and plasticizers.

Sucrose Acetate Isobutyrate \ˈsü-ˌkrōs ˈa-ə-ˌtāt – ˈbyü-tə-ˌrāt\ *n* (SAIB) A modifying extender for lacquers and finishes based on resins such as cellulose acetate butyrate, acrylics, alkyds, and polyesters.

Sucrose Octaacetate *n* $C_{12}H_{14}O_3(OOCCH3)_8$. A plasticizer for cellulosic resins and polyvinyl acetate.

Sucrose Octabenzoate \-ˈben-zə-ˌwāt\ *n* $C_{12}H_{14}O_3(OOCC_6H_5)_8$. A plasticizer for polystyrene, cellulosics, and some vinyls.

Suction \ˈsək-shən\ *n* [LL *suction-, suctio*, fr. L *sugere* to suck] (1626) Force that causes a coating to be drawn into the pores of, or adhere to, a surface because of the difference between the external and internal pressures.

Suede Fabric \ˈswād\ *n* Woven or knitted cloth finished to resemble suede leather, usually by napping, shearing, and sanding techniques.

Suede Finish See ▶ Ripple Finish.

Sulfanilic Acid \ˌsəl-fə-ˈni-lik-\ *n* [ISV *sulf-* + *-anil*ine + *-ic*] (1856) $C_6H_7NO_3S$. A crystalline acid obtained from aniline.

Sulfar Fiber *n* A manufactured fiber in which the fiber-forming substance is a long chain, synthetic polysulfide in which at least 85% of the sulfide (-S-) linkages are attached to two aromatic rings (FTC definition). The raw material is polyphenylene sulfide which is melt spun and processed into staple fibers. These are high performance fibers with excellent resistance to strong chemicals and high temperature. They show excellent strength retention in harsh environments; are flame retardant; and are non-conducting. They find use in high-temperature filter fabrics, electrical insulation, coal-fired boiler bag houses, papermaker's felt, and high-performance composites.

Sulfate \ˈsəl-ˌfāt\ *n* [F, fr. L *sulfur*] (1790) SO_4. (1) A salt or ester of sulfuric acid. (2) A bivalent group or anion characteristic of sulfuric acid and the sulfates.

Sulfated Oil *n* A newer term signifying the same type of material as sulfonated oil.

Sulfate Pin Oil See ▶ Pine Oil.

Sulfate Pitch See ▶ Tall Oil.

Sulfate Pulp *n* Paper pulp made from wood chips cooked under pressure in a solution of caustic soda and sodium sulfide. *Known as Kraft.*

Sulfate Pump See Kraft Pumps.

Sulfate Resin See ▶ Tall Oil.

Sulfate Wood Turpentine See ▶ Turpentine.

Sulfation *n* The introduction into an organic molecule of the sulfuric ester group (or its salts), $-O-SO_3H$, where the sulfur is linked through an oxygen atom to the parent molecule.

Sulfide Staining *n* (1) Discoloration of a plastic caused by the reaction of one of its constituents with a sulfide in a liquid, solid, or gas to which the plastic article has been exposed. Stabilizers based on salts of lead, cadmium, antimony, copper, or other metals sometimes react with external sulfides to form a staining metallic sulfide. (2) The formation of dark stains in a paint film, as a result of the reaction of atmospheric hydrogen sulfide with metallic compounds such as lead, mercury, or copper in the paint.

Sulfite Pulp *n* Paper pulp made from wood chips cooked under pressure in a solution of bisulfite of lime.

Sulfite Turpentine *n* This term is not in good usage, because the volatile oil recovered in the conversion of wood to pulp by the sulfite process consists chiefly of cymene, $C_{10}H_{14}$, rather than pinene and other terpenes.

Sulfonate-Carboxylate Copolymer See Polysulfonate Copoly-Mer.

Sulfonated *n* A term describing a material that has been reacted with sulfonic acid, usually to impart solubility, dyeability with cationic dyes, or other properties.

Sulfonated Castor Oil See ▶ Turkey Red Oil.

Sulfonated Oil *n* A water-dispersible or soluble surface active material obtained by treating an unsaturated or hydroxylated fatty oil, acid, or ester with an agent capable of sulfating or sulfonating it at least partially. Also known as *Sulfated Oil*.

Sulfonation *n* The introduction into an organic molecule of the sulfonic acid group (or its salts), $-SO_3H$, where the sulfur atom is joined to a carbon atom of the parent molecule.

Sulfone Polymers *n* Alternate name for polysulfone.

Sulfonic Acid *n*(1873) Any acid containing the sulfonic group (SO_3H).

Sulfonyl \ˈsəl-fə-ˌnil\ *n* (1920) An organic radical of the form $RSO_2–$, in which R may be aliphatic or aromatic. If the bond is filled by an –OH group, the compound is called a sulfonic acid. The sulfonyl chlorides are highly active reagents used in many organic syntheses.

Sulfonylazidosilane \-ˈā-zə(ˌ)dō-ˈsi-ˌlān\ *n* Any of a family of organofunctional coupling agents that, in contrast to most other ▶ Silane Coupling Agents, enter into direct chemical reaction with organic polymers. They function by insertion into carbon-hydrogen bonds, which avoids generation of free radicals and degradation of radical-sensitive polymers such as polypropylene, polyisobutylene, and polystyrene.

Sulfonyldianiline *n* (aminophenyl Sulfone) $(NH_2C_6H_4)_2SO_2$. A curing agent for epoxy resins.

Sulfur Dyes See ▶ Dyes.

Sulfuric Acid *n* (1790) H_2SO_4. A heavy corrosive oily dibasic strong acid that is colorless when pure and is a vigorous oxidizing and dehydrating agent.

Sulfurized Oil *n* An oil which has been reacted with sulfur or sulfur-containing compounds like sulfur chloride. The sulfur is believed to attach itself to the unsaturated bonds. Sulfur-treated oils set or polymerize rapidly, and they are also rather unstable on storage.

Sulfur Nitride Polymer *n* (polysulfur nitride, polythiazyl) $[-(SN)–]n$. First synthesized in 1910 but ignored for 6 decades, this covalent polymer has been restudied and found to have the physical and electrical properties of a metal. It is formed by passing the vapor of $(SN)_4$ over a catalyst that cracks it to $(SN)_2$, which is condensed on a cold surface where it spontaneously polymerizes into crystalline form. The crystals are malleable and can be cold-worked into thin sheets or fibers under pressure, having an electrical conductivity similar to that of mercury, and superconducting near 0 K.

Sulfurous Acid *n* (1790) H_2SO_3. A weak unstable dibasic acid known in solution and through its salts and used as a reducing and bleaching agent.

Sulfur Vulcanization *n* The vulcanization of a diene rubber by heating to about 150°C with sulfur. In practice, a vulcanization accelerator and activators can be used in addition to sulfur.

Sumi *n* (Japanese) Black Chinese ink.

Sumi-ye (Japanese) A print in black and white only.

Summerwood \ˈsə-mər-ˌwu̇d\ *n* (1902) The portion of the annual growth ring that is formed during the latter part of the yearly growth period. It is usually more dense and stronger mechanically than springwood.

Sunflower Oil \-ˈflau̇(-ə)r\ *n* Seed oil of *Helianthus annus*, which grows in abundance in many parts of the world. It can only be regarded as a semidrying oil, but has the useful property of being nonyellowing. Sp gr, 0.930/15°C; iodine value, 135.

Sunlight Resistance See ▶ Light Resistance.

Superabsorbent *n* A material that can absorb many times the amount of liquid ordinarily absorbed by cellulosic materials such as wood pulp, cotton, and rayon.

Super Absorbent Fibers Besides granular super absorbent polymers, ARCO Chemical developed a super absorbent fiber technology in the early 1990s. This technology was eventually sold to Camelot Absorbents (Now bankrupt). Another fiber-spinning SAP technology was developed by Courtalds & Allied Colloids in the UK in the early 1990s. These acrylic-based products, while significantly more expensive than the granular polymers, offer technical advantages in certain niche markets including cable wrap, medical devices and food packaging.

Uses include:

- Wire & Cable Waterblocking/Filtration Applications
- Spill Control
- Hot & Cold Therapy Packs
- Composites and Laminate

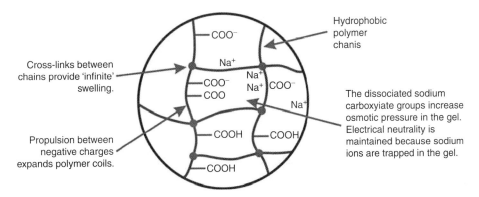

Charge repulsion between carboxyl groups in cross-linked SAP

- Diapers and Incontinence Garments
- Waste Stabilization & environmental Remediation
- Fragrance Carrier
- Wound Dressings
- Fire Protection
- Surgical pads
- Water Retention for Supplying Water to Plants
- Controlled Release of Insecticides & Herbicides
- Chemical Sensors

Sodium polyacrylate also named acrylic sodium salt polymer or simply ASAP (repeating unit: $-CH_2-CH(COONa)-$) is a polymer widely used in consumer products. Acrylate polymers generally are considered to possess an anionic charge. While sodium neutralized acrylates are the most common form used in industry, there are also other salts available including potassium, lithium and ammonium.

Applications include wide variety of industrial uses:
1. Sequestering agents in detergents. (By binding hard water elements such as Ca and Mg, the surfactants in detergents work more efficiently.)
2. Thickening agents.
3. Coatings.
4. Super absorbent polymers. These cross-linked acrylic polymers are used in baby diapers. Co-polymer versions are used in agriculture and other specialty absorbent applications.

The origins of super absorbent polymer chemistry trace back to the early 1960s when the US Department of Agriculture developed the first super absorbent polymer materials. Sodium polyacrylate belongs to a family of water loving or hydrophilic polymers. It has the ability to absorb up to 800 times its weight in distilled water. Sodium polyacrylate is a powder takes the form of a coiled chain. There are two important groups that are found on the polymer chains, carbonyl (COOH) and sodium (Na). These two groups are important to the overall absorption potential of the polymer. When the polymer is in the presence of a liquid, the sodium dissociates from the carbonyl group creating two ions, carboxyl (COO–) and sodium cation (Na+). The carboxyl groups then begin to repel each other because they have the same negative charge. As a result of the repulsion between the like charges, the sodium polyacrylate chain uncoils or swells and forms a gel substance. The action of swelling allows more liquid to associate with the polymer chain. There are four major contributors to sodium polyacrylate's ability to absorb liquids or swell. These contributors are hydrophilic chains, charge repulsion, osmosis, and cross-linked between chains. Ions in the polymer chain such as carboxyl groups (COO–) and sodium (Na+) attract water molecules, thus making the polymer hydrophilic. Charge repulsion between carboxyl groups allow the polymer to uncoil and interact with more water molecules (see figure above).

When these dry coiled molecules are placed in water, hydrogen bonding with the HOH surrounding them causes them to unfold and extend their chains as shown in figures below.

Acrylate polymers – in a dry coiled state

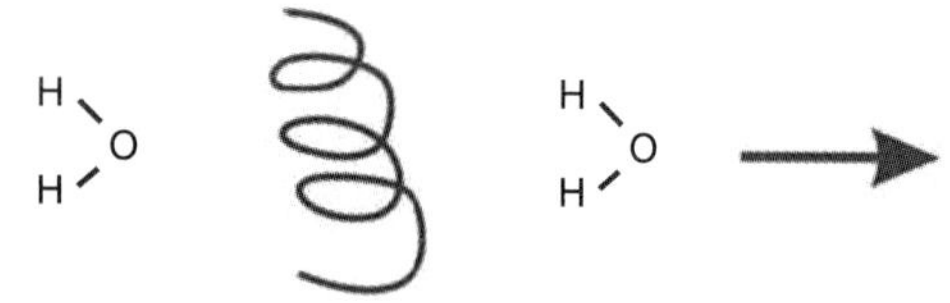

Interaction of SAP with water

When the molecules straighten out, they increase the viscosity of the surrounding liquid. That is why several types of acrylates are used as thickeners. Super absorbent chemistry requires two things: The addition of small cross-linking molecules between the polymer strands; and the partial neutralization of the carboxyl acid groups (–COOH) along the polymer backbone (–COO⁻Na⁺). Water molecules are drawn into the network across a diffusion gradient which is formed by the sodium neutralization of the polymer backbone. The polymer chains want to straighten but are constrained due to the cross-linking. Thus, the particles expand as water moves into the network.

Expansion between polymer chains with interaction with water

The water is tightly held in the network by hydrogen-bonding. Many soluble metals will also ion-exchange with the sodium along the polymer backbone and be bound (see Cross-linked polymer chains below).
A view of a crosslinks in super absorbent molecules is shown in the following figure, and the more practical polymer network in the following figure.

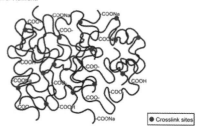

Superabsorbent Polymer Network:

View of superabsorbent polymer network with cross-linked sites

Superabsorbent Polymers Superabsorbent polymers (SAP) are now commonly made from the polymerization of acrylic acid blended with sodium hydroxide in the presence of an initiator to form a poly-acrylic acid, sodium salt (sometimes referred to as cross-linked sodium polyacrylate). This polymer, such as polyacrylamide copolymer, ethylene maleic anhydride copolymer, cross-linked carboxy-methyl-cellulose, polyvinyl alcohol copolymers, cross-linked polyethylene oxide, and starch grafted copolymer of polyacrylonitrile to name a few. Superabsorbent polymers are made using one of two primary methods; suspension polymerization or solution

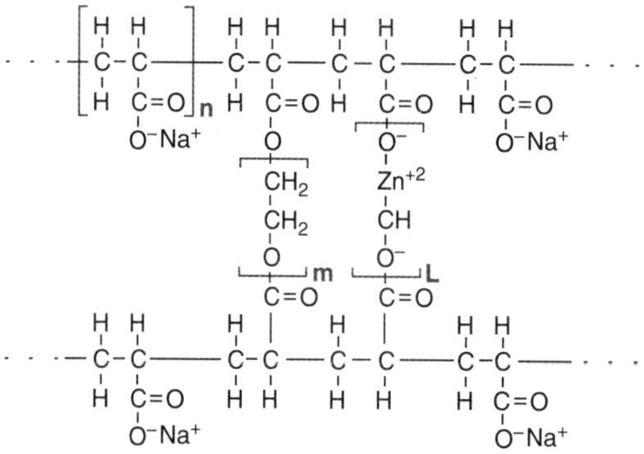

n = Number of repeated units

m = Cross linking agent such as a diol, could be diol terminated polymer of m repeat units

L = Cross linked by divalent metal, could be useful as cationic exchange resin

Cross-linked polymer chains

polymerizations. Both processes have their advantages over the other and both yield a consist quality of product. Solution based polymerization is the most common process used today for SAP manufacture. This process is efficient and generally has a lower capital cost base. The solution process uses a water based monomer solution to produce a mass of reactant polymerized gel. The polymerization's own reaction energy (exothermic) is used to drive much of the process, helping reduce manufacturing cost. The reactant polymer gel is then chopped, dried and ground to its final granule size. Any treatment to enhance performance characteristics of the SAP is usually accomplished after the final granule size is created. The suspension process is practiced by only a few companies because it offers a higher degree of production control and product engineering during polymerization step. This process suspends the water based reactant in a hydrocarbon based solvent. The net result is that the suspension polymerization creates the primary polymer particle in the reactor rather than mechanically in post-reactions stages. Performance enhancements can also be during or just after the reaction stage. The total absorbency and swelling capacity are controlled by the type and degree of crosslinking to the polymer. Low density cross-linked SAP generally has a higher absorbent capacity and swells to a larger degree. These types of SAPs also have a softer and more cohesive gel formation. High cross-link density polymers exhibit lower absorbent capacity and swell. The gel strength is firmer and can maintain particle shape even under modest pressure. The largest use of SAP is found in personal disposable hygiene products, such as baby diapers, adult protective underwear and feminine napkins. SAP was discontinued from use in tampons due to 1980s concern over a link to toxic shock syndrome. They are also used for blocking water penetration in underground poser or communications cable, horticultural water retention agents, control of spill and waste aqueous fluid, artificial snow for motion picture and stage production. The first commercial use was in 1978 for use in feminine napkins in Japan, disposable bed liners for nursing home patients in the USA.

The SDA gave the technical expertise to several USA companies for further development of the basic technology. A wide range of grating combinations was attempted including work with acrylic acid, acrylamide and polyvinyl alcohol (PVA). Copolymer Chemistry includes materials as polyacrylate/polyacrylamide copolymers originally designed for use in conditions with high electrolyte/mineral content and a need for long term stability including numerous wet/dry cycles. Uses include agricultural and horticultural. With the added strength of the acrylamide monomer, used as medical spill control, wire and cable waterblocking.

Superconducting Polymers n A polymer exhibiting electrical conductivity of $\cong 10^{20}$ S/cm.

Supercooling n The rapid cooling of a normally crystalline plastic through its crystallization temperature so it does not get a change to crystallize and it remains in the amorphous state.

Superheating n The heating of a liquid above its boiling point without boiling taking place.

Superpolymers n Polymers having molecular weights above 10,000 (date). This term was first used by Carothers, (Adams, Roger (1940) A Biography, in High Polymers: A Series of Monographs on the Chemistry, Physics and Technology of High Polymeric Substances Vol.1 Collected Papers of W.H. Carothers on High Polymeric Substances, New York, NY: Interscience Publishers, Inc. XVIII).

Superposition Principle See ▶ Boltzmann Superposition Principle.

Supersaturated Solution n An unstable solution in which the concentration of solute is greater than its solubility.

Superstruture Paint n Ship paint used on part of the ship's structure above the main deck.

Suppliers of Advanced Composite Materials Association n (SACMA) A recently formed industry group headquartered at 1600 Wilson Blvd, Suite 1008, Arlington, VA 22209.

Supralen n Poly(ethylene) (pipes), manufactured by Mannesmann, Germany.

Surah \ˈsúr-ə\ n [prob. alter. of *surat*, a cotton produced in Surat, India] (1873) A soft fabric of silk or filament polyester or acetate, usually a twill and often woven in a plaid. Surah is used for ties, mufflers, blouses, and dresses.

Surface Active *adj* (1920) Altering the properties and lowering the tension at the surface of contact between phases (soaps and wetting agents are typical surface-active substances).

Surface-Active Agents n Substances that, when used in small quantities, modifies the surface properties of liquids or solids. A surface-active agent reduces surface tension in a fluid or the interfacial tension between two immiscible fluids, such as oil and water. Surfactants are particularly useful in accomplishing the wetting or penetration of solids by aqueous liquids and serve in the manner of detergent, emulsifying, or dispersing agents.

Surface Area *n* Total area of the surface of all the particles in a mass of pigment; this is usually expressed in terms of square meters per gram.

Surface Area *n* (**B.E.T.**) The total surface area of a solid calculated by the B.E.T. (Brunauer, Emmett, Teller) equation, from nitrogen adsorption or desorption data obtained under specified conditions.

Surface Characteristics *n* All those properties attributable to surfaces, such as roughness, adsorptivity, coefficient of friction, surface energy, and chemical activity.

Surface Charge *n* The electrical charge on the surface of a substance.

Surface Conditioner *n* Preparatory coating applied to a chalked surface to bind chalk to the substrate, prior to top coating.

Surface Conductance *n* The direct-current conductance (A/V) between two electrodes in contact with a specimen of solid insulating material when the current is passing only through a thin film of moisture on the surface of the specimen.

Surface Density of Electricity *n* Quantity of electricity per unit area.

Surface Density of Magnetism *n* Quality of magnetism per unit area.

Surface Drying *n* The premature drying of the surface of a liquid coating film, so that the under portion is retarded in drying. See ▶ Drying (1) and ▶ Drying Time.

Surface Energy *n* (1) The free energy of the surfaces at an interface that arises because of differences in the tendencies of each phase to attract its own molecules. (2) The work that would be required to increase the surface area of a liquid by one unit area (3) An alternate aspect of ▶ Surface Tension. (4) Surface energy is work per area, and simply expressed as dyne/cm.

An erg is the unit of energy and work in the CGS system of units, derived from the Greek ergon meaning "work." 1 erg = 10^{-7} J as expressed in International Standards Organization (SI) units.
= 1 dyne cm
The erg is the amount of work done by a force of 1 dyne exerted for a distance of 1 cm, and surface energy expressed as,

$$\frac{\text{erg}}{\text{cm}^2} = \frac{\text{dyne cm}}{\text{cm}^2} = \frac{\text{dyne}}{\text{cm}}.$$

So, surface energy (SE) is universally expressed as dyne/cm.

Surface energy (SE) and surface tension (ST) are identical for a liquid. Critical surface tension (CST or γ_c) is measured by plotting the contact angle of a series of liquid of different known surface tensions with the contact angle on a surface and extrapolating to 0° contact angle. This plot is referred to as the following Zeisman plot to determine γ_c or Cos θ versus dyne/cm. The units of surface tension are either in dyne/cm (conventional) or mN/m (SI).

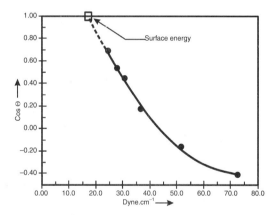

Solid surface energy data (SFE) for common polymers

Name	CAS Ref.-No.	Surface free energy (SFE) at 20°C in mN/m	Temp. coefficient SFE in mN/(m K)	Dispersive contrib. of SFE in mN/m	Polar contrib. of SFE in mN/m
Polyethylene-linear PE	9002-88-4	35.7	−0.057	35.7	0
Polyethylene-branched PE	9002-88-4	35.3	−0.067	35.3	0
Polypropylene-isotactic PP	25085-53-4	30.1	−0.058	30.1	0
Polyisobutylene PIB	9003-27-4	33.6	0.064	33.6	0
Polystyrene PS	9003-53-6	40.7	−0.072	(34.5)	(6.1)
Poly-α-methyl styrene PMS (Polyvinyltoluene PVT)	9017-21-4	39.0	−0.058	(35)	(4)
Polyvinyl fluoride PVF	24981-14-4	36.7	–	(31.2)	(5.5)

Solid surface energy data (SFE) for common polymers (Continued)

Name	CAS Ref.-No.	Surface free energy (SFE) at 20°C in mN/m	Temp. coefficient SFE in mN/(m K)	Dispersive contrib. of SFE in mN/m	Polar contrib. of SFE in mN/m
Polyvinylidene fluoride PVDF	24937-79-9	30.3	–	(23.3)	(7)
Polytrifluoroethylene P3FEt/PTrFE	24980-67-4	23.9	–	19.8	4.1
Polytetrafluoroethylene PTFE (Teflon™)	9002-84-0	20	−0.058	18.4	1.6
Polyvinylchloride PVC	9002-86-2	41.5	–	(39.5)	(2)
Polyvinylidene chloride PVDC	9002-85-1	45.0	–	(40.5)	(4.5)
Polychlorotrifluoroethylene PCTrFE	25101-45-5	30.9	−0.067	22.3	8.6
Polyvinylacetate PVA	9003-20-7	36.5	−0.066	25.1	11.4
Polymethylacrylate (Polymethacrylic acid) PMAA	25087-26-7	41.0	−0.077	29.7	10.3
Polyethylacrylate PEA	9003-32-1	37.0	−0.077	30.7	6.3
Polymethylmethacrylate PMMA	87210-32-0	41.1	−0.076	29.6	11.5
Polyethylmethacrylate PEMA	9003-42-3	35.9	−0.070	26.9	9.0
Polybutylmethacrylate PBMA	25608-33-7	31.2	−0.059	26.2	5.0
Polyisobutylmethacrylate PIBMA	9011-15-8	30.9	−0.060	26.6	4.3
Poly(t-butylmethacrylate) PtBMA	–	30.4	−0.059	26.7	3.7
Polyhexylmethacrylate PHMA	25087-17-6	30.0	−0.062	(27.0)	(3)
Polyethyleneoxide PEO	25322-68-3	42.9	−0.076	30.9	12.0
Polytetramethylene oxide PTME (Polytetrahydrofurane PTHF)	25190-06-1	31.9	−0.061	27.4	4.5
Polyethyleneterephthalate PET	25038-59-9	44.6	−0.065	(35.6)	(9)
Polyamide-6,6 PA-66	32131-17-2	46.5	−0.065	(32.5)	(14)
Polyamide-12 PA-12	24937-16-4	40.7	–	35.9	4.9
Polydimethylsiloxane PDMS	9016-00-6	19.8	−0.048	19.0	0.8
Polycarbonate PC	24936-68-3	34.2	−0.04	27.7	6.5
Polyetheretherketone PEEK	31694-16-13	42.1	–	36.2	5.9

Surface Pin Syn: ▶ Ejector-Return Pin.

Surface Preparation *n* A physical and/or chemical preparation of an adherend to render it suitable for adhesive joining.

Surfacer *n* Pigmented composition for filling minor irregularities to obtain a smooth, uniform surface preparatory to applying finish coats; usually applied over a primer and sandpapered for smoothness. Some types combine the properties and functions of both a primer and a surfacer and are called primer surfacers.

Surface Resistivity *n* The electrical resistance between two parallel electrodes in contact with the specimen surface and separated y a distance equal to the contact length of the electrodes. The resistivity is therefore the quotient of the potential gradient, in V/m, and the current per unit of electrode length, A/m. Since the four ends of the electrodes define a square, the lengths in the quotient cancel and surface resistivities are reported in "ohms per square." For reproducibility of results, specimens must be carefully cleaned and dried, and protected from contamination.

Surface Tensiometer *n* An instrument used to measure surface and interfacial tensions of liquids.

Surface Tension *n* Surface Energy. Surface energy is work per area, and simply expressed as dyne/cm.
An erg is the unit of energy and work in the CGS system of units, derived from the Greek ergon meaning "work."
1 erg = 10^{-7} J as expressed in International Standards Organization (SI) units.
= 1 dyne cm

The erg is the amount of work done by a force of 1 dyne exerted for a distance of 1 cm, and surface energy expressed as,

$$\frac{erg}{cm^2} = \frac{dyne\ cm}{cm^2} = \frac{dyne}{cm}$$

So, surface energy (SE) is universally expressed as dyne/cm.

Surface energy (SE) and surface tension (ST) are identical for a liquid. Critical surface tension (CST or γ_c) is measured by plotting the contact angle of a series of liquid of different known surface tensions with the contact angle on a surface and extrapolating to 0° contact angle. This plot is referred to as the following Zeisman plot to determine γ_c or Cos θ versus dyne/cm. The units of surface tension are either in dyne/cm (conventional) or mN/m (SI).

Surface Tension *n* (1876) (1) (free-surface energy, surface energy) The work required to increase the surface of a solid or liquid (in contact with air) by one unit of area. Surface energy and surface tension are equivalent terms, but with liquids it is possible to measure directly the surface tension. The unit still in common use is dynes per centimeter, equal to 0.001 N/m, or, in surface-energy terms, 0.001 J/m². Surface tension is measured with a *tensiometer* or by capillary rise. Surface energies of plastics are determined indirectly by observing the angles of contact of a graded series of increasingly polar liquids of known surface tension on the plastic surface (see ASTM D2578 - 09 Standard Test Method for Wetting Tension of Polyethylene and Polypropylene Films), then applying regression analysis to determine the polar and non-polar components of surface energy. The surface energy is the sum of the two components. (2) The tension exhibited by the free surface of liquids measured in dynes per centimeter. (3) The tension exhibited by the free surface of liquids, measured in dynes per centimeter. Surface tension can be expressed as the total force along a line of length l on the surface of a liquid whose surface tension is T,

$$F = lT.$$

Capillary tubes – If a liquid of density d rises a height h in a tube of internal radius r the surface tension is,

$$T = \frac{rhdg}{2}$$

The tension will be in dynes per centimeter is r and h are in centimeter, d in g per cubic centimeter and g in centimeter per square second. *Drops and bubbles* – Pressure in dynes per square centimeter due to surface tension on a drop of radius r centimeter for a liquid whose surface tension is T dynes per centimeter.

$$P = \frac{2T}{r}$$

For bubble of mean radius r centimeter,

$$P = \frac{4T}{r}$$

Surface Tension and Hansen Solubility Parameters *n*
The relationship between surface tension (surface energy) is discussed by Koenhen and Smolders (1975). An updated version of the relationship is expressed as:

Surface Tension $(dyne/cm) @ 20°C = 0.0146\ (2.28\delta_D^2 + \delta_P^2 + \delta_H^2) Mvol.^{0.2}$

δ_D = Solubility Parameter, Dispersion Forces, $MPa^{0.5}$
δ_P = Solubility Parameter, Polar Forces, $MPa^{0.5}$
δ_H = Solububility Parameter,
   Hydrogen Bonding Forces, $MPa^{0.5}$
Mvol = Molecular Volume, $m^3/mole$

Parameters used to derive the above relations are listed in the following table (data extracted from HSPiP (www.hansen-solubility.com), courtesy of Professor

Liquids, surface tension, molecular volume and Hansen Solubility Parameters

Name	ST	MVol	SP-D	SP-P	SP-H	Calculated HSP
Gamma-Butyrolactone	36.5	76.8	19	17	7.4	41.36714755
1.1.1-Trichloroethane	25.18	99.3	17	4.3	2	25.14976212
1,1,2,2-Tetrabromoethane	49.44	117	23	5.1	8.2	49.46600004
1,1,2,2-Tetrachloroethane	35.58	105	19	5.1	8.2	32.85009812
1,1,2-Trichlorotrifluoroethane	17.75	119	15	1.6	0	19.39790732
1,1-Dichloroethane	24.07	84.8	17	7.8	3	25.26622209

Liquids, surface tension, molecular volume and Hansen Solubility Parameters (Continued)

Name	ST	MVol	SP-D	SP-P	SP-H	Calculated HSP
1,1-Dichloroethylene	17.7	80.4	16	5.2	2.4	23.38676354
1,1-Dimethylhydrazine	28	76	15	5.9	11	24.67964273
1,2-Dichlorotetrafluoroethane	19.4	117	13	1.8	0	14.47662969
1,3-Benzenediol	47.6	87.5	18	8.4	21.1	46.17835877
1,3-Butanediol	37.8	89.9	17	10	21.5	44.05997806
1,3-Dimethyl-1-butanol<-Hen	22.63	127	15	3.3	12.3	27.60992955
1,4-Dioxane	32.75	85.7	19	1.8	7.4	32.30535188
1-Bromonaphthalene	44.19	140	20	3.1	4.1	39.07234186
1-Butanol	24.93	91.5	16	5.7	15.8	32.15877797
1-Decanol	28.51	192	16	4.7	10	30.39953459
1-Methylnaphthalene	40.5	139	21	0.8	4.7	39.98585774
1-Octanol	27.1	158	16	5	11.9	31.07536654
1-Pentanol	25.36	109	15.9	5.9	13.9	30.92122367
2,2,4-Trimethylpentane	18.77	166	14	0	0	18.9589125
2-Butanol	22.54	92	16	5.7	14.5	30.19050391
2-Ethoxyethyl acetate	31.8	136	16	4.7	10.6	28.59217551
2-Ethyl-1-butanol	25.06	123	16	4.3	13.5	30.35132172
2-Ethyl-1-hexanol	30	157	15.9	3.3	11.8	30.06508783
2-Furfuraldehyde	43.09	83.2	19	15	5.1	37.78674377
2-Methyl-1-propanol	22.98	92.8	15	5.7	16	30.10889934
2-Nitropropane	29.29	90.6	16	12	4.1	28.2335735
2-Octanol	26.32	159	16	4.9	11	30.84387899
2-Pyrrolidone	45.7	76.4	19.4	13.8	11.3	42.14236508
Acetaldehyde	20.5	56.5	14.7	12.5	7.9	23.99317962
Acetic acid	27.1	57.1	15	8	13.5	24.525933
Acetic anhydride	31.93	94.5	16	12	10.2	30.82739164
Acetone	23.46	74	16	10	7	25.09648287
Acetonitrile	28.66	52.6	15	18	6.1	29.75503401
Acetophenone	39.04	117	20	8.6	3.7	37.62332738
Acetyl chloride	25	71	16.2	11.2	5.8	26.74396259
Acrylonitrile	26.7	66.2	16	12.8	6.8	27.63567399
Aniline	42.12	91.5	19	5.1	10.2	36.70731713
Anisole	35.1	109	18	4.1	6.7	30.17677447
Benzaldehyde	38	102	19	7.4	5.3	35.68757925
Benzene	28.22	89.4	18	0	2	28.69207892
Benzoic acid	31.9	112	18	7	9.8	34.8467211
Benzonitrile	38.79	103	19	9	3.3	34.12163706
Benzyl alcohol	39.96	104	18	6.3	13.7	38.0527222
Benzyl chloride	37	115	19	7.2	2.7	33.63601241
Biphenyl	29.5	155	20	1	2	36.74751584
Bis(2-chloroethyl) ether	37.6	118	19	9	5.7	35.90887603
Bromobenzene	35.24	105	21	5.5	4.1	38.40778128
Bromochloromethane	32.4	67.2	17	5.7	3.5	25.67112387

Liquids, surface tension, molecular volume and Hansen Solubility Parameters (Continued)

Name	ST	MVol	SP-D	SP-P	SP-H	Calculated HSP
Bromoform	44.87	87.5	22	4.1	6.1	40.7933482
Bromotrifluoromethane	6.51	97	9.6	2.5	0	8.132685428
Butyraldehyde	24.2	88.5	15	5.3	7	21.02501906
Butyronitrile	26.92	87.9	15	12	5.1	26.29066172
Carbon disulfide	31.58	60	21	0	0.6	32.72778465
Carbon tetrachloride	26.43	97.1	18	0	0.6	27.17162966
Chlorobenzene	32.99	102	19	4.3	2	32.10911092
Chlorodifluoromethane	7.6	72.9	12	6.3	5.7	14.80613736
Chloroform	26.67	80.7	18	3.1	5.7	27.69589122
cis-Decahydronaphthalene	32.2	157	19	0	0	33.34693205
Cyclohexane	24.65	109	17	0	0.2	24.74611617
Cyclohexanol	32.92	106	17	4.1	13.5	34.02166282
Cyclohexanone	34.57	104	17.8	8.4	5.1	31.21195794
Cyclohexylamine	31.22	115	17	3.1	6.5	28.86992799
Di-(2-methoxyethyl) ether	29.5	142	15.7	6.1	6.5	26.01792403
Di-(i-butyl) ketone	23.9	177	16	3.7	4.1	26.03794578
Diacetone alcohol	31	124	16	8.2	10.8	29.73501938
Dibenzyl ether	38.2	191	19.6	3.4	5.2	39.3432669
Dichlorofluoromethane 2568?	35.6	75.4	16	3	5.7	21.8244186
Diethyl carbonate	26.44	122	15.1	6.3	3.5	22.4885061
Diethyl ether	16.65	105	15	2.9	5.1	19.61194008
Diethyl ketone	24.74	106	16	7.6	4.7	24.84849432
Diethyl phthalate	37.3	200	18	9.6	4.5	35.54874951
Diethyl sulfate	34.02	132	16	15	7.2	33.43193787
Diethyl sulfide	24.57	108	17	3.1	2	25.2443064
Diethylamine	19.85	103	15	2.3	6.1	20.87935664
Diethylene glycol	44.77	95.3	16.6	12	20.7	44.96029972
Diethylene glycol monoethyl ether	31.8	131	16	9.2	12.2	33.19206918
Diethylene glycol monomethyl ether	34.8	118	16.2	7.8	12.6	31.9682153
Diethylene glycol mono-n-butyl ether	30	171	16	7	10.6	31.34711949
Diethylenetriamine	40.5	108	16.7	7.1	14.3	34.20214862
Dimethyl phthalate	41.8	163	19	11	4.9	38.75236312
Dimethyl sulfone	60.15	75	19	19	12.3	48.21209637
Dimethyl sulfoxide	42.92	71.3	18	16	10.2	40.45615212
Dimethylformamide	36.76	77	17	14	11.3	36.08700845
Di-n-butyl phthalate	33.4	267	18	8.6	4.1	37.42800442
Di-n-butyl sebacate	30.99	337	16.7	4.5	4.1	32.45177946
Di-n-propylamine	22.31	137	15	1.4	4.1	22.2474773
Dioctyl phthalate	33.4	399	17	7	3.1	34.24747339
Dipropylene glycol (mixed isomers)	39.8	131	16	10	18.4	40.81038241
Epichlorhydrin	36.36	78.4	18.9	7.6	6.6	32.98338137
Ethanethiol	23.08	74.3	16	6.5	7.1	23.32674959
Ethanolamine	48.32	59.8	17	16	21.3	46.14612785

Liquids, surface tension, molecular volume and Hansen Solubility Parameters (Continued)

Name	ST	MVol	SP-D	SP-P	SP-H	Calculated HSP
Ethyl acetate	23.39	98.5	16	5.3	7.2	24.47104398
Ethyl bromide	23.62	74.6	16.5	8.4	2.3	24.84105879
Ethyl chloroformate	26.2	96.3	16	10	6.7	25.99379019
Ethyl formate	23.18	80.2	16	8.4	8.4	24.92356522
Ethyl lactate	28.9	115	16	7.6	12.5	31.01572878
Ethyl trans-cinnamate	38	167	18	8.2	4.1	35.85505187
Ethylbenzene	28.75	123	18	0.6	1.4	28.56925177
Ethylene carbonate	41.7	66	19	22	5.1	47.14438997
Ethylene cyanohydrin (hydracrylonitrile)	45	68.3	17	19	17.6	46.85917246
Ethylene diamine	40.77	67.3	17	8.8	17	34.74431039
Ethylene dibromide	39.55	87	19.2	3.5	8.6	34.08026365
Ethylene dichloride	31.86	79.4	19	7.4	4.1	32.30549557
Ethylene glycol	47.99	55.8	17	11	26	48.97756704
Ethylene glycol monoethyl ether	31.8	97.8	16.2	9.2	14.3	33.40608225
Ethylene glycol monomethyl ether	30.84	79.1	16.2	9.2	16.4	34.34281271
Ethylene glycol mono-n-butyl ether	26.14	131.6	16	5.1	12.3	30.39780547
Formamide	57.03	39.8	17.2	26.2	19	54.13957869
Formic acid	37.13	37.8	14.3	11.9	16.6	27.4923882
Furan	24.1	72.5	17.8	1.8	5.3	26.72699664
Furfuryl alcohol	38	87.1	17.4	7.6	15.1	35.89905777
Glycerol	63.3	73.4	17.4	12.1	29.3	60.23915987
Hexamethyl phosphoramide	33.8	175.7	18.4	8.6	11.3	41.20452441
Hexylene glycol	33.1	123	15.7	8.4	17.8	37.4110143
i-Amyl acetate	24.62	148.8	15.3	3.1	7	24.25169791
i-Butyl acetate	23.06	133.5	15.1	3.7	6.3	22.96626956
i-Butyl i-butyrate	22.6	169.8	15.1	2.9	5.9	23.67117495
i-Pentane	14.4	117.4	13.7	0	0	16.71053487
i-Propyl palmitate	30	330	16	3.9	3.7	29.41261295
Isophorone	32.3	150.5	16.6	8.2	7.4	30.7868774
m-Cresol	35.69	104.7	18	5.1	12.9	35.53278485
Mesityl oxide	28.3	115.2	16.4	6.1	6.1	26.74907086
Mesitylene	27.55	139.8	18	0.6	0.6	29.90094138
Methyl acetate	24.73	79.7	15.5	7.2	7.6	23.75402762
Methyl chloride	29.3	50.1	15.3	6.1	3.9	19.30320178
Methyl ethyl ketone	23.97	90.1	16	9	5.1	25.578026
Methyl i-amyl ketone	25	142.8	16	5.7	4.1	25.70368207
Methyl i-butyl ketone	23.64	125.8	15.3	6.1	4.1	23.26918337
Methyl oleate	31.3	337.3	16.2	3.8	4.5	30.52877828
Methyl salicylate	39.22	129	18.1	8	13.9	39.95146973
Methylal CH2(OCH3)2	21.12	88.8	15.1	1.8	8.6	22.04694552
Methylcyclohexane	23.29	128.3	16	0	1	23.24041387
Methylene dichloride	27.2	64.4	18.2	6.3	6.1	28.81526901
Methylene diiodide	50.68	80.5	22	3.9	5.5	41.6046342

Liquids, surface tension, molecular volume and Hansen Solubility Parameters (Continued)

Name	ST	MVol	SP-D	SP-P	SP-H	Calculated HSP
Morpholine	37.63	87.1	18.8	4.9	9.2	33.63820151
N,N-Dimethylacetamide	33.1	92.5	16.8	11.5	10.2	32.75058933
Naphthalene	27.98	131	19	2	5.9	34.4012868
n-Butane	14.8	96.5	14.1	0	0	17.0199894
n-Butyl acetate	24.88	132.5	15.8	3.7	6.3	24.90470603
n-Butyl lactate	33.3	149	15.8	6.5	10.2	29.29906171
n-Butylamine	23.44	99	16.2	4.5	8	25.76019411
n-Butylchloride	23.18	104.9	16.4	5.5	2	24.71992292
n-Butyric acid	26.05	110	14.9	4.1	10.6	24.4860829
n-Decane	23.37	195.9	15.7	0	0	24.3120617
n-Dodecane	24.91	228.6	16	0	0	26.04178852
n-Eicosane	28.6	363.7	16.5	0	0	30.39010118
n-Heptane	19.65	147.4	15.3	0	0	21.81212388
n-Hexadecane	27.05	294.1	16.4	0	0	28.77413642
n-Hexane	17.89	131.4	14.9	0	0	20.21655758
Nitrobenzene	43.35	102.7	20.1	8.6	4.1	38.46924808
Nitroethane	32.13	71.5	16	15.5	4.5	29.84651682
Nitromethane	36.53	54.3	15.8	18.8	5.1	31.74162408
N-Methyl-2-pyrrolidone	41	96.5	18	12.3	7.2	35.35979138
n-Nonane	22.38	179.7	15.8	0	0	24.20133853
n-Octane	21.14	163.5	15.6	0	0	23.15092486
n-Octanoic acid	29.2	159.5	15.1	3.3	8.2	24.82524534
Nonyl phenol	30	232.2	16.6	4.1	9.2	32.65697358
n-Pentane	15.49	116.2	14.5	0	0	18.68068629
n-Propyl chloride	21.3	88.8	16	7.8	2	23.94747396
n-Propylamine	21.75	83	16.9	4.9	8.6	27.29204557
o-Dichlorobenzene	26.84	112.8	19.2	6.3	3.3	34.51783461
Oleic acid	32.5	320	16	2.8	6.2	30.06103128
Oleyl alcohol	31.7	316	16.5	2.6	8	32.91367895
o-Methoxyphenol	38.6	109.5	18	8.2	13.3	37.8428395
o-Xylene	29.76	121.2	17.8	1	3.1	28.80609708
p-Diethylbenzene	29	156.9	18	0	0.6	30.58416708
Perfluorodimethylcyclohexane	13.6	217.4	12.5	0	0	15.73575305
Perfluoromethylcyclohexane	15.7	196	12.5	0	0	15.4129879
Perfluoro-n-heptane	12.9	227.3	12.1	0	0	14.87668698
Phenol	40.9	87.5	18	5.9	14.9	36.64978123
Propionitrile	26.75	70.9	15.3	14.3	5.5	27.12387175
Propylene carbonate	40.7	85	20.1	18	4.1	46.18858851
Propylene glycol	36.51	73.6	16.8	9.4	23.3	45.32627243
Pyridine	36.56	80.9	19	8.8	5.9	33.90024994
Quinoline	42.59	118	19.8	5.6	5.7	37.43425325
Stearic acid	28.6	336.5	16.4	3.3	5.5	31.54157426
Styrene	32.3	115.6	18.6	1	4.1	31.39958617

Liquids, surface tension, molecular volume and Hansen Solubility Parameters (Continued)

Name	ST	MVol	SP-D	SP-P	SP-H	Calculated HSP
Tetrachloroethylene	32.86	102.8	18.3	5.7	0	30.26932963
Tetrahydrofuran	26.4	81.7	16.8	5.7	8	26.87323903
Tetrahydronaphthalene	33.17	136.7	19.6	2	2.9	35.75920041
Toluene	27.93	106.8	18	1.4	2	28.5339916
trans-Decahydronaphthalene	29.9	159.9	18.8	0	0	33.47348951
Trichloroethylene	29.5	90.2	18	3.1	5.3	28.76111083
Trichlorofluoromethane 5041	18	92.8	15.3	2	0	20.03312616
Tridecyl alcohol'	30.8	242	16.2	3.1	9	31.08927727
Triethyl phosphate	30.22	171	16.8	11.5	9.2	36.21686575
Triethylene glycol	45.2	134.2	16	12.5	18.6	43.5403727
Trimethyl phosphate	37.8	115.8	16.7	15.9	10.2	38.65020249
Tri-n-butyl phosphate	29	274	16.3	6.3	4.3	30.71458565

Steven Abbott, University of Leeds, U K). Any parameter in the equation can be calculated using the information in this table. References: Koenhen DM, Smolders CA (1975) The determination of solubility parameters of solvents and polymers by means of correlations with other physical quantities. J Appl Polym Sci 19:1163–1179; Abbott S, Hansen CM (2008) Hansen solubility parameters in practice, complete with software, data, and examples, 1st edn.(2008), 2nd edn. (2009), 3rd edn. (2010). ISBN: 9780955122026.

Surface Treating *n* Any method of treating a plastic surface to render it more receptive to adhesives, paints, inks, lacquers, or to other surfaces in laminating processes. The two methods in widest use are ▶ Corona-Discharge Treating and ▶ Flame Treating. See also ▶ Casing, ▶ Ion Plating, and ▶ Plasma Etching.

Surfacing Mat *n* A very thin mat, usually 0.2–0.5 mm thick, of highly filamentized glass fiber used primarily to produce a smooth, strong surface on a reinforced-plastic laminate.

Surfactant \(ˌ)sər-ˈfak-tənt\ *n* [*surf*ace-*act*ive + -*ant*] (1950) Contracted from surface-active agents, these are additives which reduce surface tension and may form micelles and thereby improve wetting (wetting agents); help disperse pigments (dispersants); inhibit foam (defoamers); or emulsify (emulsifiers). conventionally, they are classified as to their charge; anionic (negative); cationic (positive); nonionic (no charge) or amphoteric (both positive or negative).

Surging *n* Any irregularity in the output rate of an extruder. Extrusion engineers have recognized two kinds: long- and short-period surging. Long-period surging has been traced to diurnal changes in electrical supply an environmental conditions, while short-period surging, which has sometimes been so serious as to interrupt flow, is attributed to many causes such as bridging of pellets in the feed throat or compression section of the screw, too abrupt transition between deep feed flights and shallow metering flights, inadequate melting capability for the rate attempted, improper profile of barrel temperatures, voltage surges because of large loads coming on or going off the line, and changing character of feedstock loaded from bags into the hopper. In tracking down and correcting the causes of surging, it is imperative for the extruder to be equipped with instruments that sense melt temperature and pressure at the head, screw speed, extrusion rate, and either screw torque or motor kilowattage, with recorders or computer data acquisition to record these quantities.

Surimono *n* (Japanese) A print, generally of small size and on thick soft paper, intended as a festival greeting or memento of some social occasion.

Surlyn® {*tradename*} Trade name for an ▶ Ionomer.

Surlyn A *n* Ionomer (copolymer from ethylene + some acrylic acid or maleic anhydride). Manufactured by DuPont, US.

SUS *n* Abbreviation for Second, Saybolt Universal. Viscosity is expressed in SUS, as determined by the Standard Method of Test for Saybolt Viscosity, ASTM D 88, and may be determined by use of the SUS conversion tables specified in ASTM Method D 2161, following determination of viscosity in accordance with the procedures specified in the Standard Method of Test Viscosity of Transparent and Opaque Liquids.

Susceptibility *n* (**Magnetic**) Measured by the ratio of the intensity of magnetization produced in a substance to the magnetizing force or intensity of field to which it is subjected. The susceptibility of a substance will be unity when unit intensity of magnetization is produced by a field of one gauss. Dimension $[e^{-1}l^{-2}t^2]$, $[\mu]$.

Suspending Agent *n* A material used in a paint to improve its resistance to the settling of pigments. See ▶ Anti-Settling Agent.

Suspension *n* A fluid medium with fine particles of any solid more or less stably dispersed therein. The particles are called the *disperse phase*, and the suspending medium is called the *continuous phase*. When the particles do not settle out and are small enough to pass through ordinary filters, the suspension is called a *colloid* or *colloidal suspension*. Solid particles suspended in air or other gas, particularly solid particles arising from burning, are called a *smoke*. In the plastics field, a suspension is essentially synonymous with Dispersion. See also ▶ Colloid and ▶ Emulsion.

Suspension Polymerization *n* (pearl, bead, or granular polymerization) A polymerization process in which the monomer, or mixture of monomers, is dispersed by mechanical agitation in a second liquid phase, usually water, in which both monomer and polymer are essentially insoluble. The monomer droplets are polymerized while maintained in dispersion by vigorous agitation. Polymerization initiators and catalysts used in the process are generally soluble in the monomer. According to the type of monomer, emulsifier, protective colloid, and other modifiers used, the resulting polymer may be in the form of pearls, beads, soft spheres, or irregular fine granules that are easily separated from the suspending medium when agitation is stopped. Suspension polymerization is used primarily for PVC, polyvinyl acetate, polymethyl methacrylate, polytetrafluoroethylene, and polystyrene.

Sustained-Pressure Test *n* In testing pipe or tubing, subjecting the specimen to constant internal pressure over a long time, up to 1000 h or more, during and after which changes in diameter, wall thickness, and appearance are noted. It is a special-purpose creep test.

Sward-Zeidler Rocker *n* See ▶ Rocker Hardness Tester.

Swarf \ˈswórf\ *n* [of Scan origin; akin to ON *svarf* file dust; akin to OE *sweorfan* to file away] (1587) The work piece cuttings, dust, oil, grain particles, etc., created by abrading action of sandpaper, etc.

Swatch \ˈswäch\ *n* [origin unknown] (1647) A piece of fabric used as a representative sample of any fabric.

Sweating *n* (1) Exudation of oily matter from a film of paint, varnish, or lacquer after the film has apparently dried. (2) On a paint or varnish film, the development of gloss on a dull or matte finish; caused by rubbing the film. (3) Development of gloss in a dry film of paint or varnish after it has been flatted down (sanding). (4) Often incorrectly used to describe condensation of moisture from humid atmospheres on relatively cold surfaces, e.g., walls.

Swedish Oil \ˈswē-dish-\ See ▶ Tall Oil.
Swedish Olein See ▶ Tall Oil.
Swedish Pine Oil See ▶ Tall Oil.
Swedish Resin See ▶ Tall Oil.
Swedish Rosin See ▶ Tall Oil.
Swedish Rosin Oil See ▶ Tall Oil.
Sweep Blast *n* See Nace No. 4 Brush-Off Blast Cleaned Surface Finish.

Swelling *n* (1) Volume expansion of a material specimen or manufactured article due to a rise in temperature (see ▶ Coefficient of Thermal Expansion) or absorption of water or other liquid. It is usually expressed as a percentage of the original volume or, sometimes, as percentage change in lineal dimensions. (2) ▶ Extrudate Swelling.

Swirl \ˈswər(-ə)l\ *n* (15c) A term applied to visual and tactile surface roughness sometimes observed in the structural-foam-molding process. It results from jetting at the gate of the mold, causing surface wrinkling as the polymer melt flows along the wall of the mold. The condition can be alleviated by measures such as reducing the filling speed and raising the temperature of melt and/or mold.

Syenite \ˈsī-ə-ˌnīt\ *n* [L *Syenites* (*lapis*) stone of Syene, fr. *Syene*, ancient city in Egyptian] (ca. 1796) An igneous rock composed primarily of alkali feldspar together with other minerals, such as hornblende.

Sylvic Acid See ▶ Abietic Acid.
Sylvic Oil See ▶ Tall Oil.

Sym- (s-) Prefix abbreviation for Symmetrical, referring to the relative location of atoms, chain groups, or substituents groups in a cyclic compound. Two examples are *sym*-trioxane, 1,3,55-$(CH_2O)_3$, and 1,3,5-trinitrobenzene (the corresponding 1,2,4-isomer in each case being unsymmetrical). The prefix is usually ignored in alphabetizing lists of compound names.

Symbiont \ˈsim-bē-ˌänt\ *n* [prob. fr. Gr, mod. of Gk *symbiount-*, *symbiōn*, pp of *symbioun*] (1887) An organism which lives in a state of symbiosis.

Symbiosis A living together of dissimilar organisms.
Symbiotic Living in that relation called symbiosis.

Sympathetic Ink *n* Inks that give markings which become invisible and can be made visible by the use of a developing solution.

Syndet \sin-'de\ *n* A contraction for the term "synthetic detergent" used loosely to signify synthetic detergents or compositions containing synthetic detergents. See ▶ Detergent.

Syndiotactic *n* Derived from the Greek words *syndio*, meaning "every other" and *tattoo*, meaning "to put in order." The term refers to polymers having alternating, different substituent groups along the chain capable of exhibiting mirror-image symmetry (see ▶ Enantiomer). The polymer structure has such groups attached to the backbone chain in an order a-b-a-b- on one side of the chain and b-a-b-a- on the other. See also ▶ Isotactic and ▶ Tacticity.

Syndiotactic Polymer *n* A tactic polymer in which the conventional base unit possesses, as a component of the main chain, a carbon atom with two different lateral substituents, these substituents being so arranged that a hypothetical observer advancing along the bonds constituting the main chain finds opposite steric configurations around these chain atoms in successive conventional base units. NOTE - The true base unit is thus twice the size of the conventional base unit. See ▶ Syndiotactic, ▶ Tactic Polymer, ▶ Atactic Polymer, ▶ Isotactic Polymer.

Syneresis \sə-'ner-ə-səs\ *n* [LL *synaeresis*, fr. Gk *synairesis*, fr. *synairein* to contact, fr. *syn-* + *harein* to take] (ca. 1577) The separation of liquid from a gel. The spontaneous exudation or squeezing out of solvent or diluents as a separate phase which may occur when a gel stands undisturbed; the gel structure slowly densifies but there is no net volume change in the system.

Synergism *n* [NL *synergismus*, fr. Gk *synergos*] (1910) A phenomenon wherein the effect of a combination of two ingredients (or other experimental factors) is greater than the sum of their individual effects, or, in the case of polymer blends or composites, then would be estimated by a simple ▶ Law of Mixtures. For example, some stabilizers and some fire retardants for plastics have a mutually reinforcing effect and are thus termed synergistic. Similarly, some plastic alloys have higher strength than either of the neat resins.

Synergistic *adj* (ca. 1847) Relating to the cooperative action of two or more discrete agencies such that their combined effect is different than the sum of the effects due to the individual agencies. See ▶ Synergism.

Synergy *n* (1660) See ▶ Synergism.

Syntactic Foam \sin-'tak-tik\ *n* (cellular mortar) A term applied to composites of tiny, hollow spheres in a resin or plastic matrix. The spheres are usually of glass, although phenolic microspheres were used in the early years of the art. The resin most used is epoxy, followed by polyesters, phenolics, and PVC. Syntactic foams of the most common type – glass microspheres in a binder of high-strength thermosetting resin – are made by mixing the spheres with the fluid resin, its curing agent, and other additives, to form a fluid mass that can be cast into molds, troweled onto a surface or incorporated into laminates. After forming, the mass is cured by heating. These foams are characterized by densities lower than those of the matrix resins, ranging from 0.57 to 0.67 g/cm^3, and very high compressive strengths. Their first applications were for deep-submergence buoys capable of withstanding depths of 6,000 m. When both gas bubbles *and* hollow glass spheres are used in the same mixture, the resulting composite has been called a *diafoam*.

Synthetic \sin-'the-tik\ *adj* [Gk *synthetikos* of composition, component, fr. *synthithenai* to put together] (1697) Substances resulting from synthesis rather than occurring naturally.

Synthetic Detergent *n* A detergent produced by chemical synthesis and comprising an organic composition other than soap. Often contracted to syndet.

Synthetic Fiber *n* (man-made fiber) A class name for various fibers (including filaments), distinguished from natural fibers such as wool and cotton, produced from fiber-forming substances which may be: (1) modified or transformed natural polymer, e.g., alginic and cellulose-based fibers such as acetates and rayon's; (2) polymers synthesized from chemical compounds, e.g., acrylic, nylon, polyester, polyurethane, polyethylene, polyvinyl, and carbon/graphite fibers; or (3) fibers of mineral origin, e.g., glass, quartz, boron, and alumina.

Synthetic Magnetite See ▶ Black Iron Oxide.

Synthetic Paint *n* A vague term which sometimes means paints containing synthetic resins rather than naturally occurring oils or gums in the vehicle. The use of this term is deprecated.

Synthetic "Papers" *n* Non-cellulosic "papers" are film-based. The main plastics being used or studied for this purpose are polystyrene, polypropylene, high-density polyethylene, and polyvinyl chloride.

Synthetic Pine Oil See ▶ Pine Oil.

Synthetic Resin *n* (1907) Complex, substantially amorphous organic semisolid or solid material built up by chemical reaction of simple molecules. See ▶ Resin, ▶ Synthetic.

Synthetic Rubber *n* An elastomer manufactured by a chemical process, as distinguished from natural rubber obtained from trees; rubber like with respect to its degree of elasticity.

Synthetic Yellow Ocher *n* Synthetic yellow iron oxide reduced with aluminum silicate and other extender pigments. See ▶ Iron Oxides, ▶ Synthetic.

Syrian Asphaltum \ˈsir-ē-ə as-ˈfól-təm, *esp British* -ˈfal-\ *n* Natural asphaltum mined near Damascus. It is not readily soluble in petroleum hydrocarbons, and its value to the paint trade is limited. *Known also as Egyptian Asphaltum.*

System \xsis-tem\ *n* [LL *systemat-*, *systema*, fr. Gk *systēmat-*, *systēma*, fr. synistanai to combine, fr. *syn-* + *histanai* to cause to stand] (1603) A portion of the universe under observation or consideration.

T

t \ˈtē\ *n* {*often capitalized, often attributive*} (before 12c) Symbol for elapsed time, for thickness, or for Student' t.

T (1) Symbol for torque. (2) SI abbreviation for Tera-.

Taber Abraser *n* An instrument used to measure abrasion resistance. Specimen on a turntable rotates under a pair of weighted abrading wheels that produce abrasion through side slip. This is a product of Taber Industries, Inc., www.taberindustries.com (Paint and coating testing manual (Gardner-Sward handbook) MNL 17, 14th edn. ASTM, Conshohocken, 1995). See also ▶ Wear Cycles.

Tab Gate *n* In injection molding, a small removable tab of approximately the same thickness as the molded item usually located perpendicularly to the item. The tab is used as a site for edge-gate location, typically on items with large flat areas.

Tablet \ˈta-blət\ *n* [ME *tablett*, fr. MF *tablete*, dim. of *table* table] (14c) A shape having opposite faces at least 4X but less than 1X greater than the other faces.

Tablet Test *n* See ▶ Flammability Tests, ▶ Methenamine Pill Test.

Tab Out *n* A spot of ink applied to paper by a finger, using tapping action to distribute the ink to approximately printing film thickness.

TAC Abbreviation for ▶ Triallyl Cyanurate.

Tacciometer *n* A device for measuring the surface stickiness or tackiness of a dried paint or varnish film. It operates on the principle that a suitably weighted piece of paper will or will not adhere to the coated surface. The instrument consists of three steel balls which are placed on the paper, and these can be loaded further as required until on removing the load and the balls, the paper shows signs of adhering. The time during which the loaded paper is left in contact with the surface under test is usually specified, because the longer the time 0 contact, the greater the tendency for the paper to adhere.

Tack \ˈtak\ *n* [ME *tak* something that attaches; akin to MD sharp point] (1574) (1) The pull-resistance exerted by a material adhering completely to two separating surfaces. In liquids, tack is a function of viscosity; in nonelastic plastic materials, tack (*tackiness*) is a function of plastic viscosity (sometimes *viscoelasticity*) and yield value. Cohesion becomes negligible since rupture occurs at very small areas. (2) Slight stickiness of the surface of a film of paint, varnish, or lacquer, apparent when the film is pressed with the finger. (3) A relative measurement of the cohesion of an ink film which is responsible for its resistance to splitting between two rapidly separating surfaces. (4) The natural adhesiveness of rubber in the raw state. Also, the property of raw or compounded rubber which causes layers of stock to cohere. It is a desirable property only when adhesion or cohesion is desired (Skeist I (ed) (1990) Handbook of adhesives. Van Nostrand Reinhold, New York). Also see ▶ ASTM (www.astm.org) for current standard method of testing.

Tack Coat *n* An application of bituminous material to an existing, relatively absorbent surface, to prevent slippage planes and to provide bond between the existing surface and the new surfacing.

Tack, Dry *n* The property of certain adhesives, particularly nonvulcanizing rubber adhesives, to adhere on contact to themselves at a stage in the evaporation of volatile constituents, even though they seem dry to the touch. Sometimes called *Aggressive Tack*.

Tack-Free *adj* Freedom from tack of a coating after suitable drying time. In some cases, coatings are tack-free after application; tack doesn't develop until a little later. See ▶ Tack and ▶ Drying Time.

Tackifier *n* A substance (e.g., rosin ester) that is added to synthetic resins or elastomeric adhesives to improve the initial tack and extend the tack range of the deposited adhesive film.

Tackiness *n* The property of being sticky or adhesive.

Tack Rag *n* Fabric impregnated with a tacky substance such as a delayed drying varnish which is used to remove dust from a surface after rubbing down and prior to further painting. Tack rag should be stored in an airtight container to conserve its tackiness.

Tack Range *n* Period of time in which an adhesive will remain in the tacky-dry condition after application to an adhered, under specified conditions of temperature and humidity.

Tacky *v* (1) That stage in the drying of a paint at which the film appears sticky when lightly touched with the finger. (2) Having a tack. Also called *Tacky Dry*.

Tacky-Dry *adj* Pertaining to the condition of an adhesive when the volatile constituents have evaporated or been absorbed sufficiently to leave it in a desired tacky state.

Tactic Block *n* In a polymer, a regular block that can be described by only one species of configurational repeating unit in a single sequential arrangement.

Tactic Block Polymer *n* A polymer whose molecules consist of Tactic Blocks connected linearly.

Tacticity \ˈtak-ti-ˌci-tē\ *n* (1) The orderliness of the succession of configurational repeating units in the main chain of a polymer molecule. (2) Any type of regular or symmetrical molecular arrangement in a polymer structure, as opposed to random positioning of substituents groups along a polymer backbone. (3) The stereospecific configuration of a polymer containing repeating asymmetric centers. An *isotactic* polymer has the same configuration (orientation in space) at each asymmetric center while a *syndiotactic* polymer has alternating configurations at sequential asymmetric centers. An *atactic* polymer has no repeating stereoregularity of asymmetric centers. See also ▶ Stereospecific (Odian GC (2004) Principles of polymerization. Wiley, New York).

Tactic Polymer *n* A polymer in which there is an ordered structure with respect to the configurations around at least one main-chain site of steric isomerism per conventional base unit. NOTE – The carbon (or other) atom in the chain at the site of the steric isomerism need not in a strict sense by an asymmetric atom, since in a chain of infinite length the two chain portions may be considered as equivalent; however, for the purpose of this definition, such atoms are referred to as asymmetric atoms. See ▶ Syndiotactic Polymer, ▶ Isotactic Polymer, and ▶ Atactic Polymer.

Taffeta \ˈta-fə-tə\ *n* [ME, fr. MF *taffetas*, fr. OIt *taffetà*, fr. Turkish *tafta*, fr. Persian *tāftah* woven] (14c) A plain-weave fabric with a fine, smooth, crisp hand and usually a lustrous appearance. Taffeta fabric usually has a fine cross rib made by using a heavier filling yarn than warp yarn. Taffetas are produced in solid colors, yarn-dyed plaids and stripes, and prints. Changeable and moiré effects are often employed. Although originally made of silk, manufactured fibers are now often used in the production of taffeta.

Tag Closed Tester *n* An instrument for determining the flashpoint of a liquid which has a viscosity of less than 45 SUS at 47.8°C (100°F), does not contain suspended solids and does not have a tendency to form a surface film while under test. Reference Standard Method of Test ASTM D 56 for Flashpoint by Tag Closed Tester.

Tail *n* (1) Highest boiling solvent fraction. (2) Elongated, somewhat pointed extension of the lower portion of the rising bubble in a bubble tube viscometer, characteristic of a varnish or resin solution that is near or approaching gelation or which has a peculiar rheological characteristic.

Tailings *n* The remains, residues, or final products produced on refining any substance. Syn: ▶ Bottoms, ▶ Foots.

Tails *n* Finger-like spray pattern.

Tak Dyeing See ▶ Kusters Dyeing Range.

Take-Off *n* The mechanism for drawing extruded or calendered material away from the extruder or calender. The most common form of extrusion take-off is a pair of endless caterpillar belts with resilient grip pads conforming to the section being extruded, driven at a speed synchronized with that of the extrudate.

Take-Up *n* (*Twist*) The change in length of a filament, yarn, or cord caused by twisting, expressed as a percentage of the original (untwisted) length.

Take-Up *n* (*Yarn-in-Fabric*) The difference in distance between two points in a yarn as it lies in a fabric and the same two points after the yarn has been removed from the fabric and straightened under specified tension, expressed as a percentage of the straightened length. In this sense, take-up is contrasted to the crimp of a yarn in a fabric, which is expressed as a percentage of the distance between the two points in the yarn as it lies in the fabric. Take-up is generally used in connection with greige fabric.

Talc *n* \talk\ *n* [MF *talc* mica, fr. ML *talk*, fr. Arabic *ṭalq*] (1610) (steatite, talcum) $Mg_3SI_4O_{10}(OH)_2$. A natural hydrous magnesium silicate, sometimes used as a filler.

Talloel See ▶ Tall Oil.

Tall Oil \ˈtäl, ˈói(ə)l\ *n* [part trans. of Gr *Tallöl*, part trans. Swedish *tallolja*, fr. tall pine + *olja* oil]. A generic name for a number of products from the manufacture of wood pulp by the alkali process sulfate or kraft process. To provide some distinction between the various products, designations are often applied in accordance with the process or composition, some of which are crude tall oil, acid refined tall oil, distilled tall oil, tall oil fatty acids and tall oil rosin. The following designations for tall oil shall be considered obsolete: crude resinous liquid, finn oil, liquid resin, liquid rosin, resin oil, sulfate pitch, sulfate resin, sulfate rosin, Swedish pine oil, Swedish resin, Swedish rosin, Swedish rosin oil, Sylvic oil, talloel, tallol, Swedish oil, fluid resin, Swedish olein.

Tall Oil, Acid Refined *n* Product obtained by treating crude tall oil in solvent solution with sulfuric acid under controlled conditions to remove dark color bodies and odoriferous materials. Removal of the solvent yields a product with lighter color and higher viscosity than crude tall oil with approximately the same fatty acids-to-rosin ratio.

Tall Oil, Crude *n* Dark brown mixture of fatty acids, rosin, and neutral materials liberated by the acidification of soap skimmings. The fatty acids are a mixture of oleic acid and linoleic acid with lesser amounts of saturated and other unsaturated fatty acids. The rosin is composed of resin acids similar to those found in gum and wood rosin. The neutral materials are composed mostly of polycyclic hydrocarbons, sterols and other high-molecular weight alcohols.

Tall Oil, Distilled *n* Class of products obtained by distilling crude tall oil in fractionating equipment under reduced pressure under such conditions that the ratio of rosin acids to fatty acids is varied over a wide range. The products which generally contain less than 90% of fatty acids are known as distilled tall oils. The fatty acids are a mixture of oleic and linoleic acids with lesser amounts of saturated and other unsaturated fatty acids. The remainder consists of rosin and neutral materials.

Tall Oil Fatty Acids *n* Class of products generally containing 90% or more fatty acid obtained by fractionization of crude tall oil; The fatty acids are a mixture of oleic and linoleic acids with lesser amounts of saturated and other unsaturated fatty acids. The remainder consists of rosin and neutral materials (Paint: pigment, drying oils, polymers, resins, naval stores, cellulosics esters, and ink vehicles, vol 3. American Society for Testing and Material, 2001).

Tall Oil Heads (Light Ends) *n* Low-boiling fractions obtained by the fractional distillation of crude tall oil under reduced pressure. The composition of these products varies over a wide range but contains palmitic, oleic, linoleic, and stearic acids together with lesser amounts of other saturated and unsaturated fatty acids. The neutral materials content is normally high. (Paint: pigment, drying oils, polymers, resins, naval stores, cellulosics esters, and ink vehicles, vol 3. American Society for Testing and Material, 2001).

Tall Oil Pitch *n* Undistilled residue from the distillation of crude tall oil. It is generally recognized that tall oil pitches contain some high-boiling esters and neutral materials with lesser amounts of rosin and fatty acids (Usmani AM (ed) (1997) Asphalt science and technology. Marcel Dekker, New York).

Tall Oil Rosin *n* Rosin remaining after the removal of substantially all of the fatty acids from tall oil by fractional distillation or other suitable means. Such rosin shall have the characteristic form, appearance and other physical and chemical properties normal for other kinds of rosin. The fatty acid content shall not exceed 5%. See ▶ Rosin.

Tall Oil Soap *n* Product formed by the saponification or neutralization of tall oil with organic or inorganic cases.

Tallol See ▶ Tall Oil.

Tan \\tan\ *n* [F, tanbark, fr. OF, fr. ML *tanum*] (1674) (1) A light or moderate yellowish brown to brownish orange. (2) Japanese. A brick-red or orange color, consisting of red oxide of lead.

Tan Delta *n* Mathematically expressed as the loss modulus divided by the storage modulus, the tangent of the phase angle between an applied stress and the strain response in a dynamic experiment. NOTE – Tan delta versus temperature curves are commonly reported in dynamic mechanical analysis (DMA) tests.

Tan of Dielectric Loss Angle *n* In a ideal condenser of geometric capacitance C_o, in which the polarization is instantaneous, the charging current $E\omega\varepsilon' C_o$ is 90° out of phase with the alternating potential. In a condenser in which absorptive polarization occurs, the current also has component $E\omega\varepsilon'' C_o$ in phase with the potential and determined by Ohm's law. This ohmic or loss current, which measures the absorption, is due to the dissipation of part of the energy of the field as heat. In vector notation, the total current is the sum of the charging current and the loss current. The angle δ between the vector for the amplitude of the total current and that for the amplitude of the charging current is the loss angle, and the tangent of this angle is the loss tangent of dielectric loss angle:

$$\tan\delta = \frac{\text{loss current}}{\text{charging current}} = \frac{\varepsilon''}{\varepsilon'}$$

where ε' is the measured dielectric constant of the dielectric material in the condenser and ε'' is the imaginary part of the dielectric constant, commonly known as the loss factor or loss index (Ku CC, Liepins R (1987) Electrical properties of polymers. Hanser Publishers, New York).

Tandem Extruder See ▶ Extruder.

Tandem Line *n* In the coil coating field, a roller coat line with two coaters capable of applying and baking two coats (i.e., primer and enamel) prior to recoiling.

Tangent Modulus *n* The slope of the curve at any point on a static stress-strain graph ($d\sigma/d\varepsilon$) expressed in pascals per unit of strain. This slope is the tangent modulus in whatever mode of stress the curve has arisen from – tension compression, or shear. [Since strain is dimensionless, the unit given for modulus is normally just stress (Pa).] (Sepe MP (1998) Dynamic mechanical analysis. Plastics Design Library, Norwich).

Tanglelaced Fabric See ▶ Spunlaced Fabric.

Tank Coating *n* Paint used for the inside of tankers.

Tank White *n* Good hiding, self-cleaning white paint for exterior metal surfaces.

Tannin \\ˈta-nən\\ *n* [F, fr. *tanner* to tan] (1802) Organic acid obtained in the form of brownish white scales from gall nuts, sumac, and other plants used in dyeing, tanning, etc.

Tan-ye *n* (Japanese) A print in which tan is the only or chief color used. Such prints, in which the tan was applied by hand, were among the earliest productions.

Tape *n* (1) A narrow, woven fabric not over 8 in. in width. (2) In slide fasteners, a strip of material, along one edge of which the bead and scoops are attached, the bead sometimes being integral with the strip. Also see ▶ Slit Tape and ▶ Nonelastic Woven Tape.

Taper *n* (1) In a ▶ Conical Transition section of an extruder screw, the vertex angle of the axial cone defined by the increasing root diameter of the screw. Compare with ▶ Helical Transition. (2) An often used Syn: ▶ Draft in molds.

Tapered Pattern *n* Elliptical-shaped spray pattern; a spray pattern with converging lines.

Taper Pin *n* A slightly conical, hard steel dowel driven into matting holes drilled into the contact faces of adjacent major machines components, after the components have been aligned, to preserve alignment. At least two pins are normally used at each such interface.

Tape Test *n* A type of adhesion test consisting of the application of an adhesive tape to a dried coating and rapidly removing the tape with a swift, jerking motion. The coating can be either scribed or unscribed, depending on the specification. The "wet adhesion" test for latex paints is performed by first wetting the paint with a specified quantity of water for the specified time and blotting off the excess surface water. The tape test is made immediately after blotting.

Tape Yarn See ▶ Slit-Film Yarn.

Tap-Out *n* Spot of ink applied to paper by a finger, using tapping action to distribute the ink to approximately printing film thickness. With experience, one can determine whether ink has the proper tack and working properties, and also can use the tap-out in color matching. also called *Pat-out*.

Tar \\ˈtär\\ *n* [ME *terr, tarr*, fr. OE *teoru*; akin to OE *trēow* tree] (before 12c) Brown or black bituminous material, liquid or semisolid in consistency, in which the predominating constituents are bitumens obtained as condensates in the destructive distillation of coal, petroleum, oilshale, wood, or other organic materials, and which yields substantial quantities of pitch when distilled.

Tar Acid *n* Phenol or its homologues either individually or blended together (Usmani AM (ed) (1997) Asphalt science and technology. Marcel Dekker, New York).

Tar and Gravel Roofing See ▶ Built-up Roofing.

Tare \\ˈtar, ˈter\\ *n* [ME, fr. MF, fr. OIt *tara*, fr. Arabic *ṭarha*, literally, that which is removed] (15c) The weight of all external and internal packing material (including bobbins, tubes, etc.) of a case, bale, or other type of container.

Tarpaulin \\tär-ˈpó-lən\\ *n* [prob. fr. ¹*tar* + -*palling*, -*pauling* (fr. *pall*] (1605) A water-resistant fabric used to protect loads or materials from the elements. Tarpaulin may be a coated fabric, a fabric with waterproof finish, or a fabric that is tightly constructed to prevent water penetration.

Tar Spirit *n* A powerful solvent of varying composition and properties, obtained by distillation of wood tar. Boiling range is extended and has been known to vary from about 70–80°C to practically 260°C. Tar spirits have sharp penetrating odors, dark colors, and a pronounced tendency to attack mild steel containers.

Tartaric Acid \\(ˌ)tär-ˈtar-ik-\\ *n* (1810) COOHCH(OH)·CH(OH)COOH. Dihydroxydicarboxylic acid used in the preparation of plasticizers. Melting point, 169°C.

Taslin® Process *n* See ▶ Texturing.

T-Bend Flexibility Test *n* Simple method for determining the flexibility of coatings by bending a coated metal test strip over itself. A panel is bent and pressed flat by means of a jig to achieve a 180° bend. Subsequent folds are equivalent to bending the panel around a rod of diameter equal to the thickness of the panel.

TBEP *n* Abbreviation for ▶ Tributoxyethyl Phosphate.

TBT *n* Abbreviation for ▶ Tetrabutyl Titanate.

TBTF *n* Abbreviation for Tributyltin Fluoride.

TBTO *n* Abbreviation for Tributyltin Oxide.

t-**Butyl Peroxy Neodecanoate** *n* A polymerization initiator for vinyl chloride.

***t*-Butyl Peroxy Pentanoate** *n* A peroxyester catalyst.

***t*-Butyl Perphthalic Acid** *n* $(CH_3)_3C \cdot O_2 \cdot COC_6H_4COOH$. A polymerization catalyst.

TCE *n* Abbreviation for ▶ Trichloroethylene.

TCEF *n* Abbreviation for ▶ Trichloroethyl Phosphate, a plasticizer.

TCP *n* Abbreviation for ▶ Tricresyl Phosphate.

TDI *n* Abbreviation for Toluene Diisocyanate, an 80–20 mixture of the 2.4- and 2.6- isomers. See also ▶ Toluene-2,4-Diisocyanate and ▶ Diisocyanate.

T-Die *n* A center-fed, slot die for extrusion of film whose horizontal cross section, together with the die adapter, resembles the letter T.

Tear See ▶ Run.

Teardrop Die See ▶ Manifold.

Tear Resistance, Strength *n* Resistance to tear shearing of a films or fabric material.

Tear Strength *n* The force or stress required to start or continue a tear in a fabric or plastic film. See ▶ Elmendorf Tear Strength.

Tea Seed Oil *n* Nondrying oil. Iodine value, 90.

Teasel Burr *n* See ▶ Napping.

Tedlar *n* Poly(vinyl fluoride), manufactured by DuPont.

Teel Oil See ▶ Sesame Oil.

Teeth *n* The resultant surface irregularities or projections formed by the breaking of filaments or strings which may form when adhesive-bonded substrates are separated.

Teflon® \ˈte-ˌflän\ *n* (1) A polymer of fluorinated ethylene. Very inert, and in the form of a film or an impregnator, used for its heat-resistant and non-sticking properties. (2) Trade name for fluorocarbon resins, including polytetrafluoroethylene, perfluoropropylene resin and copolymers, manufactured by DuPont.

Teflon FEP *n* Copolymer from tetrafluoroethylene and hexafluoropropylene, manufactured by DuPont.

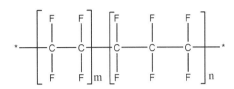

Tego *n* Phenoplast, manufactured by Resinous Products, U.S.

Tekaol *n* Linseed stand oil from which the low-polymerized (saturated) highly dispersed phase has been removed.

Telechelic Polymers *n* A polymer having purposely introduced end groups of a particular chemical type, e.g., acetal homopolymer that has been "end-capped" by treatment with acetic anhydride.

Telegraphing *n* A condition in a laminate or other type of composite construction in which irregularities, imperfections, or patterns of an inner layer are visibly transmitted to the surface. NOTE – Telegraphing is occasionally referred to as photographing. See ▶ Photographing.

Telescopic Flow *n* A picturesque name for laminar flow in a circular tube, derived from visualizing successively smaller cylindrical shells of liquid, from the tube wall toward the center, each moving faster than the next outer one, sliding like the tubes of a sectional telescope. See ▶ Laminar Flow and ▶ Poiseuille Flow.

Telomer \ˈte-lə-ˌmir\ *n* [ISV] (1940) (1) Low molecular weight polymer in which the terminal group on the end of the chain-like molecule is not the same as the side group. (2) A polymer composed of molecules having terminal groups incapable of reacting with additional monomers, under the conditions of the synthesis to form larger polymer molecules of the same chemical type. (3) An ▶ Oligomer formed by addition polymerization in the presence of excess chain-transfer agent (polymerization stopper) whose free radicals become the end groups of the telomer.

Tempera \ˈtem-pə-rə\ *n* [It *tempera*, literally, temper, fr. *temperare* to temper, fr. L] (1832) (1) A rapidly drying paint consisting of egg white (or egg yolk, or a mixture of egg white and yolk), gum, pigment, and water; especially used in painting murals. (2) A method of painting using tempera.

Temperature *n* (1) The measured relative Celsius and Fahrenheit temperature scales and the absolute temperature, Kelvin; an indication of the degree hotness; average velocity of molecules; at absolute zero, 0 K, all motion in matter stops molecular velocity is zero. (2) The property of a substance which determines the direction of heat flow into or out of the substance; heat flows from a substance of higher temperature to one of lower temperature. The temperature of a substance measures the average kinetic energy of its particles. The fundamental temperature scale is now defined by means of the equation

$$\theta(X) = 273.15°K \frac{X}{X_3}$$

where θ denotes the temperature; X the thermometric property (P, V,...); the subscript 3 refers to the triple point of water; and 273.16 K is the arbitrary fixed point for the temperature associated with the triple point of water. The ideal gas temperature θ (numerically equal to the Kelvin temperature), in particular, is defined by either of the two equations:

$$273.15° \lim_{P_3 \to o} \frac{P}{P_3}, const.V$$

$$\theta = 273.15° \lim_{P_3 \to o} \frac{V}{V_3}, const.P.$$

(Lide DR (ed) (2004) CRC handbook of chemistry and physics. CRC Press, Boca Raton).

Temperature Checking *n* Tests of finishes as applied to furniture. It is a film rupture at an angle to the run of the grain. Also referred to as *Cold Checking*.

Temperature Coefficient of Refractive Index *n* The change in refractive index *(n)* with temperature. The degree of variation of n depends on the composition of the substance and the state of aggregation, e.g., whether it is a solid or a liquid. It is usually about 100 × larger for liquids than for solids and about −0.0005/°C for liquids.

Temperature, Color See ▶ Color Temperature.

Temperature, Correlated Color See ▶ Correlated Color Temperature.

Temperature, Curing *n* The temperature to which an adhesive or an assembly is subjected to cure the adhesive. (See also ▶ Temperature, Drying and ▶ Temperature, Setting) NOTE – The temperature attained by the adhesive in the process of curing it (adhesive curing temperature) may differ from the temperature of the atmosphere surrounding the assembly (assembly curing temperature).

Temperature, Drying *n* The temperature to which an adhesive on an adherend or in an assembly or the assembly itself is subjected to dry the adhesive. (See also ▶ Temperature, Curing and ▶ Temperature, Setting) NOTE – The temperature attained by the adhesive in the process of drying it may differ from the temperature of the atmosphere surrounding the assembly (assembly drying temperature).

Temperature, Maturing *n* The temperature, as a function of time and bonding condition, which produces desired characteristics in bonded components. NOTE – The term is specific for ceramic adhesives.

Temperature of Zero Birefringence *n* The temperature at which the refractive indexes of a material are equal in two perpendicular directions (longitudinally and transversely for a fiber).

Temperature Profile *n* (1) In extrusion or injection molding, the sequence of barrel temperatures from feed opening to head, sometimes presented as a plot of temperature versus longitudinal position, hence *profile*. (2) The sequence of metal temperatures across a sheet or film die, or around a large blown-film die. (3) The sequence of temperatures across the width of a slab of newly extruded or cast plastic foam, is indicated by temperature sensors placed laterally at the same depth in the foam. (4) In analysis of Nonisothermal, laminar flow of very viscous liquids (e.g., polymer melts) within tubes and dies, the sequence of temperatures from the one sidewall through the center to the opposite sidewall at any point along the axis of flow. Such profiles have also been measured experimentally with traversing thermocouples.

Temperature Resistance Coefficient *n* The ratio of the change of resistance in a wire due to a change of temperature of 1°C to its resistance at 0°C. Dimension, $[\theta^{-1}]$.

Temperature, Setting *n* The temperature to which an adhesive or an assembly is subjected to set the adhesive. (See also ▶ Temperature, Curing and ▶ Temperature, Drying) NOTE – The temperature attained by the adhesive in the process of setting it (adhesive setting temperature) may differ from the temperature of the atmosphere surrounding the assembly (assembly setting temperature).

Template Polymerization *n* The polymerization of monomers attached in some ordered fashion to a template to produce a polymer with a precisely determined structure complementary to that of the template.

Tenacity \tə-ˈna-sə-tē\ *n* (1526) A term used in yarn and textile manufacturing to denote the strength of a yarn or filament of a given size. Numerically it is the newtons of breaking force per tex of lineal density (replacing the deprecated old unit, grams per denier). In testing tenacity, the yarn is usually pulled at the rate of 0.5 cm/s. To convert g/denier to N/tex, multiply by 0.0883. Also see ▶ Breaking Tenacity.

Tenax Poly[oxy-1,4-(2,6-diphenyl)-phenylene], manufactured by AKZO, The Netherlands.

Tencel® Registered trademark of Tencel, Inc. for their brand of cellulosic staple fiber classified as lyocell. See ▶ Lyocell Fiber.

Tensile Bar *n* (dogbone specimen, dumbbell specimen) Any of several kinds of test specimens made for use in tests of tensile stress vs elongation. There are two main types, both having ends that have two to three times the cross section of the central, or *gauge-length* section, this geometry guaranteeing that, because the force is the same everywhere along the bar, the stress will be much lower in the end sections than in the gauge section. This geometry prevents failure at the grips and confines the plastic deformation, if any, to the gauge section. The first type is of rectangular cross section perpendicular to the length, and may be machined from sheet stock or molded. These are well defined by ASTM (www.astm.org), which also specifies certain geometries for specimens taken from tubing and round-rod stock. The round-rod specimens typify the second geometry, in which the transverse cross sections are circular. The flat type, but smaller, is also called for in the ▶ Tensile-Impact Test (Shah V (1998) Handbook of plastics testing technology. Wiley, New York).

Tensile Elongation *n* The increase in distance between two gage marks that result from stressing the specimen in tension to fracture. Usually elongation is expressed as a percentage of the original gage length. NOTE – elongation is affected by specimen geometry (length, width, thickness of gage section and adjacent regions) and test procedure, such as alignment and speed of pulling. See also ▶ Elongation.

Tensile Factor *n* The empirical factor $T \times E^{1/2}$ that describes the tenacity elongation exchange relationship for a large number of manufactured fiber systems.

Tensile Heat-Distortion Temperature *n* An obsolete misnomer for ▶ Deflection Temperature.

Tensile Hysteresis Curve *n* A complex load-elongation, or stress-strain curve obtained: (1) when a specimen is successively subjected to the application of a load or stress less than that causing rupture and to the removal of the load or stress according to a predetermined procedure; or (2) when a specimen is stretched less than the breaking elongation and allowed to relax by removal of the strain according to a predetermined procedure.

Tensile Impact Energy *n* Kinetic energy dissipated on break of a specimen in a tensile impact test. In the test, one end of the specimen is attached to a swinging pendulum while another is gripped in a crosshead that travels with pendulum. The specimen is ruptured by tensile stress as the crosshead strikes an anvil and is arrested.

Tensile-Impact Test *n* An impact test that uses a pendulum striker to break a dogbone-shaped test specimen, described in ASTM D 1822 and 1822M. It differs from the Izod impact test in two important aspects: (1) the specimen is not notched and (2) it is broken in simple tension rather than bending. For these reasons, it interpretation is more straightforward.

Tensile Modulus *n* Syn: ▶ Young's Modulus. See ▶ Modulus Of Elasticity.

Tensile Product *n* The product of tensile strength and elongation at break. In a Hooke's-law material (which few if any plastics are), the tensile product is twice the energy to rupture, so is related to toughness. In a material that exhibits a long, flat region of ductile elongation at constant stress during the tensile test, the tensile product is closely equal to the *nominal* energy to break.

Tensile Properties *n* Tensile properties are: Linear density, Tenacity, Knot tenacity, Loop tenacity, Breaking strength, Tensile strength, Elongation at break, Tensile module, Work to break, Yield point, Creep, and Elasticity.

Tensile Property Tests *n* Names and designations of the methods for tensile testing of materials. Also called tensile tests. See also ▶ Tensile Properties.

Tensile Recovery Curve See ▶ Tensile Hysteresis Curve.

Tensile-Shear Strength *n* A measure of the shear strength of an adhesive bond in which two members are bonded in a ▶ Lap Joint, then pulled at both ends until the joint fails in shear. The strength is reported as the tensile force divided by the shear area (Pa). A double lap joint may be specified. Many tests of tensile-shear strength are listed among the ASTM Standards (www.astm.org).

Tensile Strain *n* The relative length deformation exhibited by a specimen subjected to a tensile force. Strain may be expressed as a fraction of the nominal gauge length or as a percentage. Also see ▶ Elongation.

Tensile Strength *n* (ca. 1864) The maximum nominal stress sustained by a test specimen being pulled from both ends, at a specified temperature and at a specified rate of stretching. When the maximum nominal stress occurs at the ▶ Yield Point it shall be designated *tensile strength at yield*. When it occurs at break, it shall be designated *tensile strength at break*. The ASTM test for plastics is D 638 (metric, D 638M). The SI unit or tensile strength is the pascal (N/m^2), but trade publications in the U.S. are still clinging to the pound (force) per square inch (psi). The strengths of commercial plastics that are neither plasticized nor fiber-reinforced range from about 14–140 MPa (2–20 kpsi). See ▶ Strength.

Tensile Strength at Break *n* the maximum load per original minimum cross-sectional area of the plastic specimen in tension within the gage length when the maximum load corresponds to the break point. NOTE – For plastics – when the maximum load corresponds to the yield point, this property is called tensile strength at yield. See also ▶ Tensile Strength.

Tensile Strength at Yield *n* The maximum load per original minimum cross-sectional area of the plastic specimen in tension within the gage length, when the maximum load corresponds to the yield point. NOTE – When maximum load corresponds to the break point, this property is called tensile strength at break.

Tensile Stress *n* The resistance to deformation developed within a specimen subjected to tension by external force. The tensile stress is commonly expressed in two ways, either as (1) the tensile strength, i.e., the force per unit cross-sectional area of the unstrained specimen, or as (2) tenacity, i.e., the force per unit linear density of the unstrained specimen. The latter is more frequently used in textile testing.

Tensile Test *n* A method of measuring the resistance of a yarn or fabric to a force tending to stretch the specimen in one direction.

Tensile Ultimate Strength *n* the maximum tensile stress subjected to the test specimen during the tensile test. The value can be identical with the tensile stress at break.

Tensile Yield Point *n* The first engineering stress in a tensile test, in which stresses and strains are determined for a material that exhibits the phenomenon of discontinuous yielding, at which an increase in strain occurs without an increase in stress. For materials that do not exhibit a yield point, yield strength serves the same purpose as yield point.

Tensile Yield Strength *n* The engineering stress determined at the intersection of the tensile stress-strain curve with a line drawn in the diagram with a slope equal to the modulus of elasticity, and offset by the specified strain. The percent offset (0.2% is the most common in USA) must be stated for values to be meaningful.

Tensiometer \ten(t)-sē-▪ä-mə-tər\ *n* [*tension*] (1912) An instrument, invented by P. L. du Noüy in 1919, for measuring ▶ Surface Tension of liquids and interfacial tensions between immiscible liquids, consisting of a horizontal, platinum-wire ring suspended from the end of a slender cantilever beam whose movement is indicated by a pointer on a circular scale. To measure surface tension, one submerges the horizontal ring in the test liquid, and then carefully raises it by turning the knob and pointer until the meniscus lifted by the ring breaks. The pointer indicates the surface tension in d/cm ($=10^{-3}$ N/m). See ▶ Surface Tensiometer and ▶ ASTM (www.astm.org).

Tenter Frame *n* A machine that dries fabric to a specified width under tension. The machine consists essentially of a pair of endless chains on horizontal tracks. The fabric is held firmly at the edges by pins or clips on the two chains that diverge as they advance through the heated chamber, adjusting the fabric to the desired width.

Tentering *v* Biaxial orientation of film or sheet by means of a ▶ Tentering Frame.

Tentering Frame *n* (tenter frame) A machine that continuously stretches, simultaneously in two perpendicular directions, a temperature-conditioned film or sheet, imparting biaxial orientation. Clamps attached to endless chains grip the sheet on both edges and, while accelerating in the direction of sheet travel, also move outward from the longitudinal centerline. Stretch ratio is about 3–4 in each direction, with about the same factor of increase in strength and modulus over those of the unoriented sheet. Tentering is usually done shortly downstream from the sheet extruder, but can also be done on film or thin sheet that has been extruded, cooled, and wound into coils for storage, then later reheated to be oriented.

Tenter Mark See ▶ Clip Mark.

Tera- (T) The SI prefix meaning $\times 10^{12}$.

Terebine *n* Form of liquid driers, originally made by combining linseed oil and natural resin with salts of lead and manganese and thinning with turpentine.

Terephthalate Polyester \▪ter- ə(f)-▪tha-▪lāt, ▪pä-lē-▪es-tər\ *n* Any polymeric ester of terephthalic acid (1,4-benzene dicarboxylic acid), but in particular the three commercially important thermoplastic resins, ▶ Polyethylene Terephthalate and ▶ Polybutylene Terephthalate.

Terephthaldehyde Resin See Polyester, Saturated.

Terephthalic Acid \▪ter-ə(f)-▪tha-lik ▪a-səd\ *n* [ISV *tere*bene, mixture of terpenes from distilled turpentine + *phthalic acid*] (1857) (TPA, *para*phthalic acid, benzene-*p*-dicarboxylic acid) $C_6H_4(COOH)_2$. White crystals or powder, used in the production of alkyd resins and thermoplastic polyesters. Mol wt, 166.13, sublimes above 300°C without melting.

Terital *n* Poly(ethylene terephthalate), manufactured by Soc. Rhodiadoce, Italy.

Terlenka *n* Poly(ethylene terephthalate), manufactured by AKU, The Netherlands.

Terluran *n* High-impact poly(styrene) (graft polymer of styrene and acrylonitrile on styrene/butadiene copolymer), manufactured by BASF, Germany.

Termination *n* (chain termination) The final phase of a polymerization in which chain growth ends through reaction of polymeric free radicals with each other or with smaller entities.

Termolecular Process *n* An elementary process in which the activated complex is formed from the simultaneous collision of three particles.

Terpene \ˈtər-ˌpēn\ *n* [ISV *terp*- (fr. Gr *Terpentin* turpentine, fr. ML *terbentina*) + *-ene*] (1873) A class of unsaturated organic compounds having the empirical formula $C_{10}H_{16}$ occurring in most essential oils and oleoresinous plants. Structurally, the unimportant terpenes and their derivatives are classified as monocyclic (dipentene), bicyclic (pinene), and acyclic (myrcene).

Terpene Resins *n* Neutral hydrocarbon resins made by polymerization of *β*- pinene. They possess all the advantages of coumarone resins, e.g., neutrality, resistance to alkalis, etc., without the disadvantages of yellowing. They have excellent compatibility.

Terpin *n* Terpinol, 4-Hydroxy-α,α,4-trimethylcyclohexanemethanol. Additional Names: *p*-menthane-1,8-diol; dipenteneglycol. Molecular Formula: $C_{10}H_{20}O_2$. Molecular Weight: 172.26. Percent Composition: C 69.72%, H 11.70%, O 18.58%. Literature references from Merck index (2001) 13th edn.: Both *cis*-and *trans*-modifications are known. The *cis*-compd is obtained most readily in the hydrated form, *cis*-terpin hydrate. Prepn of *cis*-form from oil of turpentine: Hempel (1876) Ann 180:71; Wallach (1885) Ann 230:225; Schmitt (1955) Mfg Chemist (1955) 26:350. From *d*-limonene: Sword (1925) J Chem Soc (1925) 127:1632. Prepn of *trans*-form from 1,8-cineole, α-terpineol or *cis*-terpin hydrate: Matsuura et al (1958) Bull Chem Soc Japan 31:990; Lombard R, Ambroise G (1961) Bull Soc Chim France 230. Structure of *cis*- and *trans*-forms: Baeyer Ber (1893) 26:2861.

Terpinenes *n* $C_{10}H_{16}$. Hydrocarbons of cyclic type, present in turpentines. Bp, about 179°C; sp gr, 0.865/15°C.

Terpineol *n* $C_{10}H_{17}OH$. Useful, powerful high-boiling solvent plasticizer. Bp, 210–220°C, depending upon the type, α, β, or γ.

Terpinolene *n* $C_{10}H_{16}$. Hydrocarbon solvent, Bp, 183°C; mp, 15°C; flp, 44°C (111°F).

Terpolymer \ˌtər-ˈpä-lə-mər\ *n* (1947) A polymer composed of molecules containing three chemically different types of monomers, or of the grafting of one monomer to the copolymer of two different monomers. An important commercial terpolymer is ABS resin, derived from acrylonitrile, butadiene, and styrene.

Terra Alba \ˈter-ə ˈal-bə\ *n* $CaSO_4 \cdot 2H_2O$. A finely powdered form of gypsum, used as a filler. Another name for gypsum or calcium sulfate. See ▶ Calcium Sulfate, ▶ Anhydrous.

Terra Cariosa See ▶ Rottenstone.

Terra Cotta *n* {*often attributive*} [It *terra cotta*, literally, baked earth] (1722) A fine reddish-brown clay mixed with sand and baked until hard, used for pottery, statues and the like; also the reddish-brown color of baked clay.

Terra di Siena See ▶ Raw Sienna and ▶ Iron Oxides.

Terra Ponderosa *n* Syn: ▶ Barium Sulfate.

Terrazzo \tə-ˈra-(ˌ)zō\ *n* [It, lit. terrace, perhaps fr. OP *terrassa*] (1897) Hard, marble-like flooring material consisting of 70% marble chips in cement, that is cast in place or precast, and then ground down to a smooth surface; used as a decorative surfacing on walls as well as floors. *Also called Terrazzo Concrete*. Syn: ▶ Venetian Mosaic.

Terrazzo Seal *n* A composition of alkali-resistant resins or waves in solvent of water.

Terre Verte *n* Essentially an iron silicate. It is used as a base for lakes. Syn: ▶ Green Earth.

Terry Cloth *n* A cotton or cotton-blend fabric having uncut loops on one or both sides. Made on a dobby loom with a terry arrangement or on a Jacquard loom. It is used for toweling, beach robes, etc.

***tert*-Butyl Hydroperoxide** *n* $(CH_3)_3COOH$. A highly reactive peroxy compound used as a polymerization catalyst.

tert-Butyl Perbenzoate n $C_6H_5 \cdot O_2 \cdot C(CH_3)_3$. A catalyst for the polymerization of acrylic and styrene monomers, and the curing of polyesters. Also use din the compounding of silicones and polyethylene. This peroxide has long been the workhorse in sheet molding compounds because it is stable enough for all practical purposes but is slow-reacting, requiring activation temperatures of 121–127°C unless boosted by a less stable peroxide.

tert-Butyl Permaleic Acid n $(CH_3)_3CC \cdot O_2 \cdot COCH=CHOOH$. A polymerization catalysts.

Tertiary \ ▌tər-shē- ▐er-ē\ *adj* [L *tertiarius* of or containing a third, fr. *tertius* third] (ca. 1656) In organic chemistry, denoting a functional group in which three of its original hydrogen atoms have been replaced by other groups. Triphenylcarbinol, $(C_6H_5)_3COH$, and *t*-butyl alcohol, $(CH_3)_3-COH$, are examples of tertiary alcohols. Trimethylamine, $(CH_3)_3N$, is a tertiary amine.

Tertiary Amine Value n The number of milligrams of potassium hydroxide (KOH) equivalent to the tertiary amine basicity in 1 g of sample.

Tertiary Colors n (ca. 1864) Shades that are obtained by mixing the three primary colors or by mixing one or more of the secondary colors with gray or black.

Terylene n Poly(ethylene terephthalate), manufactured by ICI, Great Britain.

TES See ▶ Styrenic Thermoplastic Elastomers.

Tesla \ ▌tes-lə\ n [Nikola *Tesla*] (1958) (T) The SI unit of magnetic-flux density, equal to one weber per square meter (1 Wb/m^2). The older, now deprecated unit, the *gauss*, is equal to 10^{-4} T.

Test Fence n An apparatus consisting of a fence strategically located in a part of the country for specific weather conditions (temperature, humidity, sunlight duration, etc.) and facing a specific direction and angle. It contains a series of exposure racks on which test panels are exposed. Coated test panels on various substrates (wood, metal, plastic, cementitious, etc.) are tested for exterior durability properties such as chalk resistance, tint retention, adhesion, racking, peeling, etc. The panels are exposed for various durations of time typical directions are north and south. Typical angles are vertical, horizontal, 5° and 45°. Examples of U.S. test fence locations include: Arizona, Delaware, Florida, New Jersey, Pennsylvania and Puerto Rico. See ▶ Exposure Rack.

Test Method n A definitive, standardized set of instructions for the identification, measurement, or evaluation of one or more qualities, characteristics, or properties of a material.

Test Pattern n Spray pattern used in adjusting spray gun.

Test Variables n Terms related to the testing of materials such as test method names.

Testing Methods See ▶ Test Method.

TETA Abbreviation for ▶ Triethylenetetramine.

Tetrabasic Lead Fumarate n $4PbO \cdot PbC_2H_2(COO)_2 \cdot 2H_2O$. A creamy-white powder used as a heat stabilizer for electrical-grade plastisols, and insulation. It is also used as a curing agent for chlorosulfonated polyethylene.

Tetrabromobisphenol A n \ ▌ter-trə- ▌brō-(▐)mō ▌bis- ▌fē- ▐nōl\ [4,4'-isopropylidene bis(2,6-dibromophenol)] $(CH_3)_2C(C_6H_2Br_2OH)_2$. (See structure at ▶ Bisphenol A.) An off-white, crystalline solid, used as a flame retardant in epoxy resins, polyesters, and polycarbonates. Whittington's Dictionary of Plastics.

Tetrabromophthalic Anhydride \ ▌ter-trə- ▌brō-(▐) mō- ▌tha-lik (▐)an- ▌hī- ▐drīd\ n $Br_4C_6-2,3-(CO)_2O$. A reactive intermediate containing 69% bromine, used as a flame retardant.

2,2′,6,6′-Tetrabromo-3,3′,5,5′-Tetramethyl-p,p′Biphenol n (TTB) An aromatic brominated flame retardant synthesized easily by a two-step process from 2,6-dimethylphenol. The unusual chemical structure of TTB enables its use as both a reactant and additive flame retardant. It has been used in high-impact polystyrene. Whittington's Dictionary of Plastics.

Tetrabutyl Titanate n \ˈter-trə-byü-tᵊl ˈtī-tᵊn-ˌāt\ (TBT, butyl titanate, titanium butylate) Ti(OC$_4$H$_9$)$_4$. A catalyst for condensation and cross-linking reactions, also used to improve the adhesion of plastic compounds to metals.

Tetrachloride n \ˈter-trə-ˈklōr-ˌīd\ n A chloride, such as carbon tetrachloride, containing four atoms of chlorine.

Tetrachlorobisphenol \ˈter-trə-ˈklō-ˈbī-ˈfē-ˌnōl\ n [4,4'-isopropylidene bis(2,6-dichlorophenol)] (CH$_3$)$_2$C (C$_6$H$_2$Cl$_2$OH)$_2$. (See structure at ▶ Bisphenol A.) A monomer for flame-retardant epoxies, polyesters, and polycarbonates.

Tetrachloroethane \ˈter-trə-ˈklō-ˈe-ˌthān, *British usually* ˈē-\ n C$_2$H$_2$Cl$_4$. Solvent, used at one time as the principal constituent of nonflammable paint remover, but now largely discontinued because of its toxicity. Bp, 147°C; vp, 11 mmHg/20°C. Syn: ▶ Carbon Tetrachloride.

Tetraethylene Glycol Dicaprylate n (C$_7$H$_{15}$COOCH$_2$CH$_2$OCH$_2$CH$_2$)$_2$O. A plasticizer for vinyl chloride polymers and copolymers.

Tetraethylene Glycol Di(2-ethylhexanoate) n (C$_8$H$_{17}$COOCH$_2$CH$_2$O—CH$_2$CH$_2$)$_2$O. A secondary plasticizer for vinyl resins and a primary plasticizer for cellulosic plastics and synthetic rubbers. In vinyls, it is used at levels of 15–20% for the total plasticizer to impart good low-temperature flexibility. In nitrocellulose lacquer, it imparts cold-check resistance.

Tetraethylene Glycol Monostearate n C$_{17}$H$_{35}$COO(CH$_2$CH$_2$O)$_4$OH. A plasticizer for ethyl cellulose and cellulose nitrate.

Tetraethyl Lead n (1923) (PbC$_2$H$_5$)$_4$. A heavy oily poisonous liquid used as an antiknock agent.

Tetrafluoroethylene n (TFE, perfluoroethylene) CF$_2$=CF$_2$ A colorless gas used as the monomer for polytetrafluoroethylene resins. Prepared by the thermal cracking at about 700°C of cholordifluoromethane, itself produced by reaction of hydrogen fluoride with chloroform. The production of polytetrafluoroethyelene and its copolymer is carried out by free radical polymerization.

Tetrafluoroethylene-Ethylene Copolymer See ▶ Poly(Ethylene-Tetrafluoroethylene).

Tetrafluoroethylene Fiber See ▶ Polytetrafluoroethylene Fiber.

Tetrafluoroethylene-Hexafluoropropylene Copolymer (FEP) See ▶ Fluorinated Ethylene-Propylene Resin.

Tetrafluoroethylene Propylene Copolymer *n* Thermosetting elastomeric polymer of tetrafluoroethylene and propylene having good chemical and heat resistance and flexibility. Used in auto parts.

Tetrahydrofuran \ˈte-trə-ˈhī-(ˌ)drō-ˈfyúr-ˌan\ *n* (THF, tetramethylene oxide) A colorless liquid obtained by the catalytic hydrogenation of furan, with the empirical formula C₄H₈O.

Tetrahydrofurfuryl Alcohol In addition to its many uses as an industrial intermediate, THF is a powerful solvent for PVC, polyvinylidene chloride, and many other polymers. It is often used as the carrier solvent for size-exclusion chromatography of polymers. Its presence in relatively small amounts increase the "bite" of vinyl printing inks, lacquers, and adhesives. THF has been polymerized to polytetramethylene ether glycol for use in the production of polyurethanes.

Tetrahydrofurfuryl Alcohol *n* C₄H₇OCH₂OH. High-boiling solvent. Used as a solvent for vinyl resins, Bp, 178°C; vp, <1 mmHg/30°C. Also known as "THFA" and *Tetrahydrofuryl*.

Tetrahydrofurfuryl Oleate *n* CH₃(CH₂)₇CH=CH(CH₂)₂C₄H₇O. A plasticizer for polystyrene, and cellulosic, acrylic, and vinyl resins. In vinyls it is used as a secondary plasticizer imparting resistance to low temperatures, and as a lubricant in stiff or highly filled calendering and extrusion compounds.

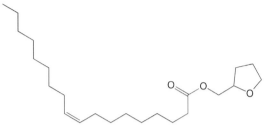

Tetrahydronaphthalene *n* (1,2,3,4-tetrahydronaphthalene, Tetralin) C₁₀H₁₂. This bicyclic, high-boiling, semiaromatic hydrocarbon, produced by the partial hydrogenation of naphthalene, is an involatile solvent for rubbers, PVC, and natural resins. Bp, 198–210°C; sp gr, 0.971/20°C; flp, 77°C (170°F); vp, <1 mmHg/30°C. It is used in alkyd compositions, chlorinated rubber finishes, and in paint removers. *Also called Tetralin.*

Tetramer \ˈte-trə-mər\ *n* (1929) A polymer formed from four molecules of a monomer and/or made up of four mer units. See ▶ Oligomer.

1,2,4,5-Tetramethylbenzene Explicit name for ▶ Durene.

Tetramethylbisphenol *n* A compound for producing polycarbonate.

1,1,3,3-Tetramethylbutyl Peroxy-2-Ethylhexanoate *n* A liquid peroxide superior to (solid) benzoyl peroxide as a catalyst for polyesters and, because it is a liquid, easier to handle.

5,5′-Tetramethylene Di(1,3,4-dioxazol-2-one) *n* (Adiponitrile carbonate, ADNC) A white crystalline solid, capable of being reacted with diols and polyols to form light-stable urethane coatings, elastomers, and foams.

Tetramethylethylenediamine *n* (TMEDA, *N,N,N′,N′*-tetramethylethylenediamine) $(CH_3)_2NCH_2CH_2N(CH_3)_2$. An anhydrous, corrosive liquid used as a catalyst for urethane foams, coatings, and elastomers, and as a curing agent.

Tetramethylene Glycol See ▶ 1,4-Butylene Glycol for epoxy resins.

Tetramethylthiuram Disulfide *n* (TMTD) $[(CH_3)_2NCS_2-]_2$. A white, crystalline powder used as a fungicide, bacteriostat, and rodent repellent in vinyl compounds, and as a secondary accelerator in rubber curing.

Tetrapolymer Syn: ▶ Quaterpolymer.

Tex *n* (1) A convenient unit of lineal density of fibers: the mass in grams of 1 km of fiber length. The SI equivalent is 10^{-6} kg/m (1 mg/m). Compare CUT, DENIER, and GREX NUMBER. (2) The system of yarn numbering based on the use of tex units. Also see ▶ Yarn Number.

Texanol® *n* 2,2,4-trimethylpentanediol monoisobutyrate. A proprietary high boiling ester alcohol water-insoluble coalescent. Flash point, 188°C (Cleveland Open Cup); density, 7.92 lb/gal; sp gr, 0.95.

Textile \ˈtek-ˌstīl\ *n* [L, fr. neuter of *textilis* woven, fr. *texere*] (1626) Originally, a woven fabric; now applied generally to any one of the following: (1) Staple fibers and filaments suitable for conversion to or use as yarns, or for the preparation of woven, knit, or nonwoven fabrics. (2) Yarns made from natural or manufactured fibers. (3) Fabrics and other manufactured products made from fibers as defined above and from yarns. (4) Garments and other articles fabricated from fibers, yarns, or fabrics when the products retain the characteristic flexibility and drape of the original fabrics (Schoeser M (2003) World textiles: a concise history, London, England. Thames and Hudson, Kadolph SJJ, Langford AL (2001) Textiles. Pearson Education, New York).

Textile Materials *n* A general term for fibers, yarn intermediates, yarn, fabrics, and products made from fabrics that retain more or less completely the strength, flexibility, and other typical properties of the original fiber or filaments. (Kadolph SJJ, Langford AL (2001) Textiles. Pearson Education, New York).

Textile Processing *n* Any mechanical operation used to translate a textile fiber or yarn to a fabric or other textile material. This includes such operations as opening, carding, spinning, plying, twisting, texturing, coning, quilling, beaming, slashing, weaving, and knitting (Kadolph SJJ, Langford AL (2001) Textiles. Pearson Education, New York; Schoeser M (2003) World textiles: a concise history. Thames and Hudson).

Texture *n* The structural quality of a surface. A term describing the surface effect of a fabric, such as dull, lustrous, wooly, stiff, soft, fine, coarse, open, or closely woven; the structural quality of a fabric.

Textured *adj* An adjective used to describe continuous filament manufactured yarns (and woven and knit fabrics made therefrom) that have been crimped or have had random loops imparted, or that have been otherwise modified to create a different surface texture. Also see ▶ Textured Yarns and ▶ Texturing.

Textured Paint See ▶ Plastic Paint.

Textured Yarns *n* Yarns that develop stretch and bulk on subsequent processing. When woven or knitted into fabric, the cover, hand, and other aesthetics of the finished fabric better resemble the properties of a fabric constructed from spun yarn. Also see ▶ Texturing. The ten types of textured yarns are: *Bulked Yarn* – Qualitative term to describe a textured yarn. A bulked yarn develops more bulk than stretch in the finished fabric. *Coil Yarn* – A textured yarn that takes on a coil or spiral configuration when further processed. A coil yarn can be either a torque yarn or a nontorque yarn. A coil yarn can be formed by the false twist or edge crimp methods. Some bilateral fibers become coiled on further processing. *Core-Bulked Yarn* – A bulky or textured yarn composed of two sets of filaments, one of which is straight to give dimensional stability and forms a core around and through which the other set is coiled or looped to give bulk. *Crinkle Yarn* – A torque-free textured yarn that is characterized by periodic wave configurations. Crinkle yarns can be formed by the stuffer box, gear crimping, or knit-de-knit methods. *Entangled Yarn* – A textured yarn of one variant that develops bulk by the air-jet texturing method. *Modified Stretch Yarn* – A stretch yarn that develops more bulk than usual but less bulk than a bulked yarn in the finished fabric. *Nontorque Yarn* – A yarn that does not rotate or kink when permitted to hang freely. A nontorque yarn may be the result of plying two equal but opposite torque yarns. *Set Yarn* – A textured yarn that is heat relaxed to reduce torque. Set yarns are not stretch yarns. *Stretch Yarn* – Qualitative term to describe a textured yarn. A stretch yarn develops more stretch than bulk in the finished fabric. *Torque Yarn* – When a torque yarn is permitted to hang freely it rotates or kinks to relieve the torque introduced into the yarn during texturing (Kadolph SJJ, Langford AL (2001) Textiles. Pearson Education, New York; Shah V (1998) Handbook of plastics testing technology. Wiley, New York).

Texture-Finished Paint See ▶ Plastic Paint.

Texturing *n* The process of crimping, imparting random loops, or otherwise modifying continuous filament yarn to increase cover, resilience, abrasion resistance, warmth, insulation, and moisture absorption or to provide a different surface texture. Texturing methods can be placed roughly into six groups. Also see ▶ Textured Yarns. *Air Jet Method* – In this method of texturing, yarn is led through the turbulent region of an air jet at a rate faster than it is drawn off on the far side of the jet. In the jet, the yarn structure is opened, loops are formed, and the structure is closed again. Some loops are locked inside and others are locked on the surface of the yarn. An example of this method is the Taslan process. (Also see ▶ Textured Yarns, Core-Bulked Yarn and Entangled Yarn.) *Edge Crimping Method* – In this method of texturing, thermoplastic yarns in a heated and stretched condition are drawn over a crimping edge and cooled. Edge-crimping machines are used to make Agilon yarns. (Also see ▶ Textured Yarns, Coil Yarn.) *False-Twist Method* – This continuous method for producing textured yarns utilizes simultaneous twisting, heat-setting, and untwisting. The yarn is taken from the supply package and fed at controlled tension through the heating unit, through a false-twist spindle or over a friction surface that is typically a stack of rotating discs called an aggregate, through a set of take up rolls, and onto a take-up package. The twist is set into the yarn by the action of the heater tube and subsequently is removed above the spindle or aggregate resulting in a group of filaments with the potential to form helical springs. Much higher processing speeds can be achieved with friction false twisting than with conventional spindle false twisting. Both stretch and bulked yarns can be produced by either process. Examples of false-twist textured yarns are Superloft®, Flufflon®, and Helanca®. (Also see ▶ Textured Yarns, Coil Yarn.) *Gear Crimping Method* – In this texturing method, yarn is fed through the meshing teeth of two gears. The yarn takes on the shape of the gear teeth. (Also see ▶ Textured Yarns, ▶ Crinkle Yarn.) *Knit-de-Knit Method* – In this method of texturing, the yarn is knit into a 2-inch diameter hose-leg, heat-set in an autoclave, and then unraveled and wound onto a final package. This texturing method produces a crinkle yarn. (Also see ▶ Textured Yarns, Crinkle Yarn.) *Stuffer Box Method* – The crimping unit consists of two feed rolls and a brass tube stuffer box. By compressing the yarn into the heated stuffer box, the individual filaments are caused to fold or bend at a sharp angle, while being simultaneously set by a heating device (Kadolph SJJ, Langford AL (2001) Textiles. Pearson Education, New York; Complete textile glossary, Celanese acetate LLC (2000) Three Park Avenue, New York).

TFE *n* Abbreviation for ▶ Tetrafluoroethylene.

Tg See Glass Transition Temperature.

TGA *n* Thermogravimetric analysis, generates a plot of "mass vs. temperature"; useful for determining temperatures of decomposition T_d or other thermogravimetric events. Abbreviation for Thermogravi-Metric Analysis.

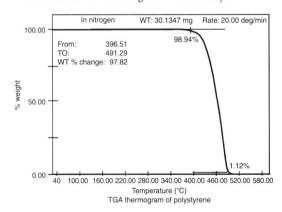

TGA thermogram of polystyrene

THEA See ▶ Tetrahydrofurfuryl Alcohol.

Thénard's Blue *n* Another name for cobalt blue. See ▶ Cobalt Blue.

Theoretical Weight *n* The weight of a part calculated from its specified dimensions and the density of the material.

Theory \thē-ə-rē\ *n* [LL *theoria*, fr. Gk *theōria*, fr. *theōrein*] (1592) A proposed explanation or justification of observed behavior made in terms of a model.

Thermal Adhesive *n* Special type of adhesive which only develops its adhesive properties when heat is applied to it, and partial liquefaction occurs.

Thermal Analysis *n* Methods of analysis that involve programmed heating of a material and observing the accompanying physical and chemical reactions, changes of state (T_g, T_m, etc.), or changes in physical or mechanical properties (Groenewoud WM (2001) Characterization of polymers by thermal analysis. Elsevier Science and Technology Books, New York). See ▶ Differential Thermal Analysis, ▶ Differential Scanning Calorimeter, ▶ Dynamic Mechanical Analyzer, and ▶ Thermogravimetric Analysis.

Thermal Black See ▶ Carbon Black.

Thermal Bonding See ▶ Thermobonding.

Thermal Capacity *n* The quantity of heat necessary to produce unit change of temperature in unit mass. It is ordinarily expressed as calories per gram per degree centigrade. Numerically equivalent to specific heat. See ▶ Heat Capacity and ▶ Specific Heat.

Thermal Capacity or Water Equivalent *n* The total quantity of heat necessary to raise any body or system unit temperature, measured as calories per degree centigrade in the cgs system. Dimension, $[m]$.

Thermal Character *n* A tactile property of a textile material. It is the difference felt in the temperature of the material and the skin of the person touching it.

Thermal Conductivity *n* (k) The basic measure of steady heat-transfer rate within solid materials (and still fluids) by atomic or molecular contact and vibration. It derives from Fourier's Law Of Heat Conduction and may be thought of as the rate of heat flow between two opposite faces of a unit cube whose other faces are perfectly insulated when the temperature at the warmer face is 1 K above that of the cooler face. The SI dimensions corresponding to this concept are $(J/S)/[m^2(K/m)]$, which reduces to $W/(m \cdot K)$. Some conversions from other units to SI are given in the Appendix. For plastics and other materials, *k* increases with rising temperature (Lide DR (ed) (2004) CRC handbook of chemistry and physics. CRC Press, Boca Raton; Ready RG (1996) Thermodynamics. Pleum Publishing Company, New York; Seanor DA (1982) Electrical conduction in polymers. Academic Press, New, Academic Press, New York).

Thermal Decomposition *n* The chemical decomposition of a plastic resulting from increasing temperature (or heat). With a given polymer, it occurs at a temperature at which some components of the material are separating, reacting together, or depolymerizing, with observable changes in micro- or macrostructure, mechanical and/or electrical properties, and, usually, reduction in molecular weight (Groenewoud WM (2001) Characterization of polymers by thermal analysis. Elsevier Science and Technology Books, New York; Ready RG (1996) Thermodynamics. Pleum Publishing Company, New York).

Thermal Diffusivity *n* (α) An important property for unsteady (transient) heat transfer, particularly in solid materials, equal to the thermal conductivity divided by the product of heat capacity and density, i.e., $\alpha = k/(C \cdot \rho)$, the SI unit is m^2/s. For most solids, a increases slowly with rising temperature. Thermal diffusivity comes into play in the heating and cooling of thermoformable sheets, in the cooling of injection moldings in the mold, and the cooling of extrudates. The most-used method for measuring thermal diffusivity of solids, the thermal-pulse method, has so far been adopted by ASTM only for synthetic carbon and graphite (Ready RG (1996) Thermodynamics. Pleum Publishing Company, New York).

Thermal Discharge *n* The introduction of water from a point source at a temperature different from the ambient temperature of the receiving waters. See ▶ Point Source.

Thermal Expansion *n* The coefficient of linear expansion or expansivity is the ratio of the change in length per degree C to the length at 0°C. The coefficient of volume expansion (for solids) is approximately three times the linear coefficient. The coefficient of volume expansion for liquids is the ratio of the change in volume per degree to the volume at 0°C. The value of the coefficient varies with temperature. The coefficient of volume expansion for a gas under constant pressure is nearly the same for all gases and temperatures and is equal to 0.00367 for 1°C. Dimension, $[\theta^{-1}]$. If l_o is the length at 0°C, α the coefficient of linear expansion, the length at t°C is,

$$l_t = lo(1 + \alpha t)$$

General formula for thermal expansion – The rate of thermal expansion varies with the temperature. The general equation giving the magnitude m_o is the magnitude at 0°C, is

$$m_t = m_o(1 + \alpha t + \beta t^2 + \gamma t^3 \ldots)$$

where α, β, γ, etc. are empirically determined coefficients. *Volume expansion* – If V represents volume and β the coefficient of expansion,

$$V_t = V_o(1 + \beta t)$$

For solids,

$$\beta = 3\alpha (\text{approximately})$$

Thermal-Expansion Coefficient See ▶ Coefficient of Thermal Expansion.

Thermal Fixation See ▶ Dyeing.

Thermal Fluid *n* Any of many types of heat-stable, noncorrosive liquids such as glycols, silicone and other oils, and the eutectic mixture of biphenyl and biphenyl oxide ("Dowtherm A") that are used to transfer heat by convection. Water, saturated steam, and glycol solutions are also important. Examples of applications in the plastics industry are jacketed molds for rotational casting, heating of calendars, injection-mold chilling, controlling the temperatures of cored extrusion screws, and maintaining temperatures of liquids in storage tanks.

Thermal-Gradient Elution *n* Method of fractionation in which a small amount of polymer, dispersed on a support, is placed on the top of a column of an inert material. The polymer is eluted from the column by extraction with solvents of increasing solvent power, as in column extraction, but, in addition, a temperature gradient is imposed along the length of the column.

Thermal Gravimetric Analysis *n* Syn: ▶ Thermogravimetric Analysis.

Thermal History *n* (heat history) The integrated product of time × temperature for a plastic, from the time it was first subjected to a high temperature to the present moment under consideration. Thermal history is an important consideration in processing (and reprocessing) heat-sensitive polymers such as rigid PVC and polypropylene.

Thermal Impulse Sealing See ▶ Impulse Sealing.

Thermally Foamed Plastic *n* A cellular plastic produced by applying heat to effect gas-generating decomposition or volatilization of a constituent (after ASTM D 883).

Thermally Stabilized See Heat Stabilized.

Thermally Stimulated Current *n* (TSC) A technique useful in studying the transitions of amorphous, polar polymers with rising temperature. A preheated sample is electrically oriented by applying a strong electric field, then chilling the polymer with the field applied. The sample is removed from the field and reheated on a temperature ramp while the current generated by the release of dipoles is tracked current peaks relate to relaxation times of molecular motions within the polymer. The technique is often carried out at a number of different preheating temperatures and the result subjected to ▶ Relaxation-Map Analysis in order to distinguish relaxations occurring at various molecular-weight levels in a typical polymer.

Thermal Neutrons *n* Neutrons slowed down by a moderator to an energy of a fraction of an electron volt – about 0.025 ev. at 15°.

Thermal Paint Coating containing small granules of metallic powder that create a resistance when current is applied and thereby heat the substrate.

Thermal Polymerization *n* A polymerization process performed solely by heating in the absence of a catalyst. Monomers such as styrene and methyl methacrylate are examples of those that an be thermally polymerized.

Thermal Properties *n* All properties of materials involving heat or changes in temperature. In section 08 of ASTM's Annual Book of Standards ("Plastics"), tests listed under "Thermal Properties" include many properties, from brittleness temperature, coefficient of expansion, deflection temperature, etc., to heat of fussion, glass-transition temperature, thermal conductivity, heat capacity, mold shrinkage, flammability, and many more.

Thermal Resistance *n* (R-valve) The ability of a material to retard the conductive passage of heat; the reciprocal of conductance. While plastics, with very few exceptions, are good thermal insulators, the property is of especial interest for cellular plastics, often used to provide thermal insulation. The R-value of a layer of thermal insulation is given by: $R = \Delta T/q$, the ratio of the temperature drop in the direction of heat flow from one surface of the layer to the other, divided by the rate of heat flow per unit of surface area (*heat flux*). English units are customary in the U.S. That is not as fundamental as property as ▶ Thermal Conductivity because the thickness of the layer is implicit in the resistance and should be separately stated.

Thermal Sealing *n* (thermal heat sealing) A method of bonding two or more layers of plastics by pressing them between heated dies or tools that are maintained at a relatively constant temperature. See also ▶ Heat Sealing.

Thermal Sensitivity *n* See ▶ Heat Sensitivity and ▶ Thermal History.

Thermal Stability *n* Syn: ▶ Heat Stability.

Thermal Stabilizer See ▶ Stabilizer.

Thermal Stress Cracking *n* (TSC) Crazing and cracking of some thermoplastics that results from overexposure to elevated temperatures.

Thermal Volatilization Analysis *n* (pyrolysis analysis, TVA) Ramp heating of a plastic with passage of the evolved volatiles through one or more chemical detectors, sometimes with intervening, controlled-temperature, vapor-condensing traps. TVA is a powerful technique when coupled with ▶ Thermogravimetric Analysis.

Thermal Welding *n* A method of bonding two or more layers of plastics by pressing them between heated dies or tools that are maintained at a relatively constant temperature.

Thermionic Emission *n* Electron or ion emission due to the temperature of the emitter. The rate of emission increases rapidly with rising temperature, and is also very sensitive to the state of the emitting surface.

Thermistor \ˈthər-ˌmis-tər\ *n* [*therm*al res*istor*] (1940) A contraction of *thermal resistor*; a semiconductor whose resistance varies sharply and reproducibly with temperature, therefore useful for temperature measurement, and sometimes used for that purpose in plastics industry.

Thermistors *n* Thermally sensitive resistors known for exhibiting a large change in resistance with only a small change in temperature.

Thermoband Welding *n* Trade name for a variant of hot-plate welding in which a metallic tape acting as a resistance element is adhered to the material to be welded. Low voltage is applied to heat the tape, and the adjacent plastic, to the plastic's melting range. Pressure may be applied to the joint while it cools.

Thermobonding *n* A technique for bonding fibers of a web with meltable powders or fibers, using infrared heating, hot air, or hot-calendering. (Also see ▶ Bonding and ▶ Bonding With Binder Fibers).

Thermochemical Equation n A chemical equation which includes an indication of the heat liberated or absorbed during the reaction.

Thermochromic *n* Changing color with changing temperature, a characteristic of special materials useful as temperature indicators.

Thermocompression Bonding *n* The joining together of two materials without an intermediate material by the application of pressure and heat in the absence of an electrical current.

Thermocouple \ˈthər-mə-ˌkə-pəl\ *n* (1890) (TC) A pair of connected, welded junctions formed by two wires of dissimilar metals such as iron and ▶ Constantan. If a temperature difference exists between the two junctions, a weak emf, 40–50 µV/K, is generated in nearly linear proportion to the ΔT. Originally, one junction was placed in an ice bath to serve as a *reference*. Then, the emf developed in the circuit was simply convertible to temperature. Modern thermocouples employ a single junction and the instrument to which the TC is connected senses its own temperature and compensates the incoming signal for the difference between that temperature and 0°C. Plain wire TCs respond very rapidly to temperature changes but those used in plastics-processing equipment are always sheathed in a sturdy protective tube, so are slower. Metal compositions of commercial TC wires are so carefully controlled that, except for exacting laboratory work, it is usually unnecessary to calibrate thermocouples. In one type of hand-held instrument, called a ▶ Pyrometer, the TC is integral with a microammeter whose needle moves across a temperature scale. Thermocouples are the most used temperature sensors in plastics processing because they are sturdy, simple, reliable, readily available, and cheap (Weast RC (ed) (1971) Handbook of chemistry and physics, 52nd edn. The Chemical Rubber Co., Boca Raton).

Thermodynamic Properties *n* A quantity that is either an attribute of the entire system or is a function of position, which is continuous and does not vary rapidly over microscopic distances, except possibility for abrupt changes at boundaries between phases of the system. Also called macroscopic properties.

Thermodynamics, Law of *n* I – *The energy of the universe is constant*. Energy is can neither created nor destroyed only changed in form. When a system gains heat, its energy increases, and when a system does work, its energy decreases. II – *The entropy of the universe is always expanding*. Entropy is a measure of disorder. In all energy changes, if no energy leaves enters or leaves the system, the potential energy of the state will be less than that of the initial state. In the process of energy exchange, some energy will dissipate as heat. A transformation whose only final result is to transform into work heat from a source which is at the same temperature is impossible (Postulate of Lord Kelvin). A transformation whose only final result is to transfer heat from a body at a given temperature to a body at a higher temperature is impossible (Postulate of Clausius). III – *A state of perfect order is a state of minimum entropy* (e.g., a perfect crystalline structure at absolute zero temperature). The British scientist and author C. P. Snow had an excellent way of stating these laws: I – You cannot win (i.e., you cannot get something for nothing, because matter and energy are conserved. II – You cannot break even (you cannot return to the same energy state, because there is constant increase in disorder). III – You cannot get out of the game (because absolute zero temperature is not attainable) (Ready RG (1996) Thermodynamics. Pleum Publishing Company, New York).

Thermodynamic Temperature See ▶ Kelvin.

Thermoelasticity *n* Rubber-like elasticity exhibited by a rigid plastic and resulting from an increase in temperature. In this state, the plastic may be formable into a different shape. Retention of the desired shape may be achieved by cooling in place after forming. With thermosetting materials, prolonged heating may be necessary to effect cure in place.

Thermoelectric Power *n* Measured by the electromotive force produced by a thermocouple for unit difference of temperature between the two junctions. It varies with the average temperature and is usually expressed in microvolts per degree C. It is customary to list the thermoelectric power of the various metals with respect to lead.

Thermoform *n* (1956) To change the shape of a plastic rod, profile, tube, or sheet by first heating it to make it pliant, next, forming the desired shape, and finally, cooling the formed shape. ▶ Sheet Thermoforming, is by far the most important class of these operations.

Thermoformability *n* The ease with which a heat-softened plastic sheet (or rod, etc.) can be given a new permanent shape, particularly by the techniques of ▶ Sheet Thermoforming. Some attempts have been made to devise tests of sheet thermoformability, but none have been widely used or adopted by ASTM as of 1992.

Thermoforming *n* (1) See ▶ Sheet Thermoforming. (2) Any process in which heat softening is used to assist in the forming or reshaping of a plastic rod, tube, bar, sheet, or profile. When performed directly following extrusion of the profile, the term *postforming* is often used.

Thermogram *n* A plot of the percent of original mass of a specimen remaining during a program of linearly rising temperature vs the specimen temperature. See ▶ Thermogravimetric Analysis.

Thermographic Nondestructive Testing *n* See ▶ Infrared Thermography.

Thermographic-Transfer Process *n* A modification of hot stamping wherein the design to be transferred is first printed (in reverse) on a film, from which it is transferred to the plastic part by means of heat and pressure.

Thermography \(▪)thər-ˈmä-grə-fē\ *n* (1840) Printing process in which the ink, while still wet on the sheet, is dusted with a resinous powder that adheres to the ink. The sheets are then put through a heating process which causes the particles of powder to fuse together with the ink, giving a raised effect to the letters which simulate steel-die engraving.

Thermogravimetric Analysis *n* (TGA) A testing procedure in which the diminishing mass of a specimen is recorded as the specimen's temperature is raised at a uniform rate (sometimes referred to as *ramped*). Typical apparatus consists of an analytical balance supporting (on a wire or rod) a platinum crucible containing the specimen, the crucible being situated inside an electric furnace, and means for recording and plotting the percent mass remaining vs temperature. Some TGA tests are conducted in air, bothers in controlled, successive atmospheres such as nitrogen in the first stage, followed by air in the second stage. Thermogravimetric curves so obtained (*thermograms*) provide the useful information regarding polymerization and pyrolysis reactions, the efficiencies of stabilizes and activators, the thermal stability of final materials, and direct analysis. An example of a TGA instrument is the Diamond TG/DTA, image courtesy of PerkinElmer, Inc. (Groenewoud WM (2001) Characterization of polymers by thermal analysis. Elsevier Science and Technology Books, New York).

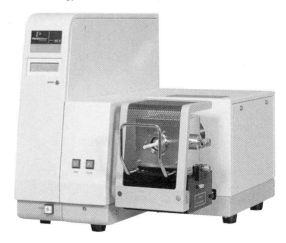

Thermolastic See Styrene-Butadiene Thermoplastic.

Thermomechanical Analysis *n* TMA or thermomechanical analysis is the measurement of material properties (coefficient of thermal expansion, modulus, etc.) by utilizing a mechanical probe or other mechanical sensor (static or oscillating) and measuring its load/position relative to the material being studied while increasing temperature. An example of a TMA instrument is the PerkinElmer Diamond TMA, image courtesy of PerkinElmer, Inc.) (Groenewoud WM (2001) Characterization of polymers by thermal analysis. Elsevier Science and Technology Books, New York).

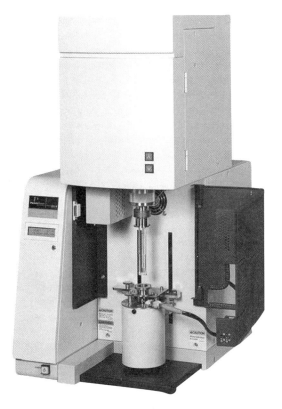

Thermomechanical Spectrum *n* A plot of a mechanical property such as tensile modulus or strength versus temperature.

Thermoplastic \ ▮thər-mə- ▮plas-tik\ *n* (1883) A resin or plastic compound that, as a finished material, is capable of being repeatedly softened by heating and hardened by cooling. Examples of thermoplastics are: acetal, acrylic, cellulosic, chlorinated polyether, fluorocarbons, polyamides (nylons), polycarbonate, polyethylene, polypropylene, polystyrene, some types of polyurethanes, and vinyl resins. (Adjective) Of an organic material, capable of being repeatedly softened and hardened by heating and cooling.

Thermoplastic Elastomers *n* (TPE) Any of a family of polymers that resemble elastomers in that they are highly resilient and can be repeatedly stretched to at least twice their initial lengths with full, rapid recovery, but are true thermoplastics and thus do not require curing or vulcanization as do most rubbers. See ▶ Elastomer for examples.

Thermoplastic Polyester *n* Any of a class of linear terephthalate polyesters that are true thermoplastics. Commercially important are ▶ Polyethylene and ▶ Polybutylene Terephthalate. U.S. sales of these resins in 1992 totaled 1.24 Tg (1.37×10^6 tons).

Thermoplastic Polyolefin *n* (TPO) Any of a group of elastomers produced by either of two processes. In one, polypropylene is melt-blended with from 15% to 85% of terpolymer elastomer, ethylenepropylene rubber, or styrene-butadiene rubber. In the other, propylene is copolymerized with ethylene-propylene elastomer in a series of reactions. The smaller elastomeric domains obtained in the latter process are claimed to provide improved properties over the blended materials.

Thermoplastic Polyurethane *n* A class of polyurethanes including rigid and elastomeric polymers that can be repeatedly made soft and pliable on heating and hard (flexible or rigid) on subsequent cooling. Also called thermoplastic urethanes, TPUR, TPU.

Thermoplastic Resins *n* Resins which remain fluid on heating and do not become infusible, and they flow whereas *thermoset* resins do not flow because they are crosslinked or set.

Thermoplastic Rubbers *n* Thermoplastic elastomers (rubbers). See ▶ Elastomer.

Thermoplastics *n* Material that can be molded and shaped when heated.

Thermoplastic Urethane See ▶ Thermoplastic Polyurethane.

Thermoset \ ▮thər-mō- ▮set\ *n* (1947) A material which will undergo or has undergone a chemical reaction by the action of heat, catalysis, ultraviolet light, etc., leading to a relatively infusible state.

Thermoset Adhesives *n* Adhesives which undergo a chemical reaction by action of heat, catalysts etc., leading to a relatively infusible state.

Thermosetting *n* Having the property of undergoing a chemical reaction by the action of heat, catalysts, ultraviolet light, etc., leading to a relatively infusible state (nonflowing).

Thermosetting Elastomer *n* A large class of polymers that can be stretched at room temperature to at least twice their original length and, after having been stretched and the stress removed, return with force to approximately their original length in a short time. To attain this elastic property the rubbers must be crosslinked or vulcanized, usually by heating in the presence of various crosslinking agents and catalysts. There are natural and synthetic rubbers. The most important synthetic rubber families are olefinic rubbers, dienic rubbers (nitrile, butadiene, neoprene), silicone rubbers, and urethane rubbers. Used often as impact modifiers/fillers in plastics.

Thermosetting Ink *n* An ink which polymerizes to a permanently solid and infusible state upon the application of heat.

Thermosetting Plastic *n* (thermoset) A resin or plastic compound that in its final state is substantially infusible and insoluble. Thermosetting resins are often liquids at some stage in their manufacture or processing, which are cured by heat, catalysis, or other chemical means. Much crosslinking occurs. After being fully cured, thermosets cannot be melted y reheating. Some plastics that are normally thermoplastic can be made unmeltable by means of crosslinking treatments or reactions. Some thermosetting plastics are alkyd, allyl, amino, epoxy, furane, phenolic, polyacrylic ester, polyester, and silicone resins.

Thermosetting Resin *n* (1) A resin that crosslinks and becomes insoluble in a solvent after curing (usually with the aid of catalysts and/or heat). (2) A resin which polymerizes to a permanently solid and infusible state upon the application of heat.

Thermosol Process See ▶ Dyeing, ▶ Thermal Fixation.

Thermospray See ▶ Flame Spray.

Thermotropic Polymer *n* Polymer that exhibits liquid crystal formation in melt form. In thermotropic polymers there must be a balance between having the necessary degree of molecular perfection to preserve the liquid crystal formation and the amount of imperfection to permit melting at workable temperatures. These polymers give high-modulus, highly oriented, extrusion products. See ▶ Liquid-Crystal Polymer.

Theta Solvent *n* A solvent, at a particular temperature, in which the polymer is at the edge of solubility and exists in the form of a statistical coil. Long-range forces between polymer molecular segments are balanced by polymer solvent interactions. At these conditions the second virial coefficient becomes zero and entropy is at its minimum (Kamide K, Dobashi T (2000) Physical chemistry of polymer solutions. Elsevier, New York; Flory PJ (1969) Statistical mechanics of chain molecules. Interscience Publishers, New York; Flory PJ (1953) Principles of polymer science. The Cornell University Press, Ithaca).

Theta Solvents *n* Solvents for a particular polymer in dilute solution, and at a temperature called the THETA TEMPERATURE that slows the polymer chains to assume their unperturbed, random-coil configurations with theoretical root-mean-square distances between chain ends (Kamide K, Dobashi T (2000) Physical chemistry of polymer solutions. Elsevier, New York; Flory PJ (1969) Statistical mechanics of chain molecules. Interscience Publishers, New York; Flory PJ (1953) Principles of polymer science. The Cornell University Press, Ithaca).

Theta State *n* A term introduced by Dr. Paul Flory to describe the condition in a polymer solution in which there is little interaction between the molecules of the solvent and those of the polymer, and in which the polymer molecules exist as statistically distributed coils (Kamide K, Dobashi T (2000) Physical chemistry of polymer solutions. Elsevier, New York; Flory PJ (1969) Statistical mechanics of chain molecules. Interscience Publishers, New York; Flory PJ (1953) Principles of polymer science. *The*, Cornell University Press, Ithaca).

Theta Temperature *n* (1) (θ, Flory temperature) See ▶ Theta Solvent. θ is also the temperature for the polymer solution at which the second virial coefficient in the equation giving molecular weight from osmotic pressure approaches zero. (2) The critical temperature of mixing for a polymer of infinitely high degree of polymerization. See ▶ Flory Temperature.

THF *n* Abbreviation for ▶ Tetrahydrofuran. CAS Registry Number: 109-99-9. Additional Names: Diethylene oxide; tetramethylene oxide. Molecular Formula: C_4H_8O. Molecular Weight: 72.11. Percent Composition: C 66.62%, H 11.18%, O 22.19%, Literature References: Prepn from 1,4-butanediol: Schmoyer LF, Case LF (1960) Nature 187:592. Manuf by catalytic hydrogenation of maleic anhydride: Gilbert, Howk, US 2772293 (1956 to du Pont); of furan: Banford, Manes, US 2846449 (1958 to du Pont); Manly, US 3021342 (1962 to Quaker Oats). Stabilization to prevent excessive peroxide formation on storage with 0.05–1.0% *p*-cresol, 0.05–0.1% hydroquinone, or less than 0.01–0.1% 4,4-thiobis(6-*tert*-butyl-*m*-cresol): Bordner, Hinegardner, US 2489260; US 2525410; Campbell, US 3029257 (1949, 1950, 1962 all to du Pont). Review of toxicology and biological effects: Moody DE (1991) Drug Chem Toxicol 14:319–342. Properties: Liquid. Ether-like odor. mp −108.5°. d_4^{20} 0.8892. bp_{760} 66°; bp_{176} 25°. Flash pt 1°F. n_D^{20} 1.4070. Dipole moment: 1.70. uv cut-off for spectro grade: 220 nm. Miscible with water, alcohols, ketones, esters, ethers, and hydrocarbons. Distil only in presence of a reducing agent, such as ferrous sulfate; peroxide explosions have occurred: Angew Chem (1956) 68:182, mp −108.5°, bp_{760} 66°; bp_{176} 25°, flash pt 1°F. Index of refraction: n_D^{20} 1.4070, density: d_4^{20} 0.8892, CAUTION: Potential symptoms of overexposure are irritation of eyes and upper respiratory system; nausea, dizziness and headache; CNS depression. See ▶ NIOSH pocket guide to chemical hazards (DHHS/NIOSH 97-140, 1997) p 302. Use: Solvent for high polymers, esp polyvinyl chloride, as reaction medium for Grignard and metal hydride reactions. In the synthesis of butyrolactone, succinic acid, 1,4-butanediol diacetate, solvent in histological techniques, may be used under Federal Food, Drug & Cosmetic Act for fabrication of articles for packaging, transporting, or storing of foods if residual amount does not exceed 1.5% of the film: Fed Reg 27:3919 (April 25, 1962) (Merck index (2001) 13th edn. Merck and Company, Whitehouse Station).

Thick \ˈthik\ *adj* [ME *thikke*, fr. OE *thicce*; akin to OHGr *dicki* thick, OIr *tiug*] (before 12c) Having relatively great consistency.

Thick-and-Thin Yarn *n* A novelty yarn of varying thickness.

Thickener *n* (1) Any material used to thicken a liquid. (2) An additive used to thicken (increase viscosity) or modify the rheology of a coating. See ▶ Thixotrope. Syn: ▶ Bodying Agent.

Thickening Agent *n* (antisag agent) A substance that increases the viscosity and/or thixotropy of fluid dispersions or solutions. Such agents are used widely in adhesives, coatings and paints to prevent flow and slumping while they are setting or drying to their final form. Examples of thickening agents are bentonite, calcium carbonates with high oil absorption, clays, chrysotile asbestos, hydrated siliceous minerals, magnesium oxide, soaps, stearates, and special organic waves.

Thick Filling See ▶ Coarse Thread.

Thickness Gauging *n* The thickness of many calendered, extruded, and cast products must be measured while they are produced in order to control the thickness within specified tolerance limits. The simplest methods called *contact gauging*, use tools such as calipers, micrometers, and rolls that physically touch the product being measured. Today these are mostly used for checking and backup while *non-contact methods* make on-line measurements and transmit signals to computers, which in turn order process adjustments to be made. Non-contact gauging devices employ nuclear radiation (see ▶ Beta-Ray Gauge), infrared radiation, ultrasound, air nozzles with means for measuring back pressure which varies with product thickness, electrical-capacitance sensors, and optical devices employing beams of light.

Thickness Variation *n*. The differences in thickness among different locations in a product of desired uniform thickness, such as film, sheet, pipe, wire coatings, laminates, bar stock. etc. In making sheet and film, one must be concerned with thickness variation in both the *machine direction* and *transverse direction*, since they usually have quite different causes and cures. The same is true of wall thickness in pipe, where the directions are axial and circumferential. (The thickness itself is radial.) Thickness variations are often of concern in molded and thermoformed products, even though major differences in thickness are there by design. In extruded products, when thickness variations are even a few percent of nominal, more material must be extruded per unit length of product in order to meet minimum thickness specifications, thus increasing unit cost and reducing profit.

Thin End See ▶ Fine End (1).

Thin-Layer Chromatography See ▶ Chromatography.

Thinner *n* (1) The portion of a paint, varnish, lacquer, or printing ink, or related product that volatilizes during the drying process. (2) Any volatile liquid used for reducing the viscosity of coating compositions or components; may consist of a simple solvent, or diluent or a mixture of solvents and diluents. (3) Solvents, diluents, low viscosity oils and vehicles added to inks to reduce their viscosity, consistency or tack.

Thinners *n* Liquids incapable of dissolving a resin but which can partly substitute for a solvent and, at the same time, reduce the viscosity of a paint, varnish, lacquer, or adhesive. See also ▶ Diluent.

Thinning Ratio *n* The amount of thinner that is recommended for a given quantity of paint.

2,2′-Thiobis(4-*t*-octylphenolate)-*N*-Butylamine Nickel *n* (Cyasorb UV 1084) An ultraviolet absorber used in polyolefins for items such as agricultural film wherein good weatherability is required.

Thiocyanogen Value \-sī- a-nə-jən val-( )yü Measure of the number of single double bonds in a substance, for example, an oil. Whereas thiocyanogen is selective in its action, adding on to isolated double bonds only, iodine is not selective and thus a combination of both thiocyanogen and iodine values provides a means of assessing quantitatively the different types of unsaturated components in a material.

Thioindigoid Maroons n (73395) These are maroon in masstone and violet in tint. They are considerably bluer than most organic maroon pigments and find application in shading durable maroon finishes. Generally speaking, thioindigoid maroons are not bake resistant enough, nor sufficiently lightfast in weak tints to be used effectively as toning agents for whites. Some varieties bleed slightly in aromatic and ketone solvents. The nonbleeding types find extensive use in automotive finishes.

Thioindigo Pigments \- in-di- gō pig-mənt\ n Vat dyes covering the orange, red, violet and brown shade ranges. They are noted for their brightness of shade and generally good fastness properties.

Thiokol® n A line of polysulfides and similar materials. See ▶ Polysulfide Rubber. Manufactured by Thiokol, U.S.

Thiourea \-yú- rē-ə\ n [NL] (1894) NH_2CSNH_2. Mp, 180°C; sp gr, 1.42. Used in the preparation of thiourea-formaldehyde resins. Known also as *Thiocarbamide*.

Thiourea-Formaldehyde Resin n An amino resin made by polycondensation of thiourea (thiocarNbamide) [$(NH_2)_2CS$] with formaldehyde.

Thiourea Resins n Resins made by the interaction of thiourea and aldehydes.

Thixotropic adj (1) A liquid or dispersion that exhibits a reduction in viscosity with *time* at constant shear stress, opposite effect is *rheopectic* (not to be confused with *pseudoplastic*, reduction of viscosity with *shear stress*). (2) A term that describe full-bodies material which undergoes a reduction in viscosity when shaken, stirred, or otherwise mechanically disturbed and which readily recovers the original full-bodied condition on standing. See ▶ Body; ▶ False Body.

Thixotrope n Additive used to impart thixotropy to a coating. See ▶ Thickener.

Thixotropic Agent n A chemical that imparts the property of Thixotropy to a solution or suspension. See also ▶ Thickening Agent.

Thixotropic Fluid See ▶ Thixotropy.

Thixotropic Paint n Paint which, while free-flowing and easy to manipulate under a brush, sets to a gel within a short time when it is allowed to remain at rest. Because of these qualities a thixotropic paint is less likely to drip from a brush then other types and can be applied in rather thicker films without running or sagging.

Thixotropy \thik- sä-trə-pē\ n [ISV *thixo*- (fr. Gk *thixis* act of touching, fr. *thinganein* to touch)] (1927) (1) A time-dependent decrease in the viscosity of a liquid subjected to shearing or stirring, followed by a gradual recovery when the action is stopped. This is a desirable property in most paints because the painter wants the pant to spread easily but not to sag or slump when brushing, rolling, or spraying is stopped. The term is often mistakenly applied to Pseudoplastic Fluids. (2) The property of a coating or other material that enables it to thicken or stiffen in a relatively short time on standing at rest but, upon isothermal agitation or manipulation to change to a very soft consistency or to a fluid of high viscosity, the process being completely reversible (gel-sol-gel transformation). Also called *False Body and Shear Thinning*.

Thomson Thermoelectric Effect n The designation of the potential gradient along a conductor which accompanies a temperature gradient. The magnitude and direction of the potential varies with the substance. The coefficient of the Thomson effect or specific heat of electricity is expressed in joules per coulomb per degree centigrade.

Thornel n Graphite yarn, originally manufactured by Union Carbide.

Thread \ thred\ n [ME *thred*, fr. OE *thrîd*; akin to OHGr *drāt* wire, OE *thrāwan* to cause to twist or turn] (before 12c) (1) A slender, strong strand or cord, especially one designed for sewing or other needlework. Most threads are made by plying and twisting yarns. A wide variety of thread types are in use today, e.g., spun cotton and spun polyester, core-spun cotton with a polyester filament core, polyester or nylon filaments (often bonded), and monofilament threads. (2) A general term for yarns used in weaving and knitting, as in "thread count" and "warp thread."

Thread Count n (1) The number of ends and picks per inch in a woven cloth. (2) The number of wales and courses per inch in a knit fabric.

Threaded-Roll Process n A high-speed method developed by Celanese for converting crimped continuous filament tow into highly bulked, uniformly spread webs of up to 108-in. widths. The webs are useful in a variety of products, such as cigarette filters, sleeping pillows, and battings.

Threading Up *v* Actions taken in starting an extrusion operation after turning on the preheated extruder at low speed and beginning the flow of feedstock. As the first extrudant emerges from the die, it is gripped and moved by hand through the various parts of the downstream equipment until each of the driven elements is itself moving the extrudate. Extrusion rate is then gradually increased to the target level.

Threadlines *n* The fiber lines of a manufactured fiber in extrusion or subsequent processes.

Thread Out See ▶ End Out.

Threadup *n* The process of directing or threading fiber or fabric through all machine positions to start or restart a process, or the configuration resulting therefrom.

Threat Plug *n* The male part of a mold that shapes an internal threat in the molding and must be unscrewed from the finished piece. Automatic unscrewing molds can do this mechanically for scores of bottle caps molded in a single shot.

Three-Bar Fabric *n* A tricot fabric made on a machine equipped with three guide bars.

Three-Center Bond *n* A bond consisting of an electron pair shared among, and bonding, three atoms; found in electron-deficient compounds.

Three-Dimensional Braid *n* (3-D braid) A recent development in building reinforcement performs for complex shapes that permits the placing of reinforcing fibers in three orthogonal (or nonorthogonal) directions so as to best support multidirectional stresses expected to act on the finished part in service.

Three-Dimensional Weaving *n* To produce three-dimensional textiles, yarns are simultaneously woven in three directions (length, width, and thickness) rather than in the conventional two. The types of structures that can be produced fall into four broad classes: (1) contoured fabrics, (2) expandable fabrics, (3) interwoven fabrics (Also see ▶ Double Weave), and (4) contoured interwoven fabrics.

Three-Plate Mold *n* An injection mold with an intermediate movable plate that permits center or offset gating of each cavity.

Three-Roll Mill See Triple Roller Mill.

Three-Roll Stack *n* (haul-off) The vertical array of three polished and cored, chrome-plated rolls that receives molten sheet from a sheet die and chills it while impressing the high finish of the rolls on the sheet itself. The position of the center roll is fixed, while the upper and lower rolls are pressed toward the center roll against adjustable stops by air cylinders exerting pressures of 300+ kPa/cm of roll width. Roll temperature are separately controlled by warm water pumped through the shell annuli at high velocity. The molten sheet passes into the nip of the two upper rolls, whose speed, relative to the mean melt velocity in the die, determines how much the sheet is drawn down. Embossed sheet can be made by replacing the center roll with an embossing roll.

Threshold Limit Value *n* (**TLV**) Refers to airborne concentrations of substances, and represents conditions under which it is believed that nearly all workers may be repeatedly exposed, day after day, without adverse effect. These values may be breathed continually for 8 h/day without harm. Because of wide variation in individual susceptibility, exposure of an occasional individual at or even below the threshold limit may not prevent discomfort, aggravation of a preexisting condition, or occupational illness. Threshold limits should be used as guides in the control of health hazards and should not be regarded as fine lines between safe and dangerous concentrations. the American Conference of Governmental Industrial Hygienists (ACGIH) adopts a list of threshold limit values (TLV) each year for more than 450 substances.

Through-Dry See ▶ Dry-Through.

Through-Drying *n* Uniform drying throughout the film as opposed to bottom-drying or top-drying.

Throughput *n* The amount of raw material processed in a specific time. This is the actual amount, not a percentage.

Throwing *n* (1) Defect of impregnating insulating varnishes used on high-speed rotating equipment, and evidenced by the ejection of particles of varnish by the centrifugal force generated. (2) A textile term referring to the act of imparting twist to a yarn, especially while plying and twisting together a number of yarns. A person doing this is called a *thruster*.

Throwing Out *n* (1) See ▶ Gas Checking. (2) Formation of a precipitate. (3) Flocculation.

Throwing Power *n* The ability of an electrodeposition coating or resin to cover an interior surface to which the current has limited access.

Throw Out *n* A precipitate or floc.

Throwster \ˈthrō-stər\ *n* (15c) A company that specializes in putting additional twist in yarn. More recently, the term also applies to a company that specializes in texturing yarns.

Thrum \ˈthrəm\ *n* [ME, fr. OE –*thrum* (in *tungthrum* ligament of the tongue), akin to OHGr *drum* fragment]

(14c) The fringe of warp yarns that remains on the loom when the woven fabric has been cut free.

Thurst Bearing *n* Of an extruder, the heavy-duty bearing upon which the rear shoulder of the screw shank pushes as it transmits the rearward force due to the head pressure. See ▶ Thurst Load. Three types of thrust bearings are in use for single-screw extruders. In order of increasing merit they are *flat-roller bearings, capered-roller bearings,* and *spherical-roller bearings*.

Thurst Load *n* In an extruder or screw-injection molder, the rear-directed force in reaction to the forward buildup of pressure in the screw, culminating in the heard pressure acting over the whole screw cross section, and equal to $\pi \cdot D^2 \cdot P/4$. This force is much less in an injection molder of the same screw diameter because, while the screw is turning, the head pressure is low and when the screw stops, the injection rams take up the thrust. Therefore, their thrust bearings can be much smaller than those of equal-size extruders.

Thus *n* Oleoresinous exudation of rosin type. Because of its relative softness it has been used to a small extent in spirit varnishes as a plasticizer.

Ti *n* Chemical symbol for the element titanium.

Ticking \ **ˈ**ti-kiŋ\ *n* [²*tick*] (1649) A durable, closely woven fabric used for covering box springs, mattresses, and pillows. Ticking may be woven in a plain, satin, or twill weave, usually with strong warp yarns and soft filling yarns.

Tie Bar *n* In a plastic molding press, one of two or four sturdy, cylindrical, steel posts that provide structural rigidity to the clamping mechanism and accurately guide platen movement.

Tie Back See ▶ Sticker (1).

Tie Coat *n* Intermediate coat used to bond different types of paint coats. Syn: ▶ Block Coat, ▶ Transition Primer, ▶ Barrier Coat.

Tiffany Finish *n* A wall decoration in which a number of translucent paints are applied in various hues in an irregular manner and modified by wiping, blending, etc.; a kind of scumbling.

Tight Cure *n* Sufficient vulcanization to give a product good tensile strength and good snap or elasticity.

Tight or Loose End *n* A taut or slack warp end caused by too much or too little tension on an individual end while weaving, by ridgy section or warp beams, by incorrect tensions in beaming or sizing, or as a result of faulty fabric design.

Tight Spot See ▶ Twit.

Tile \ **ˈ**tī(ə)l\ *n* {*often attributive*} [ME, fr. OE *tigele*, fr. L *tegula* tile; akin to L *tegere* to cover] (before 12c) A ceramic surfacing unit, usually relatively thin in relation to facial area, made from clay or a mixture of clay and other ceramic materials, called the body of the tile, having either a "glazed" or "unglazed" face and fired above red heat in the course of manufacture to a temperature sufficiently high to produce specific physical properties and characteristics.

Tile-Like Coating *n* (TLC) System applied by conventional means and intended to produce vitreous (tile-like) finishes on relatively rough masonry or other cementitious walls and ceilings. They are generally thicker, harder, and more washable than conventional paints. See also ▶ High-Build Coating.

Time, Assembly *n* The time interval between the spreading of the adhesive on the adherend and the application of pressure or heat, or both, to the assembly. NOTE – For assemblies involving multiple layers or parts, the assembly time begins with the spreading of the adhesive on the first adherend. (1) open assembly time is the time interval between the spreading of the adhesive on the adherend and the completion of assembly of the parts for bonding. (2) closed assembly time is the time interval between completion of assembly of the parts for bonding and the application of pressure or heat, or both, to the assembly.

Time, Curing *n* The period of time during which an assembly is subjected to heat or pressure, or both, to cure the adhesive. (See also ▶ Curing Time and ▶ Time, Joint Conditioning) NOTE – Further cure may take place after removal of the assembly from the conditions of heat or pressure, or both.

Time, Drying *n* The period of time during which an adhesive on an adherend or an assembly is allowed to dry with or without the application of heat or pressure, or both.

Time, Joint Conditioning *n* The time interval between the removal of the joint from the conditions of heat or pressure, or both, used to accomplish bonding and the attainment of approximately maximum bond strength. Sometimes call *Joint Aging Time*.

Time-Temperature Equivalence *n* Because an increase in temperature accelerates molecular motions, mechanical behavior of polymers at one temperature can be used to predict those at another by means of a shift factor equal to the ratio of relaxation time at the second temperature to that at the first. The principle can be used to combine measurements made at many temperatures into a single master curve for a reference temperature over many decades of time.

Time-to-Break *n* In tensile testing, the time interval during which a specimen is under prescribed conditions of tension and is absorbing the energy required to reach maximum load.

Tinctorial Strength \tiŋ(k)-ˈtōr-ē-əl-\ *n* (ca. 1864) The relative ability of a pigment or dye to impart color value to a printing ink. See ▶ Color Strength.

Tinge See ▶ Cast.

Tin Stabilizer See ▶ Organotin Stabilizer.

Tint \ˈtint\ [alter. of earlier *tinct*, fr. L *tinctus* act of dyeing, fr. *tingere* to tinge (1, *n*) The color produced by the mixture of white pigment with absorbing (generally chromatic) colorants. The color of the resulting mixture is lighter and less saturated than the color without the addition of the white. (2, *vt*) To impart or apply a tint. See ▶ Shade (4) and ▶ Tinting.

Tinter *n* Colored pigments ground in media compatible with paint vehicles, added in relatively small proportions to already prepared paints to modify their color. With the introduction of latex paints of many types, tinters have been developed which can be used both with organic solvent-thinned paints and with water-thinned paints. Such dual-purpose tinters are known as universal tinters. Also called *Stainer*.

Tinting *n* (1) Final adjustment of the color of a paint to the exact color required. (2) In lithography, a uniform discoloration of the background caused by the bleeding or washing of the pigment in the fountain solution.

Tinting (or Tint) Strength, Absorption *n* Relative change in the absorption of a standard white pigment when a specified amount of absorbing pigment, black or chromatic, is added. This is basically the common definition of tinting strength.

Tinting (or Tint) Strength, Scattering *n* Relative change in the scattering of a standard (masstone) black when a specified amount of scattering pigment, white or chromatic, is added. See ▶ Tinting Strength.

Tinting Strength *n* The relative ability of a unit quantity of colorant to alter the color of another colorant to which it is added. In popular usage, tinting strength is an index of the effectiveness with which a chromatic colorant imparts color to a standard white pigment. This definition of the term can be misleading, however. The tinting strength of a yellow, when added to a black, depends on its scattering; the tinting strength, as determined from a mixture with white, tells nothing about its behavior when mixed with colorants of low scattering, or black in the extreme case. In a mixture of pigments, the absorption strength or the scattering strength, or both, may affect its apparent strength. Therefore, see ▶ Tinting Strength, Scattering, ▶ Absorption, and ▶ Timing Strength, Scattering. In any case, tinting strength comparisons of materials of different chemical type may vary with the concentrations of colorants used, so care must be exercised in selecting relative concentrations or concentration ranged. *Also called Tint Strength and Staining Power.*

Tint-Tone *n* Color obtained when a masstone pigment is mixed with a white pigment in a vehicle system.

Tip-Sheared Carpet *n* A textured pile carpet similar to a random-sheared carpet, but with a less defined surface effect.

Tire-Builder Fabric *n* Fabric consisting of tire cord in the warp with single yarn filling at extended intervals.

Tire Construction *n* The geometry of the various layers of tire fabric in the final tire. Three constructions are commonly used. *Bias Tire–* In this construction, tire fabric is laid alternately at bias angles of 25–400° to the tread direction. An even number of layers (or piles) is used. *Radial Tire–* In a radial tire, tire fabric traverses the body of the tire at 90° to the tread direction. Atop the tire fabric are laid alternating narrow layers of fabric at low angles of 10–300° to the tread direction; the belt that is formed around the tire body restricts the movement of the body. *Bias/Belted Tire–* This tire construction combines features of the preceding two. The first layers of fabric are identical to the bias tire. The belt is added in alternating layers at 20° to the tread direction.

Tire Cord *n* A textile material used to impart the flex resistance necessary for tire reinforcement. Tire yarns of polyester, rayon, nylon, aramid, glass, or steel are twisted to 5–12 turns/in. Two or more of these twisted yarns are twisted together in the opposite direction to obtain a cabled tire cord. The twist level required depends on the material, the yarn linear density, and the particular application of the cord. Normally, tire cords are twisted to about the same degree in the S and Z directions, which means that the net effect is almost zero twist in the finished cord. Also see ▶ Tire Fabric.

Tire Fabric *n* A loose fabric woven to facilitate large-scale dipping, treating, and calendering of tire cords. Usually, 15–35 tire cords/in. of warp are woven into a tire fabric by 2–5 light filling yarns/in. In these fabrics, the strength is in the warp and the filling only holds cords in position for processing. The filling yarns are normally broken during tire molding. The warp cords are polyester, rayon, nylon, aramid, glass, or steel and range in strength from 30 to over 100 lb/cord. A 60-in. fabric would normally have a warp strength of about 7,000 lb. Such fabrics are used for tire carcasses and tire belts.

More conventional square woven fabrics are used in certain parts of a tire such as the bead, chafer, and wrapping. Also see ▶ Tire Cord.

Titanate \ˈtī-tᵊn-ˌāt\ n (1839) (1) Any various multiple oxides of titanium dioxide with other metallic oxides. (2) A titanium ester of the general formula Ti(OR)$_4$.

$$\text{O}^- - \text{Ti} \overset{\text{O}}{\underset{\text{O}^-}{}}$$

Titanate Coupler n One of a family of organo-titanium compounds first developed by Kenrich Petrochemicals in 1978 and burgeoning since then. Types available include monoalkoxy, chelate, coordinate, and quaternary salts. They form molecular bridges between organic matrices (resins) and inorganic fillers and reinforcements. Conductivities of metal-filled plastics are increased by one to four orders of magnitude, while melt viscosities are reduced by factors of 0.3–0.1.

Titania \tī-ˈtā-nē-ə\ n [NL] (1922) See ▶ Titanium Dioxide.

Titanium Brown n Brown pigment made by precipitating a mixed solution of cobalt and ferrous salts, containing a suspension of titanium hydroxide, with sodium carbonate. The mixed precipitate is washed and Calcined.

Titanium Dioxide n (1877) (titanic anhydride, titanic acid anhydride, titanium white, Titania) TiO$_2$. A white powder available in two crystalline forms, both tetragonal: *anatase* and *rutile*. Both are widely used as opacifying and brightening pigments in thermosets and thermoplastics, used alone when whites are desired or in conjunction with other pigments when tints are desired. They are essentially chemically inert, light-fast, resistant to migration and heat. The rutile form is denser and has the higher refractive indices (2.61 and 2.90 vs 2.55 and 2.45 for the anatase form), and thus has somewhat greater opacifying power for a given volume percent and particle-size distribution.

$$\text{Ti}^{+4} \quad \text{O}^{--} \quad \text{O}^{--}$$

Titanium Dioxide, Anatase n TiO$_2$. Pigment White 6 (77891). A high opacity, bright pigment of the chalking type, used as a prime pigment in paints, rubber, plastics. Prepared from the mineral, ilmenite, or rutile ore. Density, 3.8–4.1 g/cm^3 (32–34 lb/gal); O.A., 18–30; particle size, 0.3 μm, refractive index, 2.55. Syn: ▶ Titania. See also ▶ Titanium Dioxide, Rutile.

C. I. pigment (General category)	White (Hue)	6 (Consecutive number)	77891 (Chemical class)

Titanium dioxide, rutile. TiO$_2$. pigment

Titanium Dioxide, Rutile n TiO$_2$. Pigment White 6 (77891). A high-opacity, bright white pigment, nonchalking type, used as a prime pigment in paints, rubber, plastics. Prepared from the mineral, ilmenite, or rutile ore. Density, 3.9–4.2 g/cm^3 (33–35 lb/gal); O.A., 16–48; particle size, 0.2–0.3 μm, refractive index, 2.76. Syn: ▶ Titania. See also ▶ Anatase.

Titanium Green n Complex pigments based on Calcined mixtures of titanium oxide or hydroxide with suitable other metallic oxides, carbonates, etc. The other metallic compounds include those of zinc.

Titanium Lithopone n This may be made by mixing a minor proportion of titanium dioxide into lithopone, or possibly by the coprecipitation of the usual lithopone constituents in the presence of titanium hydroxide. The resultant product in the latter case is subjected to controlled calcinations. *Also known as Titanated Lithopone*.

Titanium Pigments n Titanium dioxide (TiO$_2$), rutile and anatase, are the best examples of white pigments.

Titanium Trichloride n TiCl$_3$. A catalyst for olefin polymerization.

$$\text{Cl}^- \quad \text{Ti}^{+++} \quad \text{Cl}^- \quad \text{Cl}^-$$

Titanium White n (1920) A brilliant white lead-free pigment consisting of titanium dioxide often toether with barium sulfate and zinc oxide.

Titanium Yellow See Nickel Titanate.

Titrant \ˈtī-trənt\ n (1939) The substance slowly added during a titration.

Titration \tī-ˈtrā-shən\ n (ca. 1859) The slow addition of a solution of one reactant to one of a second reactant until the equivalence point is signaled by an indicator color change or other method.

Tobacco Cloth n A thin, lightweight, open cloth used to shade and protect tobacco plants.

Tobacco Seed Oil n Seed oil obtained from *Nicotiana tabacum*. Considerable divergencies in composition of the oil have been reported. Some types contain as much as 70% linoleic acid, whereas others contain no linoleic acid and more than 54% of linoleic acid. In consequence, its constants as reported vary considerably.

Certain types have excellent drying properties, and can replace linseed oil without detriment.

Tobias Acid \tə-ˈbī-əs ˈa-səd\ *n* Intermediate used in the manufacture of dyestuffs. 2-naphthylamine-1-sulfonic antioxidant, generally regarded as safe by the Food and Drug Administration. It has been shown to be a good heat stabilizer in polyolefins, providing protection at levels around 250 ppm. Both ATP and its breakdown products are environmentally safe.

Tocopherols \tō-ˈkä-fə-ˌról\ *n* In I[ISV] (ultimo. fr. Gk *tokos* childbirth, offspring (akin to Gk *tiktein* to beget) + *pherein* to carry, bear] (1936) occurring anti-oxidants in vegetable oils.

Toe Closing *n* In knitting hosiery, this term refers to closing the toe opening. It may be knit closed, or in tube hosiery, sewn closed.

TOF *n* Abbreviation for ▶ Trioctyl Phosphate.

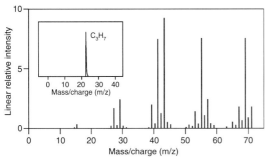

TOF-SIMS spectrogram of polypropylene specimen

Toggle Action *n* A mechanism that magnifies force exerted on a knee joint. It is used as a means of closing and locking press platens and also serves to apply pressure at the same time.

Toile \ ˈtwäl\ *n* [F, cloth, linen, fr. MF] (1794) (1) A broad term describing many simple plain weave twill fabrics, especially those made from linen. (2) Sheer cotton and linen fabrics.

Tole \ ˈtōl\ *n* {*often attributive*} [F *tôle*, fr. MF dialect *taule*, fr. L *tabula* board, tablet] (1927) Painted tin ware.

Tolerance *n* The total range of variation (usually bilateral) permitted for a size, position, or other required quantity; the upper and lower limits between which a dimension must be held.

Tolerance Interval *n* The specified allowance of variation of a dimension (or other quantity) above and below the nominal or target values in a production part or product. For a process whose average level and random variation are in control, symmetrical tolerance limits should be at least six processes standard deviations apart in order to approach 0% defective parts.

Tolerances are better understood and getting much more attention now than a few decades ago when a New York molder was asked bout tolerances on parts he was producing. His reply: "Hey, we got lotsa tolerance here! We hire people no matter what color or nationality they are."

Toluene (Tuluol) \ ˈtäl-yə-ˌwēn\ *n* [F *toluène*, fr. *tolu* balsam fr. the tropical Americal tree *Myroxylon balsamum*, fr. Sp *tolú*, fr. Santiago de *Tolú*, Colombia] (1871) (toluol, methylbenzene, methylbenzol) $H_3CC_6H_5$. A colorless, flammable liquid with a sharp, benzene-like odor, used as a solvent for cellulosics, vinyl organosols, and other resins and is used in the manufacture of coatings. Toulene is also a synthesis intermediate for polyurethanes and polyesters. The commercial product has a boiling range, 105–112°C; flp, 50°C; vp, 26 mmHg/30°C. The term "toluol" is still used commercially but is not preferred.

Toluene-2,4-Diamine *n* \-ˈdī-ə-ˌmēn\ (TDA, *m*-tolylene diamine) $H_3CC_6H_3(NH_2)_2$. A colorless, crystalline material used in the product of toluene diisocyanate, a key material in the manufacture of polyurethanes.

Toluene-2,4-Diisocyanate *n* (2,4-toulene diisocyanate, Br: tolylene diisocyanate, TDI) $H_3CC_6H_3(NCO)_2$. A water-white to pale yellow liquid with a sharp, pungent odor, produced by reacting toluene-2,4-diamine with phosgene. (Some of the 2,6-isomer is usually present.) It reacts with water to produce carbon dioxide. TDI is widely used in the production of polyurethane foams and elastomers, but is toxic and requires careful handling to keep its concentration in work-space air below the permissible threshold limits. See also ▶ Diisocyanate.

p-Toluenesulfonamide \-ˌsəl-ˈfä-nə-ˌmīd\ n (PTSA) $H_3CC_6H_4SO_2NH_2$. White leaflets, existing also in the o-form; both are used as solid plasticizers for ethyl cellulose, polyvinyl acetate, and rigid PVC.

Toluenesulfonamide Resin n Resins made by the interaction of toluene sulfonamide and formaldehyde.

p-Toluenesulfonylhydrazide n A blowing agent similar to benzene-sulfonylhydrazide, but having higher melting and decomposition temperatures.

p-Toluenesulfonylsemicarbazide n A blowing agent with a high decomposition temperature (235°C) that makes it useful for foaming plastics that are processed at high temperatures, such as high-density polyethylene, polypropylene, rigid PVC, acrylonitrile-butadiene-styrene resins, polycarbonates, and nylons.

Toluidine Blue \tə-ˈlü-ə-ˌdēn-\ n (1898) A basic thiazine dye that is related to methylene blue and is used as a biological stain.

Toluidine Reds n Pigment Red 3 (12120). Series of red dyestuffs made by diazotizing 2-nitro-p-toluidine and coupling this with β naphthol under alkaline conditions. By altering the conditions of preparation, reds of different shade, brilliance, strength, etc., are obtained.

Toluidine Toner See ▶ Toluidine Reds.

Toluidine Yellow See ▶ Hansa Yellow.

2,4-Tolulene Diisocyanate n British Syn: ▶ Toluene-2,4-Diisocyanate.

Tone \ˈtōn\ v (1660) (1) A modification of a full-strength color (mass tone) secured by blending with other colors. (2) Use of this term is to be deprecated since it is variously employed in different senses. According to the context, "hue" or "undertone" are, respectively, the preferred terms. See ▶ Hue and ▶ Undertone Color.

Toner n (1) A single organic pigment which does not contain inorganic pigment (extender), or a single inorganic pigment having maximum absorption strength for the given type of pigment. (2) Also loosely applied to pure pigment dyestuff, but this use of the term is deprecated. (3) A highly concentrated pigment and/or dye used to modify the hue or color strength of an ink. Also: the "ink" in electrostatic printing.

Tongue Tear Strength n The average force required to tear a rectangular sample with a cut in the edge at the center of the shorter side. The two tongues are gripped in a tensile tester and the force required to continue and tear is measured.

Toning Blue See ▶ Iron Blue.

Tooth n In a dry paint film, a fine texture imparted either by a proportion of relatively coarse or abrasive pigment or by the abrasives used in sanding; this texture improves the rubbing properties and also provides a good base for the adhesion of a subsequent coat of a paint or varnish.

Top n (1) A wool sliver that has been combed to straighten the fibers and to remove short fiber; an intermediate stage in the production of worsted yarn. (2) A similar untwisted strand of manufactured staple delivered by the comb or made directly from tow.

Top Coat n The coating intended to be the last coat applied in a coating system; usually applied over a primer, undercoaters, or surfacers. See ▶ Coat. Also known as Finish Coat.

Top Color n Colors used on the ground color to form a design. See ▶ Face Color.

Top Drying n Drying of a film at the top only, e.g., cobalt naphthenate is used as a top drier.

Top Dyeing *n* (1) The process of covering with an additional dye, not necessarily of the same color or class, to obtain the desired shade. (2) Fiber in top form is placed in cans and dyed in a batch-dye vessel with reverse cycling capability. An expensive process that is used primarily for fancy yarns.

Topology of Polymers \tə-ˈpä-lə-jē-\ *n* The surface texture of polymers.

Toptone See ▶ Masstone.

Torpedo *n* (1) (spreader) A streamlined, conicylindrical block, supported by three ridges (*spider*) and placed in a path of flow of the plastic pellets within the heating cylinder of a plunger-type injection molder, to spread the stock into a thin annulus, thus providing intimate contact with the heating surfaces. (2) Years ago, some extruder screws ended in smooth torpedoes only a little less in diameter than the barrels, the idea being to provide final shear mixing and improve homogeneity of the melt arriving at the die. (3) The core of an in-line pipe die supported by spider legs is sometimes called a torpedo.

Torque \ˈtórk\ *n* [F, fr. L *torques*, fr. *torquēre* to twist] (1695) A force or a combination of forces that produces or tends to produce a twisting or rotating motion. In reference to yarn, torque refers to the yarn's tendency to turn on itself, or kink, as a result of twisting.

Torque Produced by the Action of One Magnet on Another *n* The turning moment experienced by a magnet of pole strength m' and length $2l'$ placed at a distance r from another magnet of length $2l$ and pole strength m, where the center of the first magnet is on the axis (extended) of the second and the axis of the first is perpendicular to the axis of the second,

$$C = \frac{mm'll'}{r^2} = \frac{2MM'}{r^3}$$

If the first magnet is deflected through an angle θ, the expression becomes,

$$C = \frac{2MM'}{r^3}\cos\theta$$

Torque Yarns See ▶ Textured Yarns.

Torr \ˈtór\ *n* [Evangelista *Torricelli*] (1949) A unit of pressure: 1 torr = 1 mmHg = $\frac{1}{760}$ atm. The torr (for E. Torricelli, who invented the mercury barometer in 1643) was introduced a few decades ago when people grew weary of saying "millimeters of mercury," to which it was set equal. Now the torr, too, is depreciated.

Torsion *n* [LL *torsion-, torsio* torment, alter. of L *tortio*, from *torquēre* to twist] (1543) Engineering term for modes of shear stress and shear strain caused by twisting of bodies (Shah V (1998) Handbook of plastics testing technology. Wiley, New York).

Torsional Braid Analysis *n* A method of performing torsional tests on small amounts of materials in states in which they cannot support their own weight, e.g., liquid thermosetting resins. A glass braid is impregnated with a solution of the material to be tested. After evaporation of the solvent, the impregnated braid is used as a specimen in an apparatus that measures motion of the oscillating braid as it is being heated at a programmed rate in a controlled atmosphere. ASTM D 4065 provides information on torsional testing (Shah V (1998) Handbook of plastics testing technology. Wiley, New York).

Torsional Modulus *n* Shear modulus (G) as measured in a test in which the specimen is twisted. In ASTM D 1043, the test specimen is a flat rectangular bar with length about 7 times its width of 6.35 mm and thickness 1/3 the width. The apparent shear modulus is given by:

$$G = k \cdot T \cdot L / (a \cdot b^3 \cdot \phi)$$

where T is the torque exerted on the specimen, L, a, and b are its length, width and thickness, ϕ is the angle of twist, and k is a coefficient dependent on units and the ratio a/b (Shah V (1998) Handbook of plastics testing technology. Wiley, New York).

Torsional Rigidity *n* Of a fiber, wire, bar, tube, or profile shape subjected to twisting of one end relative to the other, the torque required to produce a twist of one radian. This rigidity is proportional to the shear modulus of the material and is strongly dependent on all dimensions, especially section thickness.

Torsional Test *n* A test for determining shear properties of plastics, such as shear modulus, based on measuring the torques required to twist a specimen through a prescribed arc. See ▶ ASTM, www.astm.org, or standard test methods.

Torsional Vibration See Angular Harmonic Motion.

Torsion-Braid Analyzer *n* An instrument which permits the measurement of thermomechanical properties of polymers that are undergoing structural changes during cure (Shah V (1998) Handbook of plastics testing technology. Wiley, New York).

Torsion Pendulum *n* A device for performing dynamic mechanical analysis in which a sample is deformed torsionally and allowed to oscillate in free vibration (Shah V (1998) Handbook of plastics testing technology. Wiley, New York).

Tortuosity Factor *n* (1) The distance a permeating molecule must travel to pass through a film, divided by the thickness of the film. (2) The mean length of path of fluid molecules passing through a packed bed or porous medium (such as open-cell foamed plastic) divided by the thickness of the bed in the direction of the pressure gradient (Shah V (1998) Handbook of plastics testing technology. Wiley, New York).

Total Color Difference n The perceived difference between two colors, including the differences in hue, saturation, and lightness; generally used as a value calculated from a color difference equation and designated as ΔE (McDonald, Roderick (1997) Colour physics for industry, 2nd edn. Society of Dyers and Colourists, West Yorkshire). See ▶ Color Difference and ▶ Color Difference Equations.

Total Denier n The denier of a tow before it is crimped. It is the product of the denier per filament and the number of filaments in the tow. The total denier after crimping (called crimped total denier) is higher because of the resultant increase in weight per unit length.

Total Denier (of Tow) n The product of the denier per filament times the number of filaments in a tow.

Total Reflectance See ▶ Reflectance.

Total Reflection n When light passes from any medium to one in which the velocity is greater, refraction ceases and total reflection begins at a certain critical angle of incidence of θ such that

$$\sin \theta = \frac{1}{n}$$

where n is the index of the first medium with respect to the second. If the second medium is air n has the ordinary value for the first medium. For any other second medium,

$$n = \frac{n_1}{n_2}$$

where n_1 and n_2 are the ordinary indices of refraction for the first and second medium respectively (Moller KD (2003) Optics. Springer, New York; Kokhanovsky AA (2004) Light scattering media optics. Springer, New York; Saleh BEA, Teich MC (1991) Fundamentals of photonics. Wiley, New York).

Total Shear n A concept (1956) that indicates shear mixing in extruders, but also applicable to other processes in which shear is the principal flow mode. It is the integrated product of shear rate times time over the region in which a melt is undergoing shear flow. The ratio of initial to final *striation thickness* is proportional to total shear. See ▶ Striation.

Total Solids See ▶ Solids.

Total Transmittance See ▶ Transmittance, Total.

TOTM Abbreviation for ▶ Trioctyl Trimellitate.

Touch-Dry See ▶ Set-to-Touch Time and ▶ Drying Time.

Touch-Up Painting n Application of paint on small areas of painted surfaces to repair mars, scratches, and small areas where the coating has deteriorated, in order to restore the coating to an unbroken condition.

Toughness n A term with a wide variety of meanings, no single precise mechanical definition being generally recognized, but tensile strength with pact resistance are indicator of toughness. Also, it is the measure of the ability of a sample to absorb mechanical energy without breaking, usually defined as the area underneath a stress-strain curve (Shah V (1998) Handbook of plastics testing technology. Wiley, New York). Toughness generally implies a lack of brittleness; having very substantial elongation to break accompanied by high tensile strength. One proposed definition for toughness is the energy per unit volume to break a material, equal to the area under the stress-strain curve. Toughness has also been equated to impact resistance, especially resistance to repeated impacts. Energy required to break a specimen in the ▶ Tensile-Impact Test, divided by the gauge-length volume of the specimen, is a fairly straightforward toughness measure. Also, that energy to break in tension is definitely dependent on the time scale of the test. Toughness has also been equated to resistance to abrasion, and to resistance to penetration by points and cutting with sharp tools (see ▶ ASTM and ▶ www/astm.org).

Tow \ˈtō\ n [ME, fr. OE *tow-* spinning; akin to ON *tō* tuft of wood for spinning, OE *tawian* to prepare for use] (14c) A large strand of continuous manufactured fiber filaments without definite twist, collected in loose, rope-like form, usually held together by crimp. Tow is the form that most manufactured fiber reaches before being cut into staple. It is often processed on tow-conversion machinery into tops, sliver, or yarn, or on tow-opening equipment to make webs for various uses (Wallenberger FT, Weston NE (eds) (2003) Natural fibers, plastics and composites. Springer, New York; Kadolph SJJ, Langford AL (2001) Textiles. Pearson Education, New York).

Towpreg A prepreg consisting of resin-impregnated TOW, that may be braided or woven to form a reinforced-plastic structure (Wallenberger FT, Weston NE (eds) (2003) Natural fibers, plastics and composites. Springer, New York; Kadolph SJJ, Langford AL (2001) Textiles. Pearson Education, New York; Chung DD (1994) Carbon fiber composites. Elsevier Science and Technology Books, New York).

Toxic \täk-sik\ {*combining form*} [NL, fr. L *toxicum*] Poisonous.

Toxicity *n* (1) The measure of the adverse effect exerted on the human body by a poisonous material. (2) A relative property of a chemical agent with reference to a harmful effect on some biological mechanism and the condition under which this effect occurs. The quality of being poisonous (Ashford NA, Miller CS (1997) Chemical exposures: low levels and hihg stakes. Wiley, New York).

Toxic Pollutants *n* Those pollutants, which, after discharge and upon contact with any organism, either directly from the environment or indirectly by ingestion through food chains, will cause death, disease, behavioral abnormalities, cancer, genetic mutations, physiological malfunctions or physical deformities in such organisms or their offspring (Ashford NA, Miller CS (1997) Chemical exposures: low levels and hihg stakes. Wiley, New York).

Toxic Substance *n* A substance that demonstrates the potential to induce cancer, to produce short and long term disease or bodily injury, to affect health adversely, to produce acute discomfort, or to endanger life of man or animal, resulting from exposure via the respiratory tract, skin, eye, mouth or other routes in quantities which are reasonable for experimental animals or which have been reported to have produced toxic effects in man (Ashford NA, Miller CS (1997) Chemical exposures: low levels and hihg stakes. Wiley, New York).

TPA *n* Abbreviation for Tererphthalic acid.

TPE *n* Abbreviation for Thermoplastic Elastomer.
TPO *n* Abbreviation for ▶ Thermoplastic Polyolefin.
TPP *n* Abbreviation for ▶ Triphenyl phosphate.

TPR *n* Registered trade name of Uniroyal, Inc., for a family of thermoplastic rubbers based mainly on ethylene and propylene. Grades range in hardness (Shore A scale) from 65 to 90. Processable by the usual thermoplastics methods, these materials have the properties of vulcanized rubber.

TPS *n* Abbreviation used by the British Standards Institution for "Toughened Polystyrene," equivalent to U.S. high-impact polystyrene.

TPTF *n* Abbreviation for Triphenytinfluoride. A toxicant used in anti-fouling paints.

TPU See Thermoplastic Polyurethanes.
TPUR See Thermoplastic Polyurethanes.
TPX *n* Poly(4-methyl-pentene-1). Manufactured by ICI, Great Britain. Abbreviation for Poly(4-Methylpentene-).
TR *n* Thermoplastic elastomers. Abbreviation used by British Plastics Institution for Thio Rubber. See ▶ Polysulfide Rubber.

Tracking *n* An electrical-breakdown phenomenon in polymers in which current caused by an excessive voltage difference between two conductors in contact with an insulating material gradually creates a conductive leakage path across the surface of the material by forming a carbonized track that appears as a thin, wrigly line between the electrodes (Dissado LA, Fothergill CJ (eds) (1992) Electrical degradation and breakdown of polymers. Institution of Electrical Engineering (IEE), London).

Tractive Force of a Magnet *n* If a magnet with induction B has a pole face of area A the force, is,

$$F = \frac{B^2 A}{8\pi}$$

If B and A are in cgs units, F will be in dynes.

Trade Molder *n* The British term for ▶ Custom Molder.
Trade Sales Coating See ▶ Architectural Coatings.

Trade Sales Paints *n* Coatings applied on-site at ambient conditions by the consumer using application methods such as brushing or roller coating. See also ▶ Architectural Coatings.

Traffic Marking *n* Marring or discoloration, or both, of a floor surface by traffic.

Traffic Paint *n* Paint specially formulated to withstand wear of vehicular traffic and to be highly visible at night; used to mark center lines on roadways, traffic lanes, crosswalks, etc. (Wicks ZN, Jones FN, Pappas SP (eds) (1999), 2nd edn. Organic coatings science and technology. Wiley, New York). See also ▶ Beaded Paint.

Tragacanth \ˈtra-jə-ˌkan(t)th\ *n* [MF *tragacanthe*, fr. L *tragacantha*, fr. Gk *tragakantha*, fr. *tragos* goat + *akantha* thorn] (1573 Water-soluble gum derived from *Astragalus* shrubs, and exported from Iran, Iraq, Turkey, India, Syria and neighboring countries. It is obtained both as a natural and as an artificial exudation. It resembles gum Arabic in being composed of calcium, magnesium and potassium salts. It is used as a stabilizer for emulsions (Whistler JN, BeMiller JN (eds) (1992) Industrial gums: polysaccharides and their derivatives. Elsevier Science and Technology Books, New York, New York).

Trailing Flight Face *n* (trailing flight) In an extruder screw, the rear side of any flight. The forward side is called the ▶ Leading Flight Face.

trans- {*prefix*} [L *trans-*,*tra-* across, beyond, through, so as to change, fr. *trans* across, beyond] An organic-chemistry prefix denoting an isomer in which certain atoms or groups are located on opposite sides of a plane of symmetry. Usually ignored when alphabetizing names of compounds. Compare *CIS-* (Morrison RT, Boyd RN (1992), 6th edn. Organic chemistry. Prentice Hall, Englewood Cliffs).

Transducer \-ˈdü-sər\ *n* (1924) A device that transforms the value of a physical variable into an electrical signal, usually voltage or current. Examples are: thermocouple, pressure transducer, linear variable differential transformer, (a motion transducer), tachometer generator, and force cell.

Transesterification *n* In the production of polyester from dimethyl terephthalate and ethylene glycol, the process of exchanging ethylene glycol for the methyl groups to obtain bis-*p*-hydroxyethyl terephthalate. The methanol generated in the reaction is removed as it is formed to drive the reaction to completion.

Transfer Coating *n* A process for coating fabrics such as knits, which are extremely difficult to coat directly by conventional spread-coating methods. In a typical version of the process, a layer of plastisol is cast against a silicone-treated release paper. This first layer becomes the top coat or wear layer in the final product. After gelling of the first layer a second coating of urethane solution is applied to serve as an adhesive layer that bonds the wear layer to the fabric substrate. The composite is finally heated and pressed together, and the paper is stripped away. Many variations of this process are possible, e.g., using vinyl or polyurethane foam, embossing, etc. (Pittance JC (ed) (1990) Engineering plastics and composites. SAM International, Materials Park).

Transfer Molding *n* A molding process used mainly for thermosetting resins and vulcanizable elastomers. The molding material, usually preheated, is placed in an open "pot" with a hole in its bottom atop the closed mold. The cross-sectional areas of this pot is about 15% larger than the total projected area of all cavities and runners in the mold. A ram is placed in the pot above the material. Pressure, applied by a press platen to the ram, forces the molding material through the runners and gates, and into the cavities of the heated mold. Following a heating cycle in which the material is cured or vulcanized, the press is opened and the parts are ejected. In a variation called *plunger molding*, resembling injection molding, the plunger is more a part of the press rather than of the mold, and pressure is applied to the plunger by an auxiliary hydraulic ram. The compound flows faster in plunger molding and more frictional heat is developed, so molding cycles are generally shorter than in transfer molding (Pittance JC (ed) (1990) Engineering plastics and composites. SAM International, Materials Park).

Transfer-Molding Pressure *n* The pressure applied to the cross-sectional area of the material pot or cylinder (MPa or kpsi) (Pittance JC (ed) (1990) Engineering plastics and composites. SAM International, Materials Park).

Transfer Paper *n* The temporary backing employed in decalcomania (Skeist I (ed) (1990) Handbook of adhesives. Van Nostrand Reinhold, New York).

Transfer Printing See ▶ Decalcomania (Skeist I (ed) (1990) Handbook of adhesives. Van Nostrand Reinhold, New York).

Transfer Roll *n* In the coil coating industry, the roll between the pick-up roll and applicator roll where three rolls are present in a coating operation.

Transfer Roller See Anilox Roller.

Transfer Tail A long end of yarn wound at the base of a package that permits increased warping or transfer efficiency by providing an easily accessible connecting point for the succeeding package.

Trans Isomer \-ˈī-sə-mər\ *n* Any isomer in which two identical groups are located on opposite sides of a structure (Morrison RT, Boyd RN (1992), 6th edn. Organic chemistry. Prentice Hall, Englewood Cliffs).

Transition \tran(t)-ˈsi-shən\ *n* [L *transition-, transitio*, fr. *transire*] (1551) The pronounce change in the properties of a material that occur at a certain temperature (the transition temperature) or over a range of temperatures. First order transitions is one in which a discontinuity in the intensive properties occurs. Second order transitions are associated with the onset of particular modes of molecular motion (Groenewoud WM (2001) Characterization of polymers by thermal analysis. Elsevier Science and Technology Books, New York).

Transition Element *n* A member of one of the B groups, which intervene between group IIA and IIIA in the periodic table.

Transition Metal *n* [fr. their being transitional between the more highly electropositive and the less highly electropositive elements] (1940 Any of various metallic elements (as chromium, iron, and nickel) that have valence electrons in two shells instead of only one. *Known also as Transition Element.*

Transition Point See Phase Transition Point.

Transition Primer *n* Coating compatible with primer and also with a finish coat which is not compatible with the primer. See ▶ Block Coat, ▶ Tie Coat and ▶ Barrier Coat.

Transition Section *n* (transition zone, compression section) In a metering-type screw for a single-screw extruder the section of decreasing channel volume per turn between the feed and metering sections, in which the plastic is changing state from a loosely packed bed of particles-*cum*-voids, to a void-free melt the transition may be abrupt or gradual, the latter having been found more satisfactory for nearly all thermoplastics, and may be accomplished by increasing the root diameter of the screw or by reducing the lead angle or both. Reducing the root diameter has been by far the preferred method and may be done conically or helically. See ▶ Conical Transition, ▶ Helical Transition, and ▶ Compression Ratio.

Transition Temperature *n* (1) The Glass-Transition Temperature. (2) More generally, any temperature at which a polymer exhibits an abrupt change in phase or measurable property, or at which a property's rate of change with temperature changes abruptly (second-order transition).

Translucency \-sᵊn(t)-sē\ *n* (ca. 1610) Appearance state between complete opacity and complete transparency; partially opaque.

Translucent *n* Transmission of light in such a way that image-forming rays are irregularly refracted and reflected.

Translucent Coating *n* A liquid formulation (such as varnish, shellac, or lacquer) which when dry forms a translucent film.

Transmission \tran(t)s-ˈmi-shən\ *n* [L *transmission-, transmissio*, fr. *transmittere* to transmit] (1611) Process by which radiant energy is transmitted through a material or an object.

Transmission Electron Microscopy *n* The Transmission electron microscopy is applied to observe phase domain of a size of 50–1,000Å. This and applying dyeing techniques such as oxidizing the unsaturated domain with OsO 4 and RuO 4 (Scanning electron microscopy and x-ray microanalysis (2003) Springer, New York).

Transmittance *n* (ca. 1855) (1) Of light, that fraction of the emitted light of a given wavelength which is not reflected or absorbed, but passes through a substance. See ▶ Light Transmittance. (2) The ratio of the transmitted radiant flux to the incident flux.

Transmittance, Diffuse *n* Ratio of radiant flux transmitted in all forward directions, except the undeviated direction (the specular transmittance), to incident flux.

Transmittance, Internal *n* Ratio of radiant flux reaching the exit surface of material to flux which penetrated entry surface; sometimes referred to as transmittancy.

Transmittance, Specular *n* Ratio of flux transmitted without change in image-forming state to incident flux; undeviated transmitted beam.

Transmittance, Total *n* Ratio of total flux transmitted in all forward directions to incident flux; includes both diffuse and specular transmittances.

Transmutation \ˌtran(t)s-myú-ˈtā-shən\ *n* [ME *transmutacioun*, fr. MF or L; MF *transmutation*, fr. L *transmutation-*, transmutatio, fr. *transmutare*] (14c) The transformation of one element into another.

Transparency *n* State of being transparent or completely nonscattering; attribute of located ode of appearance that permit perception of object or space through or beyond a surface or within a volume. It can also be written as, the degree to which a material or substance transmits light.

Transparent *adj* Adjective to describe a material which transmits light without diffusion or scattering.

Transparent Coating *n* A liquid formulation (such as varnish, shellac, or lacquer) which, when dry, forms a transparent film.

Transparent Inks *n* Inks which lack hiding power and permit light to pass through. They permit previous printing to show through, the two colors blending to produce a third, e.g., a transparent yellow over a blue to produce a green, where the two colors are superimposed (Printing ink handbook, National association of printing ink manufacturers (1999), 5th edn. Kluwer, London; Leach RH, Pierce RJ, Hickman EP, Mackenzie MJ, Smith HG (eds) (1993) Printing ink manual, 5th edn. Blueprint, New York).

Transparent Painting *n* Painting with transparent pigments as opposed to opaque ones, e.g., glazing, water color painting. Transparent pigment is converted to opaque pigment by the addition of white.

Transparent Plastic *n* A plastic that transmits incident light with negligible scattering and little absorption, enabling objects to be seen clearly and brightly through it. At least nine basic classes of plastics are generally regarded as being permanently transparent in thick sections; some others possess near-transparency, at least for a limited period. the nine are: acrylics (foremost in usage for transparency), cellulosics, allyl Diglycol carbonates, some epoxies, nylons, and plasticized PVCs, polycarbonate, polysulfone, crystal polystyrene, and polyphenyl Sulfone.

Transputer *n* A type of process-control architecture in which computing elements are linked not only by a systems' bus but also by additional links combining software and hardware, called "firmware" links. The gain in speed of handling control tasks (as of 11/91) is about a factor of 20, permitting the use of complex control algorithms for, say, injection molding, that could not be used with conventional microprocessor/bus systems.

Transuranic \ ˈtran(t)-syü- ˈra-nik\ *adj* (1935) Of, relating to, or being an element with an atomic number greater than that of uranium (92).

Transuranic Elements *n* Elements of atomic numbers above 92. All of them are radioactive and are products of artificial nuclear changes. All are members of the actinide group.

Transverse Direction *n* (1) In extruding sheet or film, the direction of the width, crosswise to the direction of extrusion. (2) In a uniaxially oriented plastic, either direction perpendicular to the direction of orientation (stretching). In a biaxially oriented sheet, the direction perpendicular to both axes (the plane) of orientation. (3) In a fiber-reinforced laminate the thickness direction or, in a laminate with unidirectional reinforcement, either of the two directions perpendicular to the fiber lengths.

Trapezoid Tear Tester \ ˈtra-pə- ˌzȯid\ *n* See Elmendorf Tear Tester.

Trapped End *n* An end that is unable to unwrap or unwind from the beam. Trapping of an end may be prolonged or intermittent depending upon the cause of trapping (e.g., rolled ends at the selvage, short ends, or mechanical difficulties).

Trapped-Sheet Forming *n* A process announced in 1956 for high-speed, pressure-thermoforming of thin, biaxially oriented polystyrene sheet into snap lids for dairy-product containers, so called because each circle in the sheet was edge-restrained during heating to prevent its shrinking and losing its oriented strength.

Trapping *n* Printing of one ink film on another, in multicolor wet printing. Successful trapping depends upon the relative tack and thickness of the ink film applied. See ▶ Wet Printing.

Travel \ ˈtra-vəl\ *n* Change in color as the angle of viewing a goniochromatic material, such as a metallic paint film, is changed from the perpendicular to near-grazing. *Sometimes called Flop or Flip-Flop.*

Traveler *n* A C-shaped, metal clip that revolves around the ring on a ring spinning frame. It guides the yarn onto the bobbin as twist is inserted into the yarn.

Traverse Length *n* The lateral distance between the points of reversal of the wind on a yarn package.

Traverse Ratio See ▶ Wind Ratio.

Travis *n* Vinyl acetate/vinylidene cyanide copolymer, originally manufactured by Hoechst-Celanese.

Treated Pigment *n* A pigment that has been processed during manufacturing to impart specific properties.

Treater *n* (1) Equipment for preparing resin-impregnated reinforcements, including means for passing a continuous web or strand through a resin tank, controlling the amount of resin picked up, drying and/or partly curing the resin, and rewinding the impregnated reinforcement. (2) The equipment used in ▶ Corona-Discharge Treatment.

Tree Bark *n* A term describing the rippled or wavy effect sometimes seen when a bonded fabric is stretched in the horizontal (widthwise) direction. This defect is caused by bias tensions present when two distorted or skewed fabrics are bonded.

Tree Formation *n* The generation of a tree-like void structure in a transparent plastic by electron bombardment at a point on the surface. The effect in acrylic blocks is dramatically decorative. Similar breakdown structures form in dielectrics subjected to strong electric fields, eventually penetrating the dielectric and causing a short circuit (Ku CC, Liepins R (1987) Electrical properties of polymers. Hanser Publishers, New York; Seanor DA (1982) Electrical conduction in polymers. Academic Press, New, Academic Press, New York).

Tremolite \ˈtre-mə-ˌlīt\ n [F trémolite, fr. *Tremola*, valley in Switzerland] (1799) $Ca_2Mg_5Si_8O_{22}(OH, F)_2$. A silicate mineral similar to and sometimes sold as fibrous talc. It can be used in many applications in place of asbestos as a filler.

Trevira n Polyester (fiber), manufactured by Hoechst, Germany.

Triacetate \(ˌ)trī-ˈa-sə-ˌtāt\ n [ISV] (1885) A generic term for film or fibers of cellulose acetate in which at least 92% of the hydroxyl groups are acetylated. See also ▶ Cellulose Triacetate.

Triacetate Fiber n A manufactured fiber produced from cellulose triacetate in the forms of filament yarn, staple, and tow. Cellulose triacetate fiber differs from acetate fiber in that during its manufacture the cellulose is completely acetylated whereas acetate, which is diacetate, is only partially acetylated. A fiber may be called triacetate when not less than 92% of the hydroxyl groups are acetylated. Fabrics of triacetate have higher heat resistance than acetate fabrics and can be safely ironed at higher temperatures. Triacetate fabrics that have been properly heat-set (usually after dyeing) have improved ease-of-care characteristics because of a change in the crystalline structure of the fiber. Also see ▶ Acetate Fiber (Complete textile glossary, Celanese acetate LLC (2000) Three Park Avenue, New York).

Triacetin n A type of plasticizer for acetate fibers. It is widely used to add firmness to cigarette filter rods.

Triad \ˈtrī-ˌad *also* -əd\ n [L *triad-, trias*, fr. Gk, fr. *treis* three] (1546) Group of three colors harmoniously related to each other.

Triallyl Cyanurate n (TAC, 2,4,6-triallyloxyl-1,3,5-triazine) $(CH_2=CHCH_2OC)_3N_3$. This heterocyclic compound, a solid below 110°C, is highly reactive and is used in copolymerizations with vinyl-type monomers to form ▶ Allyl Resins. It is also used to crosslink unsaturated polyesters and raise their softening temperatures.

Triallyl Phosphate n (TAP) $(CH_2=CHCH_2O)_3PO$. A monomer that can be polymerized with methyl methacrylate to produce flame-retardant copolymers.

Triangle or Polygon of Forces n If three or more forces acting on the same point are in equilibrium, the vectors representing them form, when added, a closed figure.

Triaryl Phosphate n A synthetic-ester type plasticizer derived from isopropylphenol feedstock, useful as a flame-retarding plasticizer in vinyl plastisols.

Triaxial Braid n A braided structure with axial yarns running in the longitudinal direction.

Triaxial-Braid Performing n Any braiding technique for fibrous reinforcements by which three-dimensional performs for reinforced-plastics structures are produced.

Triaxial Fabrics n Completely isotropic fabrics made in a weaving process employing three yarns at 60° angles to each other. These fabrics have no stretch or distortion in any direction. With equal sizes and number of yarns in all three directions, the fabric approaches equal strength and stiffness in all directions (Yates M Fabrics (2002) W. W. Norton and Company, New York).

Triaxial Weaving *v* Weaving in which the cloth is made from three yarns whose axes are 120° apart. When used as a reinforcing medium, the cloth yields a laminate whose properties in the plane are nearly isotropic.

Triazine **trī-ə-**\|zēn\ *n* [ISV] (1894) Any of three compounds containing a ring composed of three carbon and three nitrogen atoms.

Triazine Resin *n* (NCNS resin) A class of thermosetting polymers prepared from primary and secondary biscyanamides with pendant arylsulfonyl groups. the biscyanamides are reacted together in solutions to form soluble prepolymers by addition polymerization. By refluxing these solutions, stable laminating varnishes are obtained. Alternatively, evaporation of solvents from the solutions yields the prepolymers in powder form for molding. Laminates prepared with triazine resins have good mechanical strength at high temperatures and are relatively fire-retardant.

Tribasic \(|)trī-**|**bā-sik\ *adj* (1837) Pertaining to acids having three replaceable hydrogen atoms per molecule, e.g., phosphoric acid, the most important one; or to acid salts in which two of three available hydrogens have been replaced by metals.

Tribasic Lead Maleate *n* A yellowish-white crystalline powder, an effective heat stabilizer for vinyls and a curing agent for chlorosulfonated polyethylene (Gooch, 1993).

Tribasic Lead Sulfate *n* $3PbO \cdot PbSO_4 \cdot H_2O$. A heat stabilizer, especially for vinyl electrical-insulation compounds. It is very effective, has good electrical properties, and produces no gassing.

Triblock Polymers *n* A block co-polymer consisting of two terminal blocks of A repeating units and a central block of B units.

Tribology \trī-**|**bä-lə-jē\ *n* (1966) The engineering science that deals with friction, wear, and lubrication of surfaces sliding or rolling on one another, and the design of systems and components, such as gears and bearings, in which these actions are involved.

Tributoxyethyl Phosphate *n* $(C_4H_9OC_2H_4O)_3PO$. A primary plasticizer for cellulosic, acrylic, and vinyl resins, imparting low-temperature flexibility and flame retardance. In vinyl plastisols, small amounts of tributoxyethyl phosphate markedly reduce viscosity. However, when used alone or in high percentages, this plasticizer causes inconveniently rapid gelation. It is an organic phosphate with wide applications in floor polishes, paints, synthetic rubbers and as a deformer. It is probably best known in emulsion floor polishes as a leveling agent. Abbreviation for TBEP.

Tributyl Borate *n* (butyl borate) $(C_4H_9)_3BO_3$. A colorless liquid, used as an Antiblocking agent in plastic films and sheets.

Tributyl Citrate See ▶ Butyl Citrate.

Tributyl Phosphate *n* (TBP) $(C_4H_9O)_3PO$. A colorless liquid used as a primary plasticizer and solvent for cellulose acetate, chlorinated rubber and, in special applications, for vinyl resins. Its relatively high volatility limits its use as a plasticizer for vinyls.

Tricalcium Silicate *n* A compound which is a main constituent of Portland cement.

Tricel *n* Cellulose triacetate, manufactured by Bayer, Germany.

4,4,4-Trichloro-1,2-Butylene Oxide *n* (TCBO) Cl$_3$CCH$_2$CHOCH$_2$. This highly reactive, liquid epoxides is used for modifying polyols to achieve fire retardance in polyurethane foams.

1,1,1-Trichloroethane *n* (methyl chloroform) CH$_3$CCl$_3$. A nonflammable solvent that can be milled with resins to produce nonflammable adhesives. Less used than formerly because of the perceived need to minimize release of chlorinated compounds into the atmosphere.

Trichloroethylene *n* (TCE, trichloroethene) ClCH=CCl$_2$. Until recently, a nonflammable solvent widely used for degreasing. It is a suspected cancer agent, mutagen, and stratospheric-ozone destroyer.

Trichloroethyl Phosphate *n* C$_2$H$_4$CL$_3$PO$_4$. Plasticizer of special value in the formulation of fire-resisting lacquers, Bp, 220°C/20 mmHg.

Trichlorofluoromethane *n* (Freon 11, Genetron 11) CCl$_3$F. A chemically inert blowing agent used until recently with water in flexible polyurethane formulations to control foam density and load-bearing properties. It is also a refrigerant and former aerosol propellant. Freon 11 has been phased out of most applications since 1995, in the general push to reduce fluorocarbons in the atmosphere.

N-(Trichloromethylthio) Phthalimide *n* A bacteriostatic agent used in vinyl fabrics for hospitals and households. It stops the growth of bacteria such as *staphylococcus aureus* that cause pink staining in white and pastel-colored PVCs.

Trichlorotrifluorethane *n* (Freon 113) Cl$_2$FCCClF$_2$. A colorless, nearly odorless solvent that boils at 47°C, formerly used as a blowing agent for integral-skin polyurethane foam, but now because of its suspected action on stratospheric ozone, being phased out.

Trichromat n \ˈtrī-krō-ˌmat\ n [back-formation fr. *trichromatic*] (1929) Person requiring mixtures of three component primaries to match colors. Normal observers fall into this classification: the CIE standard observer is defined in terms of the amounts of three primary colors required to match all spectral colors. Not all trichromats are normal, however. Some observers exhibit anomalous trichromatism, requiring abnormal proportions of three primary colors for color matching.

Trichromatic *adj* (ca. 1890) Of or consisting of three colors.

Trichromatic Coefficients n Alternate term for chromaticity coordinates or coefficients. See ▶ Chromaticity Coordinates.

Tricot \ˈtrē-(ˌ)kō\ n [F, fr. *tricoter* to knit, fr. MF, to agitate, hop ultim. fr. OF *estriquier* to stroke, of Gr origin; akin to OE *strīcan* to stroke] (1872) A generic term for the most common type of warp-knit fabric. It has fine wales on the face and coursewise ribs on the back. It can be made in a plain jersey construction or in meshes, stripes, and many other designs. Tricot is usually made of triacetate, acetate, polyester, nylon, or rayon. Also see ▶ Jersey.

Tricot Beam n A metal flanged beam, commonly 42 in. in width, on which yarn is wound for use as a supply for the tricot machine.

Tricot Fabric Yield n The number of square yards per pound of greige or finished tricot fabric.

Tricot Knitting See ▶ Knitting (1).

Tricot Section See ▶ Tricot Beam.

Tricresyl Phosphate n (TCP, TCF, tritolyl phosphate) $(CH_3C_6H_4)_3PO_4$. One of the most important and earliest of all commercial plasticizers. Also suitable for cellulosics, alkyds, and polystyrene. Like all other phosphate plasticizes, TCP imparts flame retardance and fungus resistance, even when used in amounts as low as 5% of the total plasticizer content. TCP and cresyldiphenyl phosphate are the plasticizers most widely used for these properties. Bp, 430–440°C; flp, 216°C (420°F). *Known also as TCP.* Syn: ▶ Tritolyl Phosphate.

Tricresyl Phosphite n $(CH_3C_6H_4O)_3P$. A colorless liquid used as a flame retardant plasticizer and stabilizer for thermoplastics.

Tricyclohexyl Citrate n $(C_6H_{11}OOCCH_2)_2C(OH)COOC_6H_{11}$. A nontoxic plasticizer for polystyrene, cellulosics, acrylics, and vinyls.

Tridecyl Phosphite *n* $(C_{10}H_{21}O)_3P$. A colorless liquid used as a stabilizer for polyolefins and PVC.

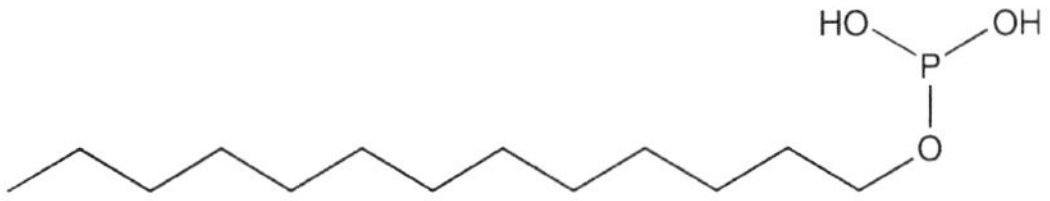

Tridimethylphenyl Phosphate *n* (trixylenyl phosphate, TXP) A plasticizer for cellulosics and vinyl compounds in which a low-density, electrical-grade, flame-retardant plasticizer is required.

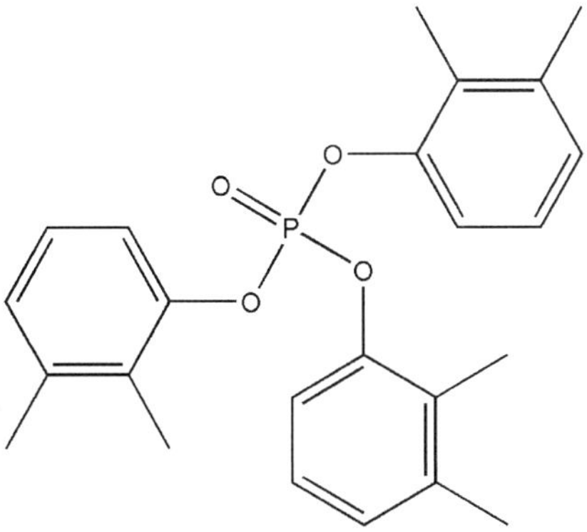

Tridirectional Laminate *n* A reinforced-plastics material having reinforcing fibers running in three principal directions. If the directions are 120° apart and there are equal percentages of fiber lying in the three directions, the laminate properties will be almost isotropic in the laminate plane.

Trienol *n* Synthetic tung oil of Swiss origin, practically indistinguishable from the natural oil.

Triethanolamine *n* 2,2,2″-Nitrilotriethanol. $N(CH_2CH_2OH)_3$. Very hygroscopic, viscous liquid with a slight ammoniacal odor. Turns brown on exposure to air and light, Used in the manufacture of surfactants and emulsions. Mol wt, 149.19; sp gr, 1.124 (9.37 lb/gal); mp, 21.2°C.

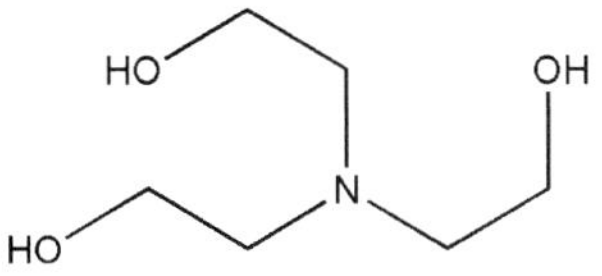

Triethyl Aluminum \(▪)trī-▪e-thəl-\ *n* (aluminum triethyl, ATE) $(C_2H_5)_3Al$. A catalyst for the polymerization of olefins.

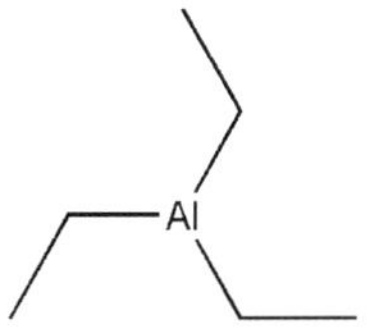

Triethyl Citrate *n* (ethyl citrate) $(C_2H_5OOCCH_2)_2C(OH)COOC_2H_5$. An ester of citric acid, used as a plasticizer for many thermoplastics including vinyls, cellulosics, and polystyrene. It has won FDA approval for use in food packaging.

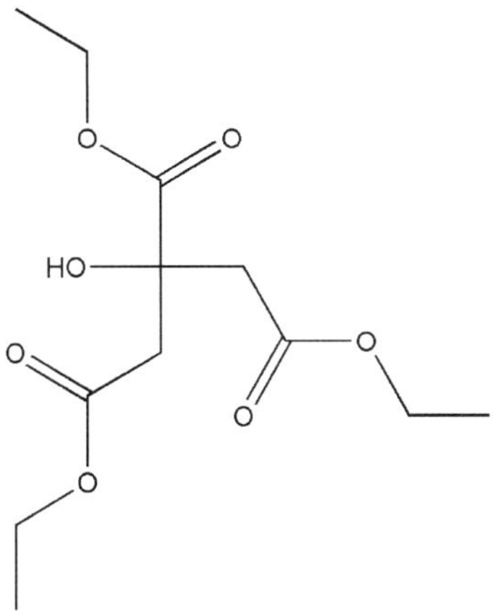

Triethylenediamine *n* (1,4-diazabicyclo-2,2,2-octane, DABCO) This tertiary diamine, whose structure is shown below, is the most widely used amine catalyst for polyurethane foams, elastomers, and coatings. It is soluble in water and polyols.

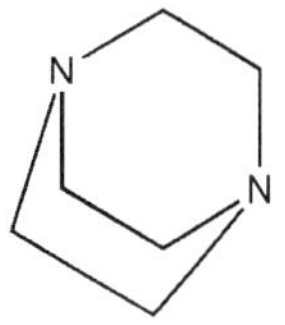

Triethylene Glycol *n* $(CH_2OCH_2CH_2OH)_2$. Polyhydric alcohol. Bp, 285°C; flp, 166°C (331°F); vp, <0.1 mmHg/20°C.

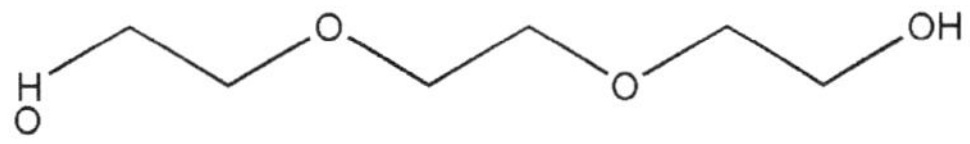

Triethylene Glycol Diacetate *n* $(-CH_2OCH_2CH_2OOCCH_3)_2$. A plasticizer for cellulosic plastics and some acrylic resins.

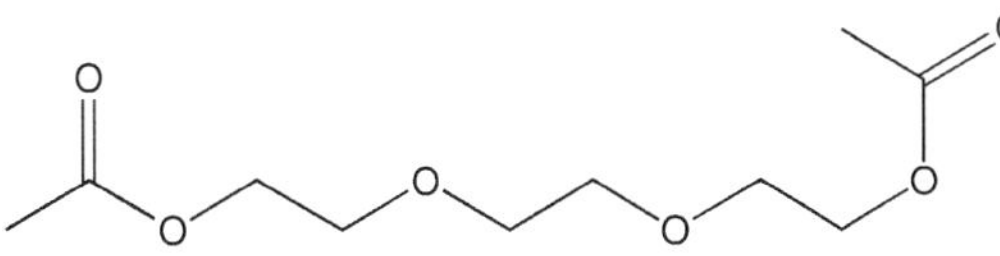

Triethylene Glycol Dibenzoate *n* $(-CH_2OCH_2CH_2OOCC_6H_5)_2$. A secondary plasticizer, partly compatible with all common thermoplastics. In most resin systems it has a tendency to crystallize and bloom at higher concentrations, which property may be used to advantage to prevent BLOCKING.

Triethylene Glycol Dicaprylate *n* (triethylene glycol trioctanoate) $(-CH_2OCH_2CH_2OOCC_7H_{15})_2$. A plasticizer for ethyl cellulose and vinyl resins, with good low-temperature flexibility.

Triethylene Glycol Di(2-ethylhexanoate) *n* $(-CH_2O CH_2CH_2OOC-C_7H_{15})_2$. A plasticizer for cellulosic plastics, polymethyl methacrylate, PVC, and vinyl chloride-vinyl acetate copolymers. In vinyls it is usually used as a secondary plasticizer at 10–25% of the total plasticizer, to impart low-temperature flexibility.

Triethylene Glycol Dipelargonate *n* $(-CH_2OCH_2 CH_2OOCC_8H_{17})_2$. A plasticizer for vinyls and cellulosics.

Triethylene Glycol Dipropionate *n* $(-CH_2OCH_2 CH_2OOCC_2H_5)_2$. A plasticizer for cellulosic resins and polymethyl methacrylate.

Triethylenetetramine *n* (TETA) $(-CH_2NHCH_2CH_2 NH_2)_2$. A viscous, yellowish liquid, TETA is used as a room temperature curing agent for epoxy resins.

Tri(2-ethylhexyl) Citrate *n* A nontoxic plasticizer for PVC.

Tri(2-ethylhexyl) Phosphate See ▶ Trioctyl Phosphate.

Triethyl Phosphate *n* (TEP) $(C_2H_5O)_3PO$. A flame-retardant plasticizer for cellulosics, acrylics, some vinyl polymers, and unsaturated polyesters.

Triglyceride \(ˌ)trī-ˈgli-sə-ˌrīd\ *n* ITS] (ca. 1860) Ester obtained by the interaction of the three reactive hydroxyl groups of glycerol with fatty or other acids.

Trihydrazide Triazine \-ˈhī-drə-ˌzīd ˈtrī-ə-ˌzēn\ *n* A heterocyclic chemical blowing agent that decomposes at 250–260°C yielding, per gram, about 175 cm^3 of gas consisting mostly of ammonia and nitrogen. It is used in foaming polypropylene, acrylonitrile-butadiene-styrene resin, nylon, and other high-melting resins.

Triisodecyl Trimellitate *n* $(C_{10}H_{21}OOC)_3C_6H_3$. An involatile plasticizer for PVC.

Triisooctyl Trimellitate *n* (TIOTM) $(C_8H_{17}OOC)_3C_6H_3$. A plasticizer for cellulosic and vinyl plastics with low volatility, high resistance to soapy-water extraction, and essentially no marring effect on lacquered surfaces.

Trim *n* (1) The visible woodwork or moldings of a room, such as the baseboards, cornices, casings, etc. (2) Any visible element, usually of metal or wood, which protects or covers joints, edges, or ends of another material; the finishing around fittings and openings, as a door trim, window trim, etc.

Trimer *n* An oligomer formed by the union of three molecules of a monomer and/or made up of three mer units. See ▶ Polymer and ▶ Oligomer.

Trimethyl Borate *n* (methyl borate, trimethoxyborine) $(CH_3O)_3B$. A colorless liquid used as a flame retardant in plastics.

Trimethylene Glycol *n* See ▶ Glycol.

Trimethylene Glycol Di-*p*-aminobenzoate *n* A diamine curing agent for polyurethanes introduced in 1976 to replace the popular but toxic MOCA®. It is very soluble in a variety of coating solvents.

2,2,4-Trimethyl-1,6-Hexane Diisocyanate *n* (TMDI) $OCNCH_2C-(CH_3)_2CH$ $(CH_3)CH_2CH_2NCO$. A branched aliphatic isocyanate used in making polyurethanes.

Trimethylolethane Tribenzoate *n* A solid plasticizer for PVC.

2,2,4-Trimethyl-1,3-Pentanediol *n* (trimethylpentanediol, TMPD) $(CH_3)_2CH(OH)C(CH_3)_2CH_2OH$.

One of the principal glycols used in making polyester resins, alkyd resins, and polyester plasticizers, containing one primary and one secondary hydroxyl group. It is made by the aldol condensation of isobutyraldehyde, yielding a water-insoluble white solid. TMPD is used in producing linear unsaturated polyesters and is particularly good for gel-coating resins.

2,2,4-Trimethyl-1,3-Pentanediol Monoisobutyrate Benzoate *n* A plasticizer for PVC imparting good stain resistance.

Trimmed Papers See ▶ Pretrimmed Papers.

Tri-*n*-butyl Aconitate *n* $C_3H_3(COOC_4H_9)_3$. A combination plasticizer and stabilizer for polyvinylidene chloride and synthetic rubbers.

Tri-*n*-butyl Citrate *n* $C_2H_4(OH)(COOC_4H_9)_3$. A nontoxic plasticizer for most thermoplastics, including cellulosics, polystyrene, and vinyls.

Tri-*n*-butyl Phosphine *n* $(C_4H_9)_3P$. A curing agent for epoxy resins, and a catalyst for vinyl and isocyanate polymerizations.

Tri(*n*-Octyl-*n*-Decyl) Trimellitate *n* (NODTM) A low-temperature plasticizer for vinyls and cellulosics. It is used in heavy-duty applications such as truck seating, film and sheeting subject to wide temperature ranges, and in baby wear.

Tri-*n*-propyl Phosphate *n* $(C_3H_7)_3PO_4$. A plasticizer and solvent. Sp gr, 1.012; bp, 135–140°C.

Trioctyl Phosphate *n* [TOP, TOF, tri-(2-ethylhexyl) phosphate] [C$_4$H$_9$CH(C$_2$H$_5$)CH$_2$O]$_3$PO. A plasticizer for PVC, imparting good low-temperature flexibility, resistance to water extraction, flame and fungus resistance, and minimum change in flexibility over a wide temperature range. It is also compatible with polyvinyl butyral, ethyl cellulose, and cellulose acetate-butyrate resins with a high butyral content.

Trioctyl Trimellitate *n* (TOTM) (C$_8$H$_{17}$OOC)$_3$C$_6$H$_3$. A primary plasticize for vinyls that hold up well at high temperatures. It combines the permanence of polymeric plasticizers with the low temperature properties of monomerics. In vinyls, it is used for auto interior parts and for wire insulation good for temperatures to 105°C. It is also used with cellulosics and acrylics.

Triol \ˈtrī-ˌȯl, -ˌōl\ *n* (1936) A term sometimes used for *trihydric alcohol*, i.e., an alcohol containing three hydroxyl (−OH) radicals.

1,3,5-Trioxane *n* (*sym*-trioxane, triformol, trioxin) $\overline{\text{CH}_2\text{OCH}_2\text{OCH}_2\text{O}}$. The stable, cyclic trimer of formaldehyde, a colorless, crystalline solid. It is easily depolymerized in the presence of acids to its monomer, or may be further polymerized to form ACETAL RESINS. This trimer should not be confused with ▶ Paraformaldehyde.

Triphenylmethane \ˌtrī-ˌfe-nᵊl-ˈme-ˌthān\ *n* [ISV] (ca. 1885) CH(C$_6$H$_5$)$_3$. A crystalline hydrocarbon that is the parent compound of many dyes.

Triphenyl Phosphate *n* (C$_6$H$_5$)$_3$PO$_4$. A crystalline powder, one of the original synthetic plasticizers for cellulose nitrate. It is also a flame retardant for vinyls, cellulosics, acrylics, and polystyrene. Bp, 410°C; mp, 50°C; flp, 220°C (428°F) Known also as *TPP*.

Triphenyl phosphate

Triple-Daylight Mold *n* A mold having four plates: feed plate, floating cavity plate, stripper plate, and moving mold plate. When the mold opens, all the plates move apart making three openings among them.

Triple-Roller Mill *n* Type of mill used for the fine grinding of pigmented compositions. The pigment agglomerates are dispersed by passage between accurately machined steel or granite cylinders, the clearance between the cylinders being capable of very delicate adjustment. The cylinders or rollers are made to operate at different speeds. The rough pigment-medium mix is fed between two rollers rotating towards each other at different speeds, and t is finally transferred to the third roller, which operates at the maximum speed. Pigment dispersion is effected by a reduction of agglomerates, controlled by the clearance between the rollers, and by a shearing action developed by the viscous medium on rollers which are rotating at different speeds. Roller mills are ideal for grinding stiff pigment pastes.

Tripoli \ˈtri-pə-lē\ *n* [F, fr. *Tripoli*, region of Africa\ (ca. 1601) An earth consisting of very friable soft schistose deposits of silica and including diatomite and kieselguhr. See ▶ Rottenstone.

Tripolite See ▶ Diatomaceous Silica.

Tris(2-Chloroethyl) Phosphate) *n* $(CH_2ClCH_2O)PO$. A plasticizer for polystyrene, cellulosics, and vinyls. It is also effective as a flame retardant for unsaturated polyesters and polyurethane foams.

Tris(2,3-Dibromopropyl) Phosphate *n* $(CH_2BrCHBr\ CH_2O)_3PO$. A flame retardant for unsaturated polyesters, polyurethane foams, and other plastics.

Tris(1,2-Dichloroisopropyl) Phosphate *n* $[CH_2ClC(CH_3)ClO]_3PO$. A flame retardant for unsaturated polyesters.

Tris(2,3-Dichloropropyl) Phosphate *n* $(CH_2ClCHCl\ CH_2O)_3PO$. A plasticizing flame retardant for many plastics including vinyls, cellulosics, acrylics, polyolefins, phenolics, polyesters, and polyurethane foams.

Triskelion Cross Section *n* A trilobal cross section in which the radiating arms are curved or bent. Also see ▶ Cross Section.

Trisnonylphenyl Phosphite *n* (TNPP) An FDA sanctioned heat stabilizer and antioxidant used in styrene-butadiene copolymers.

Tristimulus \(ˌ)trī-ˈstim-yə-ləs\ *adj* (1933) (1) Color readings based on the primary colors: yellow, magenta, and cyan. (2) Of, or consisting of, three stimuli generally used to describe components of additive mixture required to evoke a particular color sensation.

Tristimulus Colorimeter *n* Instrument used to measure quantities which can be used to obtain an approximation of tristimulus values. They are normally equipped with three (or four) special filters to obtain R, G, B values which must then be normalized to CIE magnitude. They were designed and are properly used only for measuring the color difference between two similar, nonmetameric samples (McDonald, Roderick (1997) Colour physics for industry, 2nd edn. Society of Dyers and Colourists, West Yorkshire).

Tristimulus Computation Data *n* Products of relative spectral-energy distribution of an illuminant multiplied by each of the three color matching functions; in the CIE System designated as $E_c\bar{x}, E_c\bar{y}, E_c\bar{z}$, for example, for Illuminant C and the color mixture data for the standard observer at designated wavelengths. See ▶ Tristimulus Values.

Tristimulus Integration *n* Use of tristimulus computation data to determine tristimulus values of a sample; in the CIE System, the summation of the products of the reflectance or transmittance at regular wavelength intervals by the tristimulus computation data for the same wavelengths. See ▶ Tristimulus Values and ▶ Tristimulus Integrator.

Tristimulus Integrator *n* Device for performing the tristimulus integration calculation, generally attached directly to a spectrophotometer. Several types have been used: mechanical analog type using specially cut cams giving a continuous integration, digital-mechanical type using selected ordinates (generally 100), and digital-computer type using weighted ordinates, must commonly at 10 nm intervals. The latter is most generally used today, because of the flexibility in obtaining CIE tristimulus values for any observer whose color matching functions are known, and any illuminate of defined spectral-power distribution. See ▶ Tristimulus Values.

Tristimulus Values *n*, **CIE** Amounts (in percent) of the three components necessary in a three-color additive mixture required for matching a color; in the CIE System, they are designated as X, Y, and Z. The illuminant and standard observer color matching functions used must be designated; if they are not, the assumption is made that the values are for the 1931 observer (2° field) and Illuminant C. The values obtained depend on the method of integration used and on the relationship of the nature of the sample and on the instrument design used to measure the reflectance or transmittance. Tristimulus values are not, therefore, absolute values characteristic of a sample, but relative values dependent on the method used to obtain them. Approximations of CIE tristimulus values may be obtained from measurements made on a tristimulus colorimeter, giving measurements generally normalized to 100, which must then be normalized to equivalent CIE values. The filter measurements should be properly designated as R, G, and B instead of X, Y, and Z. The calculation of tristimulus values in the CIE System is illustrated in the equation below for Y (McDonald, Roderick (1997) Colour physics for industry, 2nd edn. Society of Dyers and Colourists, West Yorkshire).

$$Y = k \int E(\lambda)\bar{y}(\lambda)R \text{ or } T(\lambda)d(\lambda)$$

where k is a normalizing factor to bring Y for a perfect white to 100.000%, and dλ is the wavelength interval used. When a large number of wavelengths is not used, the integration should be designated as a summation

$$Y = k \sum E(\lambda)\bar{y}(\lambda)R \text{ or } T(\lambda)\Delta(\lambda)$$

Tritactic Polymer *n* An isotactic or syndiotactic polymer that is also of the *cis-* or *trans* form because the molecules are unsaturated and have double bonds.

Tritium \ˈtri-tē-əm\ *n* [NL, fr. Gk *tritos* third] (1933) An isotope of hydrogen with a mass of three, structure, two neutrons, and one proton in its nucleus. Hydrogen 3, ^3_1H.

Tritolyl Phosphate *n* (**TTP**) See ▶ Tricresyl Phosphate.

Trivial Name *n* A name that is not produced by any systematic procedure of naming. Trivial names may be derived from geographic locations, names of people, or from descriptive words. Examples: "aragonite" after Aragon, Spain; "wollastonite" in honor of the English chemist, W. H. Wollaston; and "magnetite" alluding to its highly magnetic property.

Trixylenyl Phosphate *n* (TXP) See ▶ Tridimethylphenyl Phosphate.

Trolit AE *n* Cellulose ether, manufactured by Dynamit Nobel, Germany.

Trolitan Phenol-formaldehyde resin. Manufactured by Dynamit Nobel, Germany.

Trolit F *n* Cellulose nitrate. Manufactured by Dynamit Nobel, Germany.

Trolitul *n*, Poly(styrene), manufactured by Dynamit Nobel, Germany.

Trommsdorf Effect *n* The acceleration of a reaction such as polymerization which continues to increase uncontrollably without external stimulus (also referred to as gel effect). (Lenz RW (1967) Organic chemistry of high poymers. Interscience). See ▶ Autoacceleration.

Trompe l'oeil \(▪)trómp-▪ləi, trō ⁿp-líi\ *n* {*often attributive*} [F *trompe-l'oeil*, literally, deceive the eye] (1889) Literally, "fool-the-eye." A design that creates a three-dimensional illusion by means of shadow and graphic textures.

Tronal *n* High-impact poly(styrene), manufactured by Dynamit Nobel, Germany.

Trough *n* In the coil coating field, synonymous with pan.

Trovidur *n* Poly(vinyl chloride), manufactured by Dynamit Nobel, Germany.

Trovitherm Poly(vinyl chloride) (films), manufactured by Dynamit Nobel, Germany.

True Density *n* (solid density) Of a specimen of porous material, the mass of the specimen divided by (volume of the specimen le the volume of its voids), i.e., $\rho_t = M/[V(1-\varepsilon)]$, where ε is the void fraction and V is the specimen volume. Compare ▶ Apparent Density.

True Strain *n* In a tensile test, the integral of the differential increase in length divided by the length at that point in the test, i.e.,

$$True\ strain = \int_{L_O}^{L_f} \frac{dL}{L} = \ln(L_f/L_O)$$

In this equation, L_O is the original gauge length of the specimen and L_f is the gauge length at the stress-strain point of interest. True, strain is always less than nominal strain (see ▶ Strain), but the difference is small unless the strain is large. For example, a nominal strain of 0.5 (50%) corresponds to a true strain of 0.41 (41%).

True Stress *n* The quotient of force divided by true cross-sectional area in the specimen's gauge length, at any point during a tensile or compressive test. True stress is always larger than nominal stress (see ▶ Stress), but the difference is negligible at low strains. In a ductile plastic having a Poisson's ratio of 0.4, for example, if the true strain were 0.41 (41%), the cross-sectional area would be 70% of the initial area so true stress would be 1.43 times nominal.

True Tensile Strength *n* The maximum tensile stress expressed in force per unit area of the specimen at the time of rupture. Also see ▶ Tensile Strength.

Truss \▪trəs\ *n* Framework of wood or metal beams, stiffened by cross braces; used for roofs, bridges, etc.

TSC *n* Abbreviation for Thermal Stress Cracking Or Thermally Stimulated Current, either of which see.

T-Slot Die See ▶ Manifold.

TTB See 2,2′,6,6′-Tetrabromo-3,3′,5,5′-Tetramethyl-*P,P*′-Biphenol.

T-300 *n* Temperature at which the torsional modulus of an air-dried film is 300 kg/cm^2; a relative parameter of film stiffness.

Tub See ▶ Beck.

Tube *n* (1) A cylindrical holder or bobbin used as a core for a cylindrical yarn package. (2) A cylindrical yarn package.

Tubelength, Mechanical *n* The distance from the shoulder of the objective to the upper end of the drawtube. The mechanical tubelength for biological and most polarizing microscopes is 160 mm; Leitz formerly had a 170 mm tube length.

Tubelength, Optical *n* The distance from the back focal plane of the objective to the image formed by the objective (intermediate image).

Tubing *n* (1) Any of a wide range of continuous extrusions, usually having circular annular cross sections, and flexible enough to be wound on a core or coiled. Some large-diameter tubing has such thin walls that it is flattened before winding. Some stiff but coilable, small-diameter tubing, such as saran and nylon tubing used for automotive fuel lines and compressed-air service, has relatively thick walls. Unlike pipe, tubing is not usually cut to standard lengths nor is it expected to be nearly straight. With tubing and pipe, the internal diameter (ID) is usually more carefully controlled than the outside diameter (OD). (2) Cylindrical fabric made by braiding, waving, or knitting. The term *sleeving* is applied to such tubing less than 10 cm in diameter.

Tubing Die *n* A die with an annular opening used to extrude plastics tubing. The core (*mandrel*) of the die may be fitted with a water-cooled extension that aids in chilling the extrudate and bringing its internal diameter within tolerances. See also ▶ Pipe Die.

Tubular Fabric *n* A fabric woven or knit in a tube form with no seams, such as seamless pillowcases, some knit underwear fabrics, and seamless hosiery. Also see ▶ Circular-Knit Fabric.

Tubular Film See ▶ Film Blowing.

Tuck Stitch *n* A knitting stitch made when a needle receives a new yarn without losing its old loop.

Tuft \ |təft\ *n* [ME, mod. of MF *tufe*] (14c) (1) A cluster of soft yarns drawn through a fabric and projecting from the surface in the form of cut yarns or loops. (2) The portion of pile-like material that comprises a tufted fabric or carpet. Also see ▶ Tufted Fabric and ▶ Tufted Carpet (Kadolph SJJ, Langford AL (2001) Textiles. Pearson Education, New York).

Tufted Carpet *n* Carpet produced by a tufting machine instead of a loom. It is an outgrowth of hand-tufted bedspreads. Today, broadloom tufting machines produce over 90% of all domestic carpeting. Tufting machines are essentially multineedle sewing machines that push the pile yarns through a primary backing fabric and hold them in place to form loops as the needles are withdrawn. The loops are then either released for loop-pile carpets or cut for cut-pile carpets. The pile yarns may be either predyed or uncolored, in which case, the greige carpet is then piece-dyed or printed. In either case, a latex or other binding agent is applied to the backstitch to lock the tufts in place and to secure the secondary backing fabric. Formerly, all carpets were woven, either by hand or machine. The significantly greater productivity of tufting has revolutionized the carpet industry and made soft floor coverings available to the mass market (Kadolph SJJ, Langford AL (2001) Textiles. Pearson Education, New York).

Tufted Fabric *n* Cotton sheeting, lightweight duck, or other fabric decorated with fluffy tufts of multiple-ply, soft-twist cotton yarns or manufactured fiber yarns closely arranged in continuous lines or spaced at intervals to produce the type of fabric called candlewick. The tufts are inserted and cut by machine in previously woven fabric or are woven in by the loom and afterwards cut to form the tufts. They have a chenille-like softness and bulk and are erroneously called chenille. Patterns vary from simple straight lines and elaborate designs to completely covered materials resembling long pile fabrics. The may be white, solid colored, or multicolored. Tufted fabrics are used for bedspreads, bath mats, and robes, etc. (Kadolph SJJ, Langford AL (2001) Textiles. Pearson Education, New York).

Tukon See ▶ Penetrometer.

Tulle \ |tül\ *n* [F, fr. *Tulle*, France] (ca. 1818) A fine, very lightweight, machine-made net usually having a hexagonal mesh effect. Tulle is used in ballet costumes and veils.

Tumbling *n* (1) Process by which paint is applied to small articles, such as hairpins, children's building bricks, etc., which are unsuitable for coating by any of the normal methods. The articles are placed in a drum, together with a little more paint than will be sufficient to cover the total surface of all the articles, and the drum is rotated until the paint is evenly distributed. the articles are then emptied from the drum, generally onto wire trays, and the coating air-dried or baked. *Also called Rumbling, Barrelling, Barrel Finishing, Barrel Polishing.* (2) A finishing operation for small plastic articles in which gates, flash and fins are removed and/or surfaces are polished, by rotating them in a barrel together with wooden pegs, sawdust, and (sometimes) polishing compounds. The barrels are usually of octagonal shape with alternate open and closed panels the open panels covered with screen to permit fragments of removed material to fall out. Blocks of dry ice may be added to the tumbling medium to embrittle the parts and thus facilitate cleaner break-off of flash; but in this case the barrel must be closed to retard evaporation of the dry ice. (3) Process used in some paint storehouses whereby containers are repeatedly up-ended to redisperse pigments which may have settled or caked during storage of the paint.

Tumbling Agitator *n* (tumble mixer) A cylindrical or conical vessel rotating about a horizontal or inclined axis, with internal ribs that life the material and then let it tumble back into the charge. They are used mainly for blending dry materials, e.g., color concentrates and resin pellets.

Tungates *n* Metallic soaps derived from tung oil.

Tung Oil \ |tən\ *n* [part translation of Chinese (Beijing) *tóngyóu*] (1881) Drying oil obtained from *Aleurites fordii* or *Montana* now grown in many different parts of the world. The Japanese tung oil is obtained from *Aleurites cordate*. Tung oil is distinguished by the presence of a very high proportion of elaeostearic acid, which contains three conjugated double bonds, It dries rapidly, polymerizes readily on heating, "reacts" with certain types of pure phenolic resins, possesses outstanding water resistance, and webs or frosts on drying unless properly heat-treated. Sp gr, 0.939–0.943/15°C; refractive index, 1.486–1.522; iodine value, 145–175; saponification value, 189–195. *Known also as China Wood Oil, Wood Oil, and Mu Oil.*

Tung or Chinawood Oil *n* A vegetable oil having exceptional drying properties and chemical resistance (Paint: pigment, drying oils, polymers, resins, naval stores, cellulosics esters, and ink vehicles vol 3. American Society for Testing and Material, 2001).

Tungsten \ |təŋ-stən\ *n* [Sw, fr. *tung* heavy + *sten* stone] (1796) (W) A hard, dense metallic element that has occasionally been used in powder form as a plastics filler to increase density. Tungsten carbide extremely hard and widely used in metal-cutting tools, has also been used to impart abrasion resistance in plastics.

Tungstomolybdic Pigments See ▶ Phosphomolybdic Pigments.

Tunnel Gate *n* Syn: ▶ Submarine Gate.

Tunnel Test See ▶ Flammability Tests.

Tunneling *n* A condition occurring in incompletely bonded laminates, characterized by release of longitudinal portions of the substrate and deformation of these portions to form tunnel-like structures.

Turbid \ˈtər-bəd\ *adj* [L *turbidus* confused, turbid, fr. *turba* confusion, crowd; prob. fr. Gk *tyrbē* confusion] (1626) Characteristic of a liquid or solid having a relatively great amount of nonsetting floc, gels, suspended matter, particles, droplets, or other insoluble or separated matter, even though the liquid is translucent and transmits at least a little light.

Turbidity *n* (τ) Parameter used to express the intensity of light scattering. It is defined as,

$$\tau = (1/l)[\log(I_0/I)]$$

where I is the intensity of a beam of light, of initial intensity I_0, after passing through a length *l* of the scattering medium. It is related to the Raleigh ratio (R_θ) by $\tau = (16/3)\ R_\theta$.

Turbulence *n* Deviation from streamline or telescopic flow. Turbulent flow occurs at relatively high rates of shear and indicates the existence of eddy currents in sheared material.

Turbulent Flow *n* Flow for which the stream ▶ Reynolds Number exceeds about 2,100–4,000. Polymer solutions, particularly dilute ones, can easily experience turbulent flow but polymer melts rarely (if ever) do because of their extremely high viscosities.

Turkey Red *n* [*Turkey*] (1789) Red oxide of iron obtained by calcinations at relatively low temperatures. Hydroxide or carbonate of iron is precipitated from a solution of an iron salt, and the precipitate is washed before calcinations. Soft, brilliant pale reddish brown pigments are obtained. Opacity is high.

Turkey Red Oil *n* Sulfonated castor oil. The term is sometimes applied to the sodium salt of this product. the oil is chiefly the reaction product of sulfuric acid with the hydroxyl groups of castor oil, but some secondary reactions also occur. Turkey red oil is used as a wetting agent.

Turkey Umber *n* Type of umber derived from Cyprus and originally exported via Turkey. See ▶ Umber.

Turn *n* The distance parallel to the axis of a yarn or rope in which a strand makes one complete spiral. Also see ▶ Twist.

Turnbull's Blue See ▶ Iron Blue.

Turned Over Edge *n* A curled selvage.

Turnkey *n* An adjective describing a complete manufacturing system or facility, such as a polymerization plant of sheet-extrusion line, delivered to the customer in ready-to-operate condition.

Turpentine \ˈtər-pən-ˌtīn\ *n* [ME *terbentyne*, *turpentyne*, fr. MF & ML; MF *terbentine*, *tourbentine*, fr. ML *terbentina*, fr. L *terebinthina*, feminine of *terebinthinus* of terebinth, fr. *terebinthus* terebinth, fr. Gk *terebinthos*] (14c) A colorless, volatile oil distilled from the products of certain pine trees and consisting primarily of a complex mixture of terpene hydrocarbons of the general formula $C_{10}H_{16}$. Turpentine was formerly extensively used in paints and varnishes but has now been largely replaced by mineral spirits or white spirit. Four kinds of turpentine are now recognized: (1) *Gum Turpentine* or *Bum Spirits*. Obtained by distilling the crude exuded gum or oleoresin collected from living trees. (2) *Steam-distilled Wood Turpentine*. Obtained from the oleoresin within the wood of pine stumps or cuttings, either by direct steaming of the mechanically disintegrated wood or after solvent extraction of the oleoresin from the wood. (3) *Sulfate Wood Turpentine*. Recovered during the conversion of wood to paper pulp by the sulfate processes. (4) *Destructively Distilled Wood Turpentine*. Obtained by fractionation of certain oils recovered by condensing the vapors formed during the destructive distillation of pine wood. *Also known as Turps*.

Turpentine, Oil of See ▶ Oil of Turpentine.

Turpentine, Spirits of *n* (ca. 1792) See ▶ Turpentine.

Turpentine Substitute *n* This term is not recommended, because its use to describe a paint thinner of mineral spirits origin is illegal under the Federal Naval Stores Act. Syn: ▶ Sub Spirits and ▶ Sub Turps.

Turps See ▶ Turpentine.

Tuscan Red *n* Red pigment made from a mixture of iron oxide and an alizarine dye.

Tusche \ˈtüsh, ˈtü-shə\ *n* [Gr, fr. *tuschen* to lay on color, fr. F *toucher*, lit., to touch, fr. OF *tuchier*] (1885) A lithographic drawing or painting material of the same nature as lithographic ink. Also used as a resistant in the etching process.

Twaddell Hydrometer *n* Form of technical hydrometer used for measuring the specific gravity of liquids. It does not give a direct reading, but the specific gravity is calculated from the following simple equation:

$T° = 200(d−1)$, where $T°$ is the reading in degrees Twaddell, and d is the required sp gr.

Tweed \ˈtwēd\ *n* [prob short for Scots *tweedling*, *twidling* twilled cloth] (1841) An irregular, soft, flexible, unfinished, shaggy wool or wool-blend fabric made with a 2/2 twill weave. Tweeds are used in all types of coat fabrics and suitings.

Twilight Vision *n* Rod vision by the dark-adapted eye in dim light.

Twill Weave *n* A cloth weave in which the warp yarn runs alternately over two fill yarns and under one fill yarn. The twos and ones may be staggered with adjacent or regularly spaced warp yarns to produce a diagonal effect. In Dutch twill, wires of different sizes are used to produce a very dense screen with tortuous passages (light does not pass directly through) and an effective ▶ Mesh Number of 1,200 or more. Twilled screens have sometimes been used in fiber extrusion and twill-woven cloths are used in some reinforced-plastics structures.

Twilo Process *n* A spinning process in which yarn is made by binding fibers with an adhesive, then removing the adhesive after the yarn is made into fabric.

Twine \ˈtwīn\ *n* [ME *twin*, fr. OE *twīn*, akin to MD *twijn* twine, OE *twā* two] (before 12c) (1) A plied yarn made form medium-twist single yarns with ply twist in the opposite direction. (2) A single-strand yarn, usually 3 or 4 mm in diameter, made of hard fibers, such as henequen, sisal, abaca, or chromium, and sufficiently stiff to perform satisfactorily on a mechanical grain binder.

Twinning *n* A movement of planes of atoms in the crystal lattice parallel to a specific (twinning) plane so that the lattice is divided into two symmetrical parts which are differently oriented. The amount of movement of each plane is proportional to its distance from the twinning plane.

Twin-Plane *n* The common composition plane of a crystal. It always coincides with a possible crystallographic face.

Twin-Screw Extruder See ▶ Extruder, Twin-Screw.

Twin-Shell Forming *n* A high-speed thermoforming process for producing bottles and other hollow objects. Two thermoplastic sheets from separate roll-unwind stands are conveyed through heating apparatus, then positioned between facing halves of vacuum-forming molds. After closing the molds, vacuum is drawn on each half to simultaneously for the articles and seal the edges. When molds are arranged on a pair of endless conveyors, the process becomes continuous, sheets being fed into one end and a continuous web of formed products emerging from the other end, ready for separation from the waste portion of the sheets. The trim is ground and returned to the sheet-extrusion operation.

Twist *n* (1) A textile term, the number of turns (360°) per unit length that a multifilament yarn, staple yarn, or other structure is turned or twisted around its longitudinal axis into a stable structure. (2) The unintended, progressive spiraling seen in some protruded products.

Twist Bleed See ▶ Twit.

Twist Direction *n* The direction of twist in yarns and other textile strands is indicated by the capital letters S and Z. Yarn has S-twist if when it is held vertically, the spirals around its central axis slope in the same direction as the middle portion of the letter S, and Z-twist if they slope in the same direction as the middle portion of the letter Z. When two or more yarns, either single or plied, are twisted together, the letters S and Z are used in a similar manner to indicate the direction of the last twist inserted.

Twisting *n* (1) The process of combining filaments into yarn by twisting them together or combining two or more parallel singles yarns (spun or filament) into plied yarns or cords. Cables are made by twisting plied yarns or cords. Twisting is also employed to increase strength, smoothness, and uniformity, or to obtain novelty effects in yarn. (2) A very high level of twist is added to single or plied yarns to make crepe yarns. This operation generally is called creping or throwing. (3) The process of adding twist to a filament yarn to hold the filaments together for ease in subsequent textile processing, etc.

Twist Multiplier *n* The ratio of turns per inch to the square root of the yarn count.

Twist Setting *n* A process for fixing twist in yarns to deaden torque and eliminate kinking during further processing. There are several methods that use steam to condition the packages of yarns.

Twit *n* A short section of real twist in false-twist yarn that prevents crimp development and hence causes a pinhole effect in fabric. Also called twist bleed or tight spot.

Twitchell Reagent *n* Reagent employed for the splitting or hydrolysis of oils and fats, and made by the interaction of sulfuric acid, oleic acid and naphthalene.

Two-Color Molding See ▶ Double-Shot Molding.

Two-Compartment Coating *n* Crosslinking systems which must be stored in separate containers before use. Otherwise, they would react and form a useless gel. *Also called Two-pot Coating.*

Two-Component Gun *n* Spray gun having two separate fluid sources leading to the spray had.

Two-Component Spray Gun *n* A spray head with fittings for attachment to two separate feed lines, each carrying one component of a reactive resin mixture, e.g., as in urethane foam-in-place molding or polyester gel-coating.

Two-For-One Twister *n* A twister that inserts twist at a rate of twice the spindle speed. For example, at a spindle speed of 2,000 rpm, 4,000 turns/min are inserted in the yarn.

Two-Level Mold *n* (double-decker mold) An injection mold having two layers of cavities, used in making low-mass articles with large projected areas, so as to make best use of both plasticating and clamping capacities.

Two-Pot (or Two Part) *n* Systems Inks or coatings in which two reactive components are mixed together only at press time.

Two-Pot Coating See ▶ Two-Compartment Coating.

Two-Roller Mills *n* These are used for purposes similar to the triple-roller mills, except that only two rollers are involved.

Two-Roll Mill See ▶ Roll Mill.

Two-Shot Injection Molding *n* Confusingly, this term has been used in the literature for two processes that are distinctly different. One is described under ▶ Double-Shot Molding. In the other process, one first injects a metered amount of one material into a single-cavity mold. As this material just begins to chill against the cold mold surfaces, a second material is injected. This fills the interior and forces the first material outward to the cavity surfaces. The second polymer, usually a reclaimed material forms the interior of the finished article, while the virgin material first injected forms the outer shell and surface of the article.

Two-Shot Molding See ▶ Double-Shot Molding.

TXP *n* Abbreviation for ▶ Trixylenyl Phosphate.

Tylose *n* Cellulose ether. Manufactured by Kalle, Germany.

Tympan \ˈtim-pən\ *n* [in one sense, fr. ME, fr. OE *timpana*, fr L *tympanum*; in other senses, fr. ML & L *tympanum*] (before 12c) The roller opposite the printing roller on a rotary press over which the paper web passes.

Tympan Sheet *n* A sheet of paper or cloth placed between the impression surface (platen or cylinder press) and the paper to be printed.

Tyndall Effect *n* The scattering of a beam of light by a colloid.

Tynex *n,* Nylon 6,6 manufactured by DuPont, U.S.

Type 8 Nylon *n* Not to be confused with ▶ Nylon 8. Type 8 nylon is a chemically treated nylon 6/6. It is thermoplastic, light yellow, with a leathery flexibility and excellent resistance to common solvents and abrasion. It has been used to impart abrasion resistance to the denim knees of jeans and overalls by impregnation of the cloth, and as an abrasion- and solvent-resistant coating for work gloves.

Type High *n* 0.918 in.; the standard in letterpress printing.

Type J Thermocouple *n* A ▶ Thermocouple made up of one or two welded junctions of iron and ▶ Constantan wires and widely used in measuring temperatures in plastics-processing equipment.

Type K Thermocouple *n* A Thermocouple made up of one or two welded junctions of ▶ Chromel and Alumel wires, widely used in measuring high temperatures in oxidizing atmospheres, and enjoying considerable use in plastics processing.

Type P Thermocouple *n* A Thermocouple made up of one or two welded junctions of copper and ▶ Constantan wires, used more in the measurement of low temperatures than high ones.

Typography \tī-ˈpä-grə-fē\ *n* [ML *typographia*, fr. Gk *typos* impression, cast + *-graphia* -graphy] (1610) Processes of making an impression using a raised, pigmented surface, from type, line etching, halftones, or rubber stamps. See ▶ Letterpress.

U

U \ˈyü\ *n* (1) Chemical symbol for the element uranium. (2) In heat-transfer engineering symbol for ▶ Overall Conductance.

Ubbelohde Viscometer (1) (Cannon-Ubbelohde viscometer) An instrument made of Pyrex glass and consisting of an upper reservoir that drains through a marked capillary to a lower, vented chamber and thence to a second reservoir. The time taken by the test liquid to drain through the capillary is proportional to the viscosity (with slight corrections). This is one of several similar types used to measure viscosities of polymer solutions (see ASTM, www.astm.org) See also ▶ Viscosity. (2) Capillary viscometer for measurement of viscosity molecular weight of polymers in solution; advantage over Ostwald-Fenske and Cannon-Fenske is that the measurement is independent of the amount of solution in the viscometer and measurement at a series of concentrations can easily be made by successive dilution; also used for measuring intrinsic viscosity (Kamide K, Dobashi T (2000) Physical chemistry of polymer solutions. Elsevier, New York).

Uchiwa-ye *n* [Japanese] A print in the shape of a fan.

UF Abbreviation for Urea-Formaldedyge Resin. See ▶ Amino Resin.

UHMWPE *n* Abbreviation for ▶ Ultra-High-Molecular-Weight Polyethylene.

Ultimate Elongation \ˈəl-tə-mət (ˌ)ē-ˌlón-ˈgā-shən\ In a tensile test, the nominal elongation at rupture. See ▶ Elongation (Shah V (1998) Handbook of plastics testing technology. Wiley, New York).

Ultimate Strength *n* The maximum nominal stress a material can withstand when subjected to an applied tensile, compressive, or shear load. If the mode of loading is not specified, it is assumed to be tensile. In materials that exhibit a definite yield strength, ultimate strength will usually mean the nominal stress at break, which can be less than the maximum (Shah V (1998) Handbook of plastics testing technology. Wiley, New York).

Ultimate Tensile Strength See ▶ Tensile Strength.

Ultimate Tensile Stress See ▶ Tensile Strength.

Ultracentrifugation \-ˌsen-trə-fyú-ˈgā-shən\ *vt* (1930) Rotating a fluid in a centrifuge in excess of about 15,000 rpm; useful for separating components from mixtures of solids, liquids and gases; and useful for ultra-cleaning or filtering solids from solutions (Gooch JW (1997) Analysis and deformulation of polymeric materials. Plenum Press, New York; Krause A, Lange A, Ezrin M (1988) Plastics analysis guide: chemical and instrumental methods. Oxford University Press, Oxford).

Ultracentrifuge \-ˈsen-trə-ˌfyüj\ *n* (1924) A small centrifuge capable of rotational speeds to 100,000 rpm, creating sedimentation forces up to a million times gravity. At lower speeds and long sedimentation times, an equilibrium radial distribution of larger and smaller molecules is attained. As the centrifuge spins, a narrow, collimated light beam is passed through a cell containing a polymer solution. By measuring refractive index or light absorption at different radii (i.e., depths in the cell), and using complex data-reduction methods, the number- and weight-average molecular weights of the polymer can be estimated and a graph of the molecular-weight distribution drawn. A second method uses the same techniques to measure sedimentation velocities at much higher speeds and over shorter time periods. Ultracentrifugation has been most useful with biological polymers, which tend to be all one size (*monodisperse*), making the data analysis less complicated (Krause A, Lange A, Ezrin M (1988) Plastics analysis guide: chemical and instrumental methods. Oxford University Press, Oxford).

Jan W. Gooch, *Encyclopedic Dictionary of Polymers*, DOI 10.1007/978-1-4419-6247-8,
© Springer Science+Business Media LLC 2011

Ultra High Molecular Weight n Capillary viscometer for measurement of viscosity molecular weight of polymers in solution; advantage over Ostwald-Fenske and Cannon-Fenske is that the measurement is independent is independent of the amount of solution in the viscometer and measurement at a series of concentrations can easily be made by successive dilution; also used for measuring intrinsic viscosity.

Ultra-High-Molecular-Weight Polyethylene n (UHMWPE) Any polyethylene having an average molecular weight (Mw) in the range from 1 million to 5 million grams per mole. Density is about 0.94 g/cm^3. These materials, like polytetrafluoroethylene, do not truly melt and are processed by compression and sintering, and related methods. Small amounts of PE resins having somewhat lower molecular weight may be added as processing aids.

Ultra-Low-Density-Polyethylene n (ULDPE) Any linear polyethylene with density less than 0.90 g/cm^3, and possibly as low as 0.86 g/cm^3. ULDPE films have better optical properties and better resistance to puncture, impact, and tearing than conventional linear low-density PEs.

Ultramarine Blue \-mə-ˈrēn\ n [ML *ultramarinus* coming fr. beyond the sea, fr. L *ultra-* + *mare* sea] (1598) $Na_6Al_6Si_6O_{24}S_2$ (light); $Na_7Al_6Si_6O_{24}S_3$ (Medium); $Na_8Al_6Si_6O_{24}S_4$ (Dark). (1) A natural type of mineral origin. It is also known as genuine ultramarine and *lapis lazuli*. (2) The synthetic types are produced in several shades by heating sulfur, clay, soda ash and a reducing agent. They have good lightfastness and good alkali resistance, but poor opacity and poor tinting strength. Density, 2.2–2.7 g/cm^3 (18.6–20.2 lb/gal); O.A., 25–39; particle size, 0.8–2.0 μm. (3) A natural or synthetic inorganic blue pigment occasionally used for printing inks. Syn: Brillain Ultra-Marine, Factious Ultramarine, French Blue, French Ultramarine, Guimet's Blue, Huemann's Blue, Laundry Blue, Oriental Blue, and Permanent Blue.

Ultramarine Blue Pigment n A pigment family comprising a complex of double silicates of sodium and aluminum in combination with sodium polysulfide. They produce bright, clean tones even in combination with white pigments, and are resistant to the high temperatures employed in processing thermoplastics ((1996) Kirk-Othmer encyclopedia of chemical technology: pigments-powders. Wiley, New York).

Ultramarine Green n Ultramarines cover a color range from pink to violet to blue and green. The pinks, violets, and greens are especially weak colors, and the green is no longer an item of commerce ((1996) Kirk-Othmer encyclopedia of chemical technology: pigments-powders. Wiley, New York).

Ultramarine Violet n $H_2Na_{46}Al_6Si_6O_{24}S_2$. Pigment Violet 15 (77007). Ultramarine Violet is produced by mixing ultramarine blue with ammonium chloride and heating at about 150°C for several hours. It is chemically and physically similar to ultramarine blue ((1996) Kirk-Othmer encyclopedia of chemical technology: pigments-powders. Wiley, New York).

Ultramide A n Nylon 6,6, manufactured by BASF, Germany.

Ultramide B n Nylon 6, manufactured by BASF, Germany.

Ultramides Nylon 6,10, manufactured by BASF, Germany.

Ultrapas n Melamine-formaldehyde resin, manufactured by Dynamit Nobel, Germany.

Ultrasonic n \-ˈsä-nik\ *adj* (1923) Vibrations above the hearing range of humans.

Ultrasonic Assembly n A process for using ultrasonic energy for assembling polymer parts.

Ultrasonic Cleaning n A method used for thoroughly cleaning molded plastics for electrical components and mechanical parts. A piezoelectric transducer (e.g., a crystal of barium titanate), mounted on the side or bottom of a cleaning tank, is excited by an ultrasonic generator to produce high-frequency vibrations in the cleaning medium. These vibrations cause intense cavitations in the liquid and dislodge contaminants from crevices, and even from blind holes, that normal cleaning method would not remove.

Ultrasonic C-Scan n A nondestructive inspection technique for reinforced plastics in which the energy absorbed from a short ultrasonic pulse is measured. This is quantitatively different for sample containing delaminations, voids, or too little or too much reinforcement, than for solid samples of the correct composition.

Ultrasonic Degating n A degating method used for small plastic parts produced by a family mold. The molding-machine operator removes the runner system and attached parts from the mold and loads, the branched structure into the degating machine, where the runner makes contact with an ultrasonic horn. High-frequency-sound vibrations are transmitted through the runners to the narrow gates, causing them to melt, and the pats drop through holes into a sorting system.

Ultrasonic Frequency n A sound frequency above the limit of human audibility, approximately 18 kHz. Most ultrasonic devices operate well above this level

(Giambattista A, Richardson R, Richardson RC, Richardson B (2003) College physics. McGraw Hill Science, New York).

Ultrasonic Inserting *n* A method of incorporating metallic inserts into plastics articles by means of ultrasonic heating. A plain cylindrical hole or well is molded in the plastic article by means of a core pin, the hold diameter being slightly less than that of the insert. Ultrasonic vibration and light pressure are applied as the metal part is being inserted, melting the plastic within a small radial distance of the inside hole surface. The displaced melt flows into the knurls, flutes or undercuts in the insert's outer surface and freezes, locking the insert into position.

Ultrasonic Prepreg Cutting *n* A recently introduced method for cutting Prepregs that uses an intense and extremely narrow beam of ultrasonic energy.

Ultrasonic Sealing *n* Same for sealing films, etc. (Skeist I (ed) (1990) Handbook of adhesives. Van Nostrand Reinhold, New York).

Ultrasonic Staking *n* The process of forming a head on a protruding peg in a plastic article for the purpose of holding a surrounding part in position (see ▶ Staking), utilizing ultrasonic heating and pressure to melt the tip of the protrusion and form it into a head. The process is very fast, usually taking only a fraction of a second.

Ultrasonic Welding *n* (Ultrasonic sealing) A method of welding or sealing thermoplastics in which heating is accomplished with vibratory mechanical pressure at ultrasonic frequencies (20–40 kHz). Electrical energy is converted to ultrasonic vibration by a transducer, directed to the area to be welded by a horn, and localized heat is generated by friction and impact at the surfaces to be joined. Other parts of the assembly are not heated. The process is most effective for rigid and semirigid plastics, since the energy is rapidly dissipated in soft, flexible materials. It will work with dissimilar plastics if they are melt-compatible and melt in the same temperature range (Skeist I (ed) (1990) Handbook of adhesives. Van Nostrand Reinhold, New York).

Ultraviolet *n* \ˌəl-trə-ˈvī-(ə-)lət\ adj (1840) (UV) The region of the electromagnetic spectrum between the X-ray region and the violet end of the visible-light range, including wavelengths from about 3 to 200 nm. Photons of radiation in the UV region have sufficient energy to initiate some chemical reactions and to degrade many neat resins. The term "ultraviolet light" is incorrect because light refers only to visible radiant energy. See ▶ Light. (The term "light" is limited to the visible region of the spectrum.)

Ultraviolet Absorber *n* Substance which absorbs ultraviolet radiation more readily than the coating material in which it is dissolved or dispersed, and which transforms the ultraviolet energy into longer wavelength energy which is relatively harmless to the coating. While some pigments are relatively nontransparent to ultraviolet radiation, the term is usually used to describe compounds which are physically dissolved in the coating an do mot affect transparency to visible light. These UV absorbers are a class of stabilizers which have intense absorption up to 350–370 nm, but are transparent in the visible. Examples of these are benzotriazoles and 2-hydroxy-benzophenones. They are widely used to stabilize exterior varnishes against UV degradation.

Ultraviolet Absorbers *n* Absorbers for ultraviolet energy which are usually strong nucleophilic agents and decompose peroxides ionically by S_N2 reactions; e.g., benzonephenones. {G UV-Absorber m, F absorbeur d'UV, absorbeur m, S absorbente de UV, absorbente m, I assorbitore UV, assorbitore m}

Ultraviolet Curing (Or UV Curing) *n* (1) Conversion of a wet coating or printing ink film to a solid film by the use of ultraviolet radiation. (2) The process by which certain polymers or coatings are cured, with the aid of a photoinitiator, by exposure to ultraviolet radiation. One such polymer system is the oligomer tris(2-hydroxymethyl)isocyanurate triacrylate with an initiator and an acrylic monomer such as 2-phenoxyethyl acrylate in ratios from ten to 100 parts per 100 parts of the triacrylate (USP 4,812,489).

Ultraviolet Degradation *n* One of the most serious degradation threats to plastics being used outdoors. Changes in plastics such as crazing, chalking, dulling of the surface, discoloration, and fading; changes in electrical properties; lowering of strength and toughness; and even disintegration, have been caused by exposure to UV radiation, particularly the longer wavelengths near the visible violet. The presence of oxygen can exacerbate the process. Chain scission is the major mechanism.

Ultraviolet Light (Deprecated) See ▶ Ultraviolet.

Ultraviolet Printing (UV printing, UV-curing decoration) The process of printing or decorating with inks that cure rapidly be exposure to ultraviolet light. The inks are solvent-free, thus avoiding problems with air-pollution regulations, and contain a UV-sensitive catalyst that cures the ink in as little as 1 s. Mercury-vapor lamps can be used as the source of UV light. For polymers that are subject to degradation by UV-induced oxidation, equipment has been designed to blanket the exposed area with nitrogen.

Ultraviolet Resistance *n* Ability to retain strength and resist deterioration on exposure to sunlight (Zaiko GE (ed) (1995) Degradation and stabilization of polymers. Nova Science, New York).

Ultraviolet Spectrophotometry *n* A method of chemical analysis similar to Infrared Spectrophotometry, except that the spectrum is obtained with ultraviolet light. It is somewhat less sensitive than the IR method for polymer analysis, but is useful for identifying and measuring plasticizers and antioxidants. An example of a UV spectrum of pyridine is shown in figure below (Willard HH, Merritt LL, Dean JA (1974) Instrumental methods of analysis. D. Van Nostrand, New York).

Ultraviolet Spectroscopy *n* Spectroscopic analysis using the ultraviolet (uv) wavelengths (<400 nm) and useful for detecting unsaturated chemical groups, conjugation, etc.

Ultraviolet Stabilizer *n* An additive that protects plastics against ▶ Ultraviolet Degradation, and that may accomplish its protection in various ways. An additive that preferentially absorbs UV radiation and dissipates the associated energy in a harmless manner is sometimes called an *ultraviolet absorber* or *ultraviolet screening agent*. Additives that do not actually absorb UV radiation but protect the polymer in some other manner are called *ultraviolet stabilizers* or other names indicative of the mechanism of stabilization. For example, products that remove the energy absorbed by the polymer before photochemical degradation can occur are called *energy-transfer agents* or *excited-state quenchers*. Other modes of UV stabilization are singlet-oxygen quenching, free-radical scavenging, and hydroperoxide decomposition. Classes of such stabilizers in use are the benzophenones, benzotriazoles, substituted acrylates, aryl esters, and compounds containing nickel or cobalt. Still another group consists of dispersed pigments, such as the very effective carbon black, which, of course, color the plastics and render them opaque.

Umber \ˈəm-bər\ *n* [prob. fr. obs. E, shade, color, fr. ME *umbre* shade, shadow, fr. MF, fr. L *umbra*] (1568) Pigment Brown 7. A naturally occurring brown earth containing ferric oxide, together with silica, alumina, manganese oxides and lime. Raw umber pigment is umber which has been ground; burnt umber pigment is umber calcined at a low temperature and ground.

Uncertainty Principle *n* (1929) A principle in quantum mechanics which states: it is impossible to discern simultaneously and with high accuracy both the position and the momentum of a particle (as an electron). Also known as Heisenberg Uncertainty Principle (Whitten KW, Davis RE, Davis E, Peck LM, Stanley GG (2003) General chemistry. Brookes/Cole, New York).

Uncrimping Energy See ▶ Crimp Energy.

Undercoat \ˈən-dər ˌkōt\ *n* (1648) (1) A coat of paint applied on new wood, or over a primer, or over

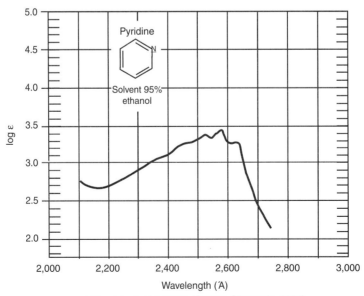

UV spectrum of pyridine. (Source: Silverstein, 1974)

a previous coat of paint; improves the seal and serves as a base for the topcoat, for which it provides better adhesion. (2) Any paint which acts as a base for enamel. (3) Any primer which is colored.

Undercoating \- ▪kō-tiŋ\ *n* (1922) A bituminous coating sprayed on the underside of automobiles to minimize rusting.

Undercure \- ▪kyúr\ A condition or degree of cure that is less than optimum, i.e., when insufficient time and/or temperature has been allowed for adequate cure; may be evidenced by tackiness, longness, or inferior physical properties.

Undercut \- ▪kət\ *n* (1859) A lateral indentation in a molded part (or protuberance in a mold) that tends to impede withdrawal of a molded part from the mold. Articles of flexible materials such as plasticized vinyls can often be removed without difficulty from molds with severe undercuts, but undercuts must either be avoided in rigid materials or, where they must be part of the design, the mold must have movable parts that withdraw (*side draws*) before the part is to be ejected. Slight undercuts are sometimes deliberately designed into one half of a mold to cause a part to remain in that half when the mold opens. See also ▶ Side-Draw Pin.

Undercutting *n* The action of etching solution on the sides of the resist of photomechanically prepared plates.

Underhue See ▶ Undertone Color.

Undertone \- ▪tōn\ *n* (1806) The color of a thin film of ink as seen on a white background. The appearance of an ink when viewed by light transmitted through the film.

Undertone Color *n* Color of a thin layer of pigment-vehicle mixture applied at incomplete hiding on a white background; sometimes referred to as cast or underhue. The term is sometimes used to refer to the color resulting when a pigment is mixed with white, but this should more properly be called tint-tone.

Underwater Pelletizing *n* A system, used mostly with 15-cm-diameter and larger, high-output extruders in resin-manufacturing plants, in which a circular, heated dire plate containing up to several hundred strand holes is enclosed within a water-tight casing and sprayed with water, while a rapidly spinning fly-knife slices the emerging strands of melt into lengths about equal to their diameters and flings the warm globs into the water spray. The water falling to the bottom carries the pellets out of the casing and onto a dewatering screen and drying conveyor. A delicate balance must be reached among several requirements: (1) cutting the pellets cleanly and chilling them instantly so that they do not grow "tails" nor form doubles and larger clusters; (2) maintaining a uniform metal temperature of the die over its entire face so that melt doesn't freeze in the holes and the extrusion rate is the same in all; (3) matching the knife speed with the extrusion rate to get the desired pellet length; and (4) draining the emerging pellets in a way that uses their remaining sensible heat to fully dry them but cools them sufficiently so that they do not soften and fuse together in the collecting bin. Surface tension in the molten particles rounds them into nearly spherical or ellipsoidal final shapes.

Undrawn Tow See ▶ Drawn Tow.

Undrawn Yarn *n* Extruded yarn (filaments), the component molecules of which are substantially unoriented. Undrawn yarn exhibits predominantly plastic flow in the initial stages of stretching and represents an intermediate stage in the production of a manufactured yarn.

Undulose Extinction *n* Nonuniform extinction of a substance between crossed polars. The areas of complete extinction move progressively with a fanlike motion across the surface of the substance as the stage is rotated.

Uneven Dyeing *n* A fabric dyeing that shows variations in shade resulting from incorrect processing or dyeing methods or from use of faulty materials.

Uneven Shrinkage A wavy, warp wise condition in the fabric that prevents it from lying flat on a horizontal surface.

Uneven Surface *n* An irregular surface characterized by nonuniformity in the physical configuration of the yarns or fibers making up the surface of the fabric.

Uneven Yarn *n* A yarn that varies in diameter to an abnormal degree.

Unfinished Worsted *n* A worsted fabric with a relatively soft hand and a light nap.

Uniaxial \ ▪yü-nē- ▪ak-sē-əl\ *adj* (ca. 1828) Of one primary axis and refractive index such as a drawn polymer fiber (e.g., drawn polyamide).

Uniaxial Crystals *n* Anisotropic crystals in the tetragonal and hexagonal systems having one unique crystallographic direction and either two (tetragonal) for three (hexagonal) directions which are alike and perpendicular to the unique direction (Rhodes G (1999) Crystallography made crystal clear: a guide for users of macromolecular models. Elsevier Science and Technology Books, New York).

Uniaxial Load *n* A condition whereby a test sample or structural member is stressed in only one direction.

Uniaxial Orientation *n* An orientation process that stretches the product in only one direction, as in manufacture of staple fiber, monofilaments, and melt-cast film.

Uniaxial Strain *n* Tensile or compressive strain in a single direction – the usual testing mode – and typically in the length direction of a test specimen or structural member. See also ▶ Strain and ▶ True Strain.

Uniaxial Stress *n* Tensile or compressive stress in a single direction, usually the lengthwise direction of a test specimen or structural member. See also ▶ Stress and ▶ True Stress.

Unicellular Plastic \ ˌyü-ni-ˈsel-yə-lər-\ A term that has sometimes been used for ▶ Closed-Cell Foamed Plastic.

Unidirectional Fabric \ ˌyü-ni-də-ˈrek-shnəl-\ A fabric having reinforcing fibers in only one direction.

Unidirectional Laminate \-ˈla-mə-ˌnāt\ A reinforced-plastic structure in which substantially all of the fibers are parallel. The modulus of elasticity (E) and strength of such a laminate in the direction o reinforcement will be somewhat more than the product of the volume fraction of reinforcing fiber times its corresponding properties. See ▶ Law Of Mixtures. However, properties in the transverse directions are essentially those of the matrix resin. See also BI- and ▶ Tridirectional Laminate.

Uniform Chromaticity Coordinates *n* Chromaticity coordinates yielding an approximately uniform chromaticity diagram for colors of equiluminosity. See ▶ Uniform Chromaticity Scale Diagram (McDonald R (1997) Colour physics for industry, 2nd edn. Society of Dyers and Colourists, West Yorkshire).

Uniform Chromaticity Scale Diagram *n* Any one of a variety of transformations of the CIE chromaticity diagram to a diagram on which all pairs of just noticeably different colors of equal luminance are represented by pairs of points separated by nearly equal distances.

Uniform Chromaticity Spacing See ▶ Uniform Chromaticity Scale Diagram.

Uniform Circular Motion *n* If *r* is the radius of a circle, *v* the linear speed in the arc, ω the angular velocity and *T* the period or time of one revolution,

$$\omega = \frac{v}{r} = \frac{2\pi}{T}$$

The acceleration toward the center is

$$a = \frac{v^2}{r} = \omega^2 r = \frac{4\pi^2 r}{T^2}$$

The centrifugal force for a mass *m*,

$$F = \frac{mv^2}{r} = m\omega^2 r = \frac{4\pi^2 mr}{T^2}$$

In the above equations ω will be in radians per second and *a* in centimeters per second per second if *r* is in centimeter, *v* in centimeter per second and *T* in second. *F* will be in dynes if mass is in grams and other units as above. *Application to the solar system* – If *M* is the mass of the sun, *G* the constant of gravitation, *P* is the period of the planet and *r* the distance of the planet from the sun, then the mass of the sun

$$M = \frac{4\pi^2 r^3}{GP^2}, (G = 6.670 \times 10^{-8} \text{ for cgs units}).$$

If *P* is the period and *r* the distance of a satellite revolving around the planet, the above expression for *M* gives the mass of the planet. The formula is written on the assumption that the orbit of the planet or satellite is circular, which is only approximately true (Giambattista A, Richardson R, Richardson RC, Richardson B (2003) College physics. McGraw Hill Science, New York).

Uniform Color Scale *n* A scale wherein the units of color difference that are judged to be equal have equal scale differences (McDonald R (1997) Colour physics for industry, 2nd edn. Society of Dyers and Colourists, West Yorkshire).

Uniform Color Space Three-dimensional space wherein all pairs of colors, judged to be equally different, are separated by nearly equal distances.

Uniform Lightness Scale *n* A scale wherein the units of lightness that are visually judged to be equal have equal scale values, and the lightness differences judged to be equal are represented by equal scale differences.

Uniformly Accelerated Rectilinear Motion *n* If v_o is the initial velocity, v_t, the velocity after time *t*, the acceleration,

$$a = \frac{v_t - v_o}{t}$$

The velocity after time *t*,

$$v_t = v_o + at$$

Space passed over in time *t*,

$$s = v_o t + \frac{1}{2}at^2$$

Velocity after passing over space *s*,

$$v = \sqrt{v_o^2 + 2as}$$

Space passed over in the *n*th second

$$s = v_o + \frac{1}{2}a(2n - 1).$$

In the above and following similar equations the values of the space, velocity, and acceleration must be substituted in the same system. For space in centimeter, velocity will be in centimeter per second and acceleration in centimeter per second per second (Giambattista A, Richardson R, Richardson RC, Richardson B (2003) College physics. McGraw Hill Science, New York).

Unimolecular Process *n* An elementary process in which the probability of collisional deactivation of the activated complex greatly exceeds the probability of its decomposing to form products.

Union Cloth *n* A term describing a fabric woven from two or more types of yarn. For example, a union cloth may have a cotton warp and a wool filling. Also see ▶ Combination Fabric.

Unit \ˈyü-nət\ *n* [back-formation fr. *unity*] (1570) Specific magnitude of a quantity, set apart by appropriate definition, which is to serve as a basis of comparison or measurement for other quantities of the same nature.

Unit, Angstrom \ˈyü-nət ˈaŋ-strəm\ See ▶ Angstrom Unit.

Unit Cell *n* (1915) The basic unit for describing the arrangement of atoms in a crystal. The smallest parallelepiped which can generate the crystal lattice by repeated translations along the axes of the lattice.

Unit Elongation Syn: ▶ Elongation.

Unit Mold *n* A mold designed for quick changing of interchangeable cavities or cavity parts.

Unit of Measurement *n* Systematic and non-systematic units for measuring physical quantities, including metric ad US pound-inch systems. Also called units.

Unit of Time *n* The fundamental invariable unit of time is the ephemeris second, which is defined as 1/31,556,925.9747 of the tropical year for 1900 January $0^d 12^h$ ephemeris time. The ephemeris day is 86,400 ephemeris seconds.

Unloading Valve *n* A valve that limits the maximum pressure in a hydraulic line (or other fluid space) to a desired value by diverting the flow of fluid from a pump to a bypass line. See also ▶ Rupture Disk.

Unopened Staple *n* Staple fiber in bunches or clusters in the bale in such a condition that it will not process smoothly through carding and subsequent operations in the spun-yarn plant.

Unperturbed Dimensions *n* The dimension of a polymer coil in dilute solution at the theta temperature. Under these conditions the long-range interactions between segments of the polymer chains, causing the chain to contract, are just balance by solvation forces.

Unpolarized Light *n* A bundle of light rays having a common propagation direction but different vibration directions.

Unrelaxed Yarn *n* See ▶ Relaxed Yarn.

Uns- *adj* (*unsym-*) Abbreviation for Unsymmetrical, a prefix denoting unsymmetrical disposition of substituents of organic compounds with respect to the carbon skeleton or a functional group. It is usually ignored in alphabetization of compound names.

Unsaponifiable Matter *n* That portion of fats and resins which does not react with alkali to form a soap.

Unsaturated Compound *n* An organic compound having one or more incidences of two or three bonds between two adjacent atoms, usually carbon or nitrogen atoms, and capable of adding other atoms at such points to reduce it to a single bond, thus becoming saturated. Multiple unsaturation is common, as in dienes, aromatics, and oils and fats.

Unsaturated Hydrocarbon *n* A hydrocarbon with one or more multiple bonds.

Unsaturated Polyester See Polyester Unsaturated.

Unsaturated Solution *n* A solution in which the concentration of solute is less than its solubility.

Unsaturation, Analysis *n* Measurement of unsaturated bonds and conjugation; e.g., ultraviolet spectroscopy.

Unwind Unit *n* (Unwind) (1) In plastics coating and film laminating, a stand or a driven machine holding a roll of the substrate to be coated or laminated and supplying the web to the coating equipment at the rate and tension needed. (2) In molding from continuous prepreg, a similar units as in (1) but sometimes including additional functions.

Unzipping *n* The fast reversal of polymerization, with release of monomer that can occur in addition homopolymers once a stable end group has been removed. Copolymerization helps to minimize unzipping (Odian GC (2004) Principles of polymerization. Wiley, New York).

UP *n* Abbreviation used by the British Standards Institution for Unsaturated Polyester.

Up-and-Down Method *n* (Staircase method) A testing protocol sometimes used to estimate the average value of a property that, with limited resources for testing, must be tested at discrete levels of experimental factors, such as temperature or impact energy, that are strong determinants of the property. One specimen is tested at level A of the factor and passes or fails. If passes, the factor intensity is raised to level B, significantly higher than A, and the test is repeated with a new specimen. (If it fails at level A, intensity is lowered.) If it passes at level B, intensity is again raised, to level C. When it eventually fails, the level is reduced one level, and so on, until the allotted number of specimens has been tested. This procedure has been used in determining ▶ Brittleness Temperature and in several impact tests. The method is fairly efficient at determining the average sought; but at the expense of standard deviation.

Updraft Metier *n* A dry spinning machine in which the air flow within the drying cabinet is countercurrent to the yarn path (upward).

Upper Critical Solution Temperature *n* The maximum temperature for phase separation of polymer-solvent solutions to occur (Flory-Huggins theory); also, phase separation occurs when the temperature is raised until a lower critical solution temperature is reached, the phenomenon is explained by the free-volume theories of polymer solutions (Kamide K, Dobashi T (2000) Physical chemistry of polymer solutions. Elsevier, New York).

Upper Newtonian Viscosity *n* (μ_∞) The coefficient of viscosity of a fluid at very high shear rates, where Newtonian behavior is observed, although the fluid is non Newtonian at lower shear rates. This is often true for polymer melts that have chains that disentangle at shear rates above a critical shear rate. See, for example, ▶ Powell-Eyring Model.

Upstroke Press *n* A hydraulic press in which the main ram is situated below the moving table, pressure being applied by an upward movement of the ram.

Uptwister *n* A machine used for twisting yarns in an upward path from a rotating vertical supply package to a horizontal take-up package. Used for spun yarns and to a small extent for adding twist to some filament yarns.

Uptwisting *n* The process of twisting yarn on the uptwister. The yarn to be twisted, which has been wound on a balanced support package, is placed on a revolving spindle. The yarn form the revolving supply package is fed upward through a gathering eye or guide, over a stop motion and a tension bar or bars, through a traversing guide, and onto the revolving collecting package.

Urea \yü-ˈrē-ə\ *n* [NL, fr. F *urée*, fr. *urine*] (1806) NH_2CONH_2. Mp, 132°C; sp gr, 1.323. A white, crystalline powder derived from the decomposition of ammonium carbonate. It is used in the production of urea-formaldehyde resins. Syn: ▶ Carbamide.

Urea-Formaldehyde *n* (1928) A thermosetting synthetic resin made by condensing urea with formaldehyde (Odian GC (2004) Principles of polymerization. Wiley, New York).

Urea-Formaldehyde Foam *n* A foam produced by combining a urea-formaldehyde resin with a detergent-type foaming agent under pressure. Upon release of pressure, a foam of about the consistency of shaving cream emerges and cold-cures within 2–4 h. The foam is of low density, is noncombustible, and dries within 1–2 days. The dried foam has some resiliency, good thermal-insulation qualities, and is sound-absorbent. Although not recommended for continuous exposure to temperatures above 100°C, the material does not decompose and release gases until heated to a much higher temperature.

Urea-Formaldehyde Polymers *n* The cheapest synthetic polymer used as bonding elements between faces and core of a laminated structures (Odian GC (2004) Principles of polymerization. Wiley, New York).

Urea-Formaldehyde Resins *n* Any of a group of resins formed by the interaction of urea and formaldehyde under conditions that include heat and pH control (Odian GC (2004) Principles of polymerization. Wiley, New York). See ▶ Amino Resin.

Urea Plastic See ▶ Amino Resin.

Urea Resins *n* A synthetic resin made from urea and an aldehyde.

Urethane \ˈyür-ə-ˌthān\ *n* [F *uréthane*, fr. ur- ^{1}ur- + *éth-* eth- + *ane*] (1838) (1) (ethyl carbamate) $H_2NCOOC_2H_5$. This compound may be thought of as urea in which one $-NH_2$ group has been replaced by an ethoxy group, $-OC_2H_5$. Curiously, urethane itself has no direct application in making polyurethanes. (2) A compound of the general structure RHNCOOR', formed by reaction of an alcohol with an isocyanate, which is (1), above with one of the amino hydrogens replaced by R and the ethyl of ethoxy broadened to include the other alkyl or aryl radicals. (3) A chain unit in polyurethanes, -RHNCOOR'-, which is formed by the reaction of a diol and an isocyanate. (4) Shorthand substitute for ▶ Polyurethane. Mp, 48–50°C; bp, 182–184°C. Syn: ▶ Ethyl Carbamate.

Urethane Coating *n* Coating vehicles containing a polyisocyanate monomer reacted in such a manner as to yield polymers containing any ratio, proportion or combination of urethane linkages, active isocyanate groups or polyisocyanate monomer. The reaction products may contain excess isocyanate groups available for further reaction at time of application or may contain essentially no free isocyanate as supplied. ASTM (www.astm.org) has designated five types of urethane coatings. *Type I* is a one-component system modified with a drying oil such as linseed or soya, which reacts with oxygen from the air to affect cure, used in wood finishes. *Type II* is based on isocyanate-terminated prepolymers in a solvent that dries by evaporation and cures by reaction with moisture in the air. It is used for coating wood, rubber, and leather. *Type III* is based on a blocked isocyanate, a polyester, a curing agent, and a suitable solvent. The applied coating is heated to effect curing. It is used for wire covering and industrial finishes. *Type IV* is a two-component system, one having a prepolymers made from a diisocyanate and a polyol, the other being a catalyst such as a tertiary amine. It is used for heavy-duty industrial finishes with good resistance to chemicals, abrasion, and corrosion. *Type V* is also a two-component system, one being a polyisocyanate (usually an adduct of a diisocyanate and trimethylolpropane) and the other being a polyol, in solvents that evaporate after application of the coating. The reaction proceeds at ambient temperatures without the aid of a catalyst. It is used as a high-performance industrial coating (Klempner D, Frisch KC (2001) Advances in urethane science and technology. Rapra Technology, Shropshire; Carley JF (ed) (1993) Whittington's dictionary of plastics. Technomic, Lancaster).

Urethane Foams *n* Urethane in the form of rigid foam displays superior thermal insulating qualities, hard degree of hardness, mar resistance, flexibility, and good chemical resistance, when used as a coating resin. Urethane foams consist of about two thirds of polyol. This allows urethane foams to produce pyrolysis products similar to those of diol or triol. See ▶ Polyurethane Foam.

Urethane-Imide Modified Foam See Polyurethane-Imide Modified Foam.

Urethane Plastic See ▶ Polyurethane and ▶ Polyurethane Foam.

Urethane Polymers See ▶ Polyurethanes.

Urethane Resins See ▶ Polyurethanes.

Urethane Thermoplastic Elastomers *n* Block polyether or polyester polyurethanes containing soft and hard segments. Have good tensile strength, elongation, adhesion, and a broad hardness and service temperature ranges, but decreased moisture resistance and processibility. Processed by extrusion, injection molding, film blowing, and coating. Used in tubing, packaging film, adhesives, medical devises, conveyor belts, auto parts, and cable jackets. Also called TPU.

Urshiol *n* Hydroxy acid of aromatic type present in *Rhus vernicifera*, the basis of Japanese lacquer.

Urushi Japanese word for lacquer.

Urushiol \yú-ˈrü-shē-ˌȯl, ú-ˈ, -ˌōl\ *n* [ISV, fr. Japanese *urushi* lacquer + ISV *ˈ-ol*] (1908) A mixture of pyrocatechol derivatives with saturated or unsaturated side chains of 15 or 17 carbon atoms that is an oily toxic irritant principle present in poison ivy and some related plants (genus *Rhus*) and in oriental lacquers derived from such plants (Langenheim JH (2003) Plant resins: chemistry, evolution ecology and ethnobotany. Timber Press, Portland; (2001) Paint: pigment, drying oils, polymers, resins, naval stores, cellulosics esters, and ink vehicles, vol 3. American Society for Testing and Material, Conshohocken).

Urushi-ye *n* (Japanese) A print for which lacquer is used to intensify the color. The term is generally employed to describe only the only the early hand-colored prints in which lacquers, colors and metallic rust were applied to the printed black outline (Gair A (1996) Artist's manual. Chronicle Books LLC, San Francisco; Mayer R, Sheehan S (1991) Artist's handbook of materials and techniques. Viking Adult, New York).

Urylon *n* Poly(nonamethylene urea), manufactured by Toya, Japan.

Usable Life See ▶ Pot Life.

Useful Life *n* The length of time the coating is expected to remain in service.

U.S. Gallon See ▶ Gallon, U.S.

Uster Tester *n* An instrument that provides a continuous measurement of the variation in weight per unit length of sliver, roving, and yarn.

UTS See ▶ Tensile Strength.

u, v *n* Chromaticity coordinates in one of the CIE Uniform Color Spacings (1964).

UV *n* Abbreviation for Ultraviolet Radiation. See ▶ Ultraviolet.

UV Absorbers *n* Polymer additives that absorb light in the UV region or that trap radicals produced in fiber during photooxidation. They provide stabilization against actinic degradation. Some critical applications include geotextiles, recreational surface polymers and fibers, tenting tarpaulins, etc. (Fouassier J-P (1995) Photoinitiation, photopolymerization and photocuring. Hanser-Gardner, New York).

UV Stabilizers *n* Additives which do not actually absorb UV radiation but protect the polymer in some other manner are called ultraviolet stabilizers (Zaiko GE (ed) (1995) Degradation and stabilization of polymers. Nova Science, New York).

V

v \ˈvē\ *n* (15c) (1) Symbol for velocity.

V (1) SI abbreviation for ▶ Volt. (2) Chemical symbol for the element vanadium. (3) Symbol for system volume.

Vacancy \ˈvā-kən(t)-sē\ *n* (1599) A lattice point in a crystal at which a particle is missing.

Vacations See ▶ Holidays.

Vacuum Bag Molding See ▶ Bag Molding.

Vacuum Calibration Syn: Vacuum Sizing.

Vacuum Casting *n* A method used for casting fluid thermosetting resins to avoid inclusions of air bubbles. The mold is placed in a vacuum chamber and filled with resin from an external hopper. Vacuum is applied to pull out bubbles, held until they have all risen to the surface, then released. Curing follows.

Vacuum Filter Filtering system in which filtrate is removed by application of a vacuum.

Vacuum Forming *n* (straight vacuum forming) Process in which heated thermoplastic sheets are converted to other configurations by causing them to flow into molds with application of a vacuum. The simplest, original technique of ▶ Sheet Thermoforming.

Vacuum Impregnating *n* The process of impregnating electrical components by subjecting the parts to a moderate vacuum to remove air and other volatiles, introducing the impregnant to penetrate the parts, then releasing the vacuum and curing. Epoxy, phenolic, and polyester resins are often used. See also ▶ Potting and ▶ Encapsulation.

Vacuum Metalizing *n* (1) A decorating process used to make plastic objects resemble shiny metals by depositing very small thicknesses of metals on plastics and films. The process consists of cleaning the surface of the object to be coated followed by placing it in a vacuum chamber where the metal is vaporized or cathode-sputtered to produce vapor that condenses on the surface of the object. A thin film of lacquer or other coating is usually applied to the metallized surface to protect the luster of the coating. (Madox, D. M., *Handbook of Physical Vapor Deposition (PVD) Processing*, Noyes Data Corporation, New York, 1998).

Vacuum Molding *n* This type of molding is used to mold fiberglass-reinforced plastics. The method allows molding without high temperatures and pressures. In this technique, layers of reinforced media are applied to the mold by hand. Resin is either sprayed or brushed on after each layer is positioned. The flexible sheet (usually cellophane or polyvinyl acetate) is placed over the lay up. Joints and seals are sealed and a vacuum causes the bag to collapse over the face of the product not contacting the mold. The resultant pressure tends to eliminate voids and forces out any excess resin or entrapped air. (Madox, D. M., *Handbook of Physical Vapor Deposition (PVD) Processing*, Noyes Data Corporation, New York, 1998)

Vacuum-Pressure Impregnation *n* Method for the commercial impregnation of electrical equipment with insulating varnish. It consists in first subjecting the equipment to a vacuum in order to remove as much occluded air as possible, permitting the varnish to enter also under vacuum to replace the air, and finally reducing the vacuum and applying pressure in order to assist the somewhat viscous varnish through the narrow interstices left between the fine windings, etc.

Vacuum Venting *n* The drawing of a vacuum on the cavity of an injected (or other) mold in order to eliminate molding defects such as short shots, voids and, particularly, burned spots (*dieseling*). The vacuum may be drawn by means of tubes leading to vents to sharp corners, blind holds, etc. In one implementation of the concept, the entire mold is enclosed in a vacuum-tight box with a parting line coplanar with that of the mold, its mating surfaces sealed by O-rings.

Valance \ˈva-lən(t)s, ˈvā\ *n* [ME *vallance*, perhaps fr. *Valence*, France] (15c) (1) In ionic chemical bonding, the property of an element that is measured by the number of atoms of hydrogen (or its equivalent) that one atom of the element can hold in combination if negative, or can displace in a reaction if positive. Many elements have more than one valence, corresponding to lower and higher states of oxidation or reduction. In covalent bonding, valance number of outermost-shell electrons that an element has available for sharing with other elements. (2) Short drapery at top of a window. A decorative frame used to conceal the top of curtains and fixtures. (Goldberg, D. E., *Fundamentals of Chemistry*, McGraw-Hill Science/Engineering/Math, New York, 2003)

Valence Electrons *n* Electrons which are gained, lost, or shared in a chemical reaction.

Valence Shell *n* The shell of electrons with the highest principal quantum number *n* in an atom.

Valerian \və-ˈlir-ē-ən\ *n*[ME, fr. MF or ML; MF *valeriane*, fr. ML*vaeriana*] Any of a genus (*Valeriana* of the family Valerianaceae, the valerian family) of perennial herbs many of which possess medicinal properties. (Langenheim, J. H., *Plant Resins: Chemistry, Evolution*

Ecology and Ethnobotany, Timber Press, Portland, OR, 2003)

Valeric Acid \və-ˈlir-ik´\ *n* Pentanoic Acid. [*valerian; fr.* Its occurrence in the roof of valerian] (1857) Any of four isomeric fatty acids or a mixture of these; a liquid acid of disagreeable odor obtained from valerian or made synthetically and used in organic synthesis.

Valley Printing *n* (inlay printing) A printing process for flat plastic surfaces in which ink is applied to the raised portions of an embossing roll that simultaneously embosses the plastic surface and deposits ink in the valleys of the embossed surface. The process is similar to flexographic heating in that both print from raised portions of a cylinder.

Value a \ ˈval-(ˌ)yü\ *n* [ME, fr. MF, fr. (assumed) VL *valuta*, fr. feminine of *valutus*, pp of L *valēre* to be worth, be strong] (14c) (color value) The lightness of a color. A color may be classified as equivalent to some member of a series of shades ranging from black (the zero-value member) to white. The other two fundamental characterizers of color are *hue* and *saturation*.

Value Extrusion *n* An extrusion operation in which melt pressure and, to a lesser extent, throughput are controlled by an adjustable value. For example, when a screen pack is used to remove foreign matter from the melt stream, a valve may be inserted between the screen pack and the die

Value, Munsell See ▶ Munsell Value.

Vanadium \və-ˈnā-dē-əm\ *n* [NL, fr. ON *Vanadīs* Freya] (1833) A grayish malleable ductile metallic element found combined in minerals and used especially to form alloys.

Vanadium Driers *n* Group of rather dark-colored driers, based on soluble vanadium salts, which are fairly strong surface-drying catalysts.

Van der Waals Equation *np* and Free Volume Effects *n* \ ˈvan-dər-ˌwólz-\ The van der Waals Equation is expressed as:

$$\left(p + \frac{a}{v^2}\right)(v - b) = RT$$

It makes allowance both for the volume occupied by the molecules and for the attractive force between the molecules; *b* is the effective volume of molecules in one mole of gas; and *a* is a measure of the attractive force between the molecules. For values of *R*, *a*, and *b* see a table of Van der Waal's constants for gasses in *CRC Handbook of Chemistry and Physics*, Lide, D. R., ed., CRC Press, Boca Raton, Fl, 2004 Version.

Van der Waals Forces *np* [Johannes D. *van der Waals* † 1923 Dutch physicist] (1926) (secondary valence force, intermolecular force) An attractive force, much weaker than primary covalence bonds, between molecules of a substance in which all the primary valences are saturated. They are believed to arise mainly from the *dispersion effect*, in which temporary dipoles induce other dipoles in phase with themselves. The primary van der Waals forces are *dipole-dipole* (polar molecules) and *London forces* (nonpolar molecules). These forces are attributed to the attractions between molecules and from noncovalent bonds (Goldberg, D. E., *Fundamentals of Chemistry*, McGraw-Hill Science/Engineering/Math, New York, 2003).

Vandyke Brown \van-ˈdīk ˈbraún\ *n* [fr. its use by the painter Vandyke] (ca. 1850) (1) A very dark, deep-brown natural earth pigment consisting essentially or organic matter obtained from peat or lignite found chiefly in parts of Germany; its principal uses are in artists' colors, stains and graining. *Also called Cassel Brown and Cassel Earth.* (2) A synthetic pigment of similar color.

Van't Hoff's Principle \vänt-ˈhóf\ *np* [Jaobus Hen.dri.cus *van't Hoff*, 1852–1911, D physical chemist] If the temperature of interacting substances in equilibrium is raised, the equilibrium concentrations of the reaction are changed so that the products of that reaction which absorb heat are increased in quantity, or if the temperature for such as equilibrium is lowered, the products which evolve heat in their formation are increased in amounts (*Handbook of Chemistry and Physics*, Lide, D. R., ed., CRC Press, 2004; *Handbook of Chemistry and Physics*, ed. R. C. Weast, The Chemical Rubber Co., Boca Raton, FL, 52nd Ed.)

Vapor \ ˈvā-pər\ *n* [ME *vapour*, fr. MF *vapeur*, fr. L *vapor* steam, vapor] (14c) As most frequently used, the term vapor means a substance that, although present in the gaseous phase, generally has a stable liquid or solid state at ambient temperature. *Gas*, on the other hand is used for substances that do not have stable liquid or solid states at ambient conditions. Thus, we speak of *water vapor* but *oxygen* and *nitrogen gases* in the atmosphere.

Vapor Barrier *n* (ca. 1941) A moisture-impervious layer or coating (such as special paint, or a membrane on roofing felt or on building paper) that prevents the passage of moisture of vapor into a material or structure.

Vapor Degreaser *n* An apparatus in which metal surfaces are cleaned by solvent vapors.

Vapor Degreasing *n* A process of removing grease from parts and equipment components by suspending them with a closed chamber over a pool of boiling, nonflammable solvent such as a mixture of chlorofluorinated hydrocarbons.

Vapor-Liquid Chromatography See ▶ Chromatography.

Vapor Permeability *n* That characteristic of a material which permits the passage of a vapor or gas. The permeability is measured under carefully specified conditions, such as total pressure, partial pressure on the two sides of the specimen, temperature, and relative humidity. As the fibers of a material such as paper have such a high affinity for water (vapor), vapor permeability should not be confused with air permeability or porosity.

Vapor Pressure *n* (1875) The pressure of the vapor phase when a solid or liquid is in equilibrium with its vapor. Vapor pressure increases exponentially with absolute temperature at a rate that is unique for each pure substance and closely related to its heat of vaporization. In homologous organic compounds, vapor pressure at any temperature decreases with increasing molecular weight. The term has little meaning for plastics because of high polymers decompose before evaporating. The relative vapor pressure density of solvents is given by the following equation:

$$d_{vp} = \frac{M_s}{M_{air}}$$

Where M_s = the molecular weight solvent, and M_{air} = the molecular weight of air. The vapor density increases with molecular weight of solvent. The Clausius-Clapeyron equation (Wypych, 2001) gives the relationship between molecular weight of solvent and vapor pressure:

$$\frac{d \ln p}{dT} = \frac{M\Delta}{RT^2}$$

Where p = vapor pressure, T = temperature, M = molecular weight of solvent, Λ = heat of vaporization and R = gas constant. Vapor pressure at any temperature can be estimated by use of the Antoine equation Wypych, 2001):

$$\log p = A - \frac{B}{C + T}$$

Where p = vapor pressure of solvent and A, B and C are constants. Vapor pressure of a solvent increases with temperature. A, B and C constants and vapor pressures of many pure substances at a temperature (*CRC Handbook of Chemistry and Physics*, Lide, D. R., ed., CRC Press, Boca Raton, Fl, 2004 Versiond; *Handbook of Solvents*, Wypych, G., ed., Chemtec Publishing, New York, 2001) and plotted "log P vs. 1/T" where P is vapor pressure mm Hg) of a solvent, T is temperature (K) and developed the equation,

$$\log P = m\frac{1}{T} + b$$

and

$$m = \frac{-\Delta H_{vap}}{2.303R}$$

Where m = molar heat of vaporization, ΔH_{vap} = heat to vaporize one mole of solvent or liquid to from an ideal gas at one atmosphere pressure, and R is the ideal gas constant. This form of the Clausius-Claperon equation is useful for determining ΔH_{vap} when the vapor pressure is known at two different temperatures; the vapor pressure at a given temperature if ΔH_{vap} is known and if the vapor pressure is known at another temperature; and the temperature at which a liquid has a given vapor pressure if ΔH_{vap} and the vapor pressure at one temperature are known. Barrow (1973) expressed the Clausius-Claperon equation as,

$$\frac{d(\ln P)}{dT} = \frac{\Delta H_{vap}}{RT^2}$$

and

$$\frac{d(\ln P)}{d(1/T)} = -\frac{\Delta H_{vap}}{R}$$

and the integrated form

$$\log P = -\frac{\Delta H_{vap}}{2.303R}\frac{1}{T} + \text{constant}$$

Mixtures of miscible solvents have limited predictability using Raoult's Law:

$$p_{12} = m_1 p_1 + (1 - m_1)p$$

Where p_{12} = vapor pressure of mixture, m_1 = molar fraction of component 1, and p_1, p_2 = vapor pressures of the components 1 and 2.

Vapor-Pressure Lowering *n* One of the colligative properties of a solution and the basis of a method for determination of the molecular mass of a solute. For a dilute solutions, the solvent vapor pressure lowering is determined by $(p_0 - p)/p_0 = x_2$, where p_0 and p are the vapor pressures of the pure solvent and the solution respectively and x_2 is the mole fraction of the solute.

Vapor-Pressure Osmometer *n* An instrument for determining number-average molecular weight of a polymer utilizing the "vapor pressure versus molecular weight relationship."

Vapor Transmission *n* If the vapor is not otherwise identified, this phase is understood to mean Water-vapor Transmission Rate.

Varnish \▮vär-nish\ *n* [ME *vernisch*, fr. MF *vernis*, fr. OIt or ML; OIt *vernice*, fr. ML *veronic-*, *veronix* sandarac] (14c) (1) A liquid composition which is converted to a transparent solid film after application as a thin layer. *Bituminous* – A dark colored varnish containing bituminous ingredients. The varnish may be either of the oil or spirit type. *Oil* – A varnish which contains resin and drying oil as the basic film-forming ingredients and is converted to a solid film primarily by chemical reaction. *Spar* – A varnish for exterior surfaces. The name originates from its use on spars of ships. *Spirit* – A varnish which is converted to a solid film, primarily by solvent evaporation. (2) In printing ink technology, a broad term including fluid compositions comprising one or more of the following: oils, resins, solvents, driers and waxes; used either as a vehicle or to cover surfaces. See ▶ Oil. (Raaf, J. J., *Dictionary of Paint, Varnish and Lacquer Terms*, English, Language Services, Inc., New York, 1967; *Paint,Coatings Dictionary*, Compiled by Definitions Committee of the Federation of Societies for Coatings Technology, 1978)

Varnish Remover See ▶ Paint and ▶ Varnish Remover.

Varnish Stain *n* A varnish which is colored with a transparent material, leaving a colored coating on the surface; has less penetrating power than a true stain. See ▶ Stain.

Varnish Tree *n* (1758) Any of various trees yielding a milky juice from which in some cases varnish or lacquer is prepared.

Vat Dyes *n* Those dyes which are insoluble in water, dilute acids and alkaline solutions. Exposure of material dyed with such a color base produces the original colored compound by oxidation. See ▶ Dyes.

Vegetable Black *n* Originally, a sooty product, obtained by burning vegetable oils with restricted air. Most vegetable black today is obtained by burning coal-tar oil. See ▶ Lampblack.

Vegetable Drying Oil See ▶ Drying Oil.

Vegetable Fiber *n* A textile fiber of vegetable origin, such as cotton, kapok, jute, ramie, and flax. (*Natural Fibers, Plastics and Composites*, Wallenberger, F. T., and Weston, N. E., eds., Springer-Verlag, New York, 2003; Kadolph, S. J. J., and Langford, A. L., *Textiles*, Pearson Education, New York, 2001)

Vegetable Oil *n* (1797) An oil extracted from vegetable matter; especially castor, linseed, safflower, soya, and tung oil; used in paints and plastics. ()

Vehicle \▮vē-ə-kəl *also* ▮vē-▮hi-kəl\ *n* [F *véhicule*, fr. L *vehiculum* carriage, conveyance, fr. *vehere* to carry] (1612) The liquid portion of an ink that holds and carries the pigment and provides workability, and drying properties, and binds the pigment to the substrate after the ink has dried. (*Coatings Technology Handbook*, Tracton, A. A., ed., Taylor and Francis, Inc. New York, 2005)

Veil \▮vā(ə)l\ A thin mat of very fine, relatively long fibers used at the outermost layer of a composite in order to improve surface appearance and smoothness. (Murphy, J., *Reinforced Plastics Handbook*, Elsevier Science and Technology Books, New York, 1998)

Veiling *n* (1) Formation of a cobweb attern. (2) Curtaining or sagging. (Murphy, J., *Reinforced Plastics Handbook*, Elsevier Science and Technology Books, New York, 1998)

Velocity \və-▮lä-sə-tē\ *n* [MF *velocité*, fr. L *velocitat-, velocitas* fr. *veloc-, velox* quick; prob. akin to L *vegēre* to enliven] (ca. 1550) Time rate of motion in a fixed direction. Cgs units – one centimeter per second. If *s* is space passed over in time *t*, the velocity (*Handbook of Chemistry and Physics*, ed. R. C. Weast, The Chemical Rubber Co., 52nd Ed., Boca Raton, FL).

$$\bar{v} = \frac{s}{t}$$

(Giambattista, A., Richardson, R., Richardson, R. C., and Richardson, B., *College Physics*, McGraw Hill Science, New York, 2003)

Velocity Gradient *n* The change *dv* in relative velocity *v* between parallel planes with respect to the change *dr* in perpendicular distance *r* throughout the depth of material being sheared. Velocity gradient has the same dimensions as rate of shear, which is reciprocal seconds. (Munson, B. R., Young, D. F., and Okiishi, T. H., *Fundamentals of Fluid Mechanics*, John Wiley and Sons, New York, 2005)

Velocity of a Compressional Wave *n* The velocity of a compressional wave in an elastic medium, in terms of elasticity *E* (bulk modulus) and density d,

$$V = \sqrt{\frac{E}{d}}$$

For the velocity of sound in air, where *p* is the pressure and *d* the density,

$$V = \sqrt{\frac{1.4p}{d}}$$

(*Handbook of Chemistry and Physics*, ed. R. C. Weast, The Chemical Rubber Co., 52nd Ed., Boca Raton, FL).

Velocity of a Transverse Wave *n* (in a stretched cord). If T is the tension of the cord and m the mass per unit length,

$$V = \sqrt{\frac{T}{m}}$$

Velocity of a Wave *n* The velocity of propagation in terms of wavelength λ and the period T or frequency n is,

$$V = \frac{\lambda}{T} = n\lambda$$

Velocity of Efflux of a Liquid *n* If h is the distance from the opening to the free surface of the liquid, the velocity of efflux is

$$V = \sqrt{2gh}$$

The above is the theoretical discharge velocity disregarding friction and the shape of orifice. For water issuing through a circular opening with sharp edges of area, A, the volume discharged per second is given approximately by,

$$Q = 0.62A\sqrt{2gh}$$

Velocity of Sound, Variation with Temperature *n* The velocity in meters per sec at any temperature t in °C is given approximately by

$$V = V_o\sqrt{1 + \frac{t}{273}}$$
$$V = 331.5 + .607t$$

The *variation with humidity* is given by the equation

$$V_d V_h \sqrt{1 - \frac{e}{p}\left(\frac{\gamma\omega}{\gamma a} - \frac{5}{8}\right)}$$

where v_d is the velocity in dry air, V_h that in air at barometric pressure p in which the pressure of water vapor is e. $\gamma\omega$ and γa are the specific heat ratios for water vapor and for air respectively. (*Handbook of Chemistry and Physics*, ed. R. C. Weast, The Chemical Rubber Co., Boca Raton, FL, 52nd Ed.)

Velocity of Water Waves *n* If the depth h is small compared with the wavelength, the velocity,

$$V = \sqrt{gh}$$

In deep water for a wavelength λ,

$$V = \sqrt{\frac{g\lambda}{2\pi}}$$

If the wavelength is very small less than about 1.6 cm, the velocity increases as the wavelength decreases and is expressed by the following,

$$V = \sqrt{\frac{2\pi T}{\lambda d} + \frac{g\lambda}{2\pi}}$$

where T is the surface tension and d the density of the liquid V will be given in cm per sec if h and λ are in cm, g in cm per sec^2, T in dynes per cm and d in g per cm^3. (*Handbook of Chemistry and Physics*, 52nd Ed., R. C. Weast, ed., The Chemical Rubber Co., Boca Raton, FL)

Velocity Profile *n* A profile of the fluid velocity in a stream at various points along a coordinate direction perpendicular to the flow and in which direction the velocity is changing most sharply. This graph has the form of a parabola for a Newtonian liquid in laminar flow through a circular tube. For pseudoplastic liquids (polymer melts), the curve is a parabola of higher degree, usually 2.25 to 4 (instead of 2.0), rising more rapidly near the tube wall and flattening near the center. For a pure drag flow between parallel surfaces, the profile is linear regardless of the type of fluid. (Shenoy, A., V., *Thermoplastics Melt Rheology and Processing*, Marcel Dekker, 1996)

Velour \və-lúr\ *n* [F *velours* velvet, velour, fr. MF *velours*, *velour*, fr. OF *velous*, fr. L *villosus* shaggy, fr. *villus* shaggy hair] (ca. 1706) (1) Generally, a soft, closely woven fabric with a short, thick pile, weighting about 10 to 20 ounces per yard and made in a plain or satin weave. Velour is usually made of cotton or wool, or with a cotton warp in wool, silk, or mohair velour. It is also made in blends of spun manufactured fiber and wool. Velours are used for coats, draperies, upholstery, powder puffs, and other pile items. (2) A felt with velvet-like texture used for men's and women's hats (*Complete Textile Glossary*, Celanese Corporation, Three Park Avenue, New York, NY)

Velvet Carpet *n* A woven carpet in which the pile ends are lifted over wires that are inserted in the same manner as the filling and that cut the pile as they are withdrawn.

Velveteen \vel-və-tēn\ *n* (1776) A fabric with a low filling pile made by cutting an extra set of filling yarns woven in a float formation and bound to the back of the material at intervals by weaving over and under one or more warp ends. (Kadolph, S. J. J., and Langford, A. L., *Textiles*, Pearson Education, New York, 2001)

Velvet Fabric *n* A warp-pile woven fabric with short, dense cut pile that produces a rich fabric appearance and soft texture. Two methods are used for weaving velvets. In the double-cloth method, two fabrics are woven face to face with the pile ends interlocking. A reciprocating knife cuts through these pile ends to produce two separate pieces of velvet. In the second method, pile ends are lifted over cutting wires that are inserted with the filling and that are withdrawn to cut the pile. (Kadolph, S. J. J., and Langford, A. L., *Textiles*, Pearson Education, New York, 2001)

Veneer \və-ˈnir\ *n* [Gr *Furnier*, fr. *furnieren* to veneer, fr. F *fournir* to finish, equip] (1702) Thin finishing or surface layer of fine wood, laminated plastic, formica or the like, bonded to a substrate. (Harris, C. M., *Dictionary of Architecture and Construction*, McGraw-Hill Co., 2005; Hoadley, R. B., *Understanding Wood*, The Taunton Press, Newtown, CT, 2000)

Veneer Plywood *n* Plywood which is faced with a decorative wood veneer.

Venetian Glass \və-ˈnē-shən-\ *n* {often capitalized V} (ca. 1845) An often colored glassware made at Murano near Venice of a soda-lime metal and typically elaborately decorated with gilt, enamel, or engraving.

Venetian Mosaic \-mō-ˈzā-ik\ See ▶ Terrazzo.

Venetian Red *n* (ca. 1753) Chemically prepared oxide of iron, red pigment, made by calcining hydrated lime and ferrous sulfate, varies in tinting strength according to the amount of ferric oxide.

Venetian Window *n* A window with one large fixed central pane and smaller panes at each side.

Venice Turpentine *n* Same as larch turpentine, the oleoresin of the European larch tree (*Larix europea* or *Larix deciduas*). The term is also now used widely to describe the clear yellow liquid portion of pine oleoresin, or a synthetic product of similar composition made by dissolving rosin in a terpene solvent.

Vent See ▶ Air Vent.

Vented Extruder See ▶ Extruder, Vented.

Venturi Cooling Ring *n* A design of air-cooling ring for blown-film extrusion in which a slot around the inside of the ring and near the bottom injects air vertically upward at high velocity. This not only cools the bubble but also, by lowering air pressure between the bubble and the annular jet, helps to quickly expand the bubble and stabilize its position and movement.

Verdigris \ˈvər-də-ˌgrēs\ *n* [ME *vertegrez*, fr. OF *vert de Grice*, literally, green of Greece] (14c) Cu $(C_2H_3O_2)_2 \cdot 2Cu(OH)_2$. (1) The dibasic acetate of copper which is a greenish blue, crystalline powder with an acetic odor. Used as a pigment. (2) The blue or green corrosion products which form the patina on copper, brass, or bronze upon weathering. Syn: ▶ Aerugo.

Verditer Basic copper carbonate.

Vergeboard See ▶ Bargeboard.

Vermiculite \(ˌ)vər-ˈmi-kyə-ˌlīt\ *n* [L *vermiculus* little worm] (1824) Hydrated magnesium-aluminum-iron silicate capable of expanding six to twenty times the volume of the unexpanded mineral when heated to about 1100°C; the platelets exhibit an active curling movement when heated, hence the name. The expanded material is sometimes used as a density-reducing filler in plastics.

Vermiculus See ▶ Mercuric Sulfide.

Vermillion *n* (**Vermilion**) \vər-ˈmil-yən\ *n* [ME *vermilioun*, fr. OF *vermeillon*, fr. *vermeil*, adj., bright red, vermilion, fr. LL *vermiculus* kermes, fr. L, little worm] (13c) A red mineral pigment consisting of a sulfide of mercury. See ▶ Mercuric Sulfide.

Versamid® *n* Group of "polymerized" vegetable oils whose ester groups are converted with di- and tramlines. Manufactured by General Mills, U.S.

Vert Emeraude *n* (**Dull**) See ▶ Chromium Oxide Green.

Vertical Extruder *n* An extruder arranged so that the barrel is vertical and extrusion is usually downward or upward.

Vertical Flame Test See ▶ Flammability Tests.

Vertical Grain See ▶ Edge Grain.

Vertical Pattern *n* A spray pattern whose longest dimension is vertical.

Very-Low-Density Polyethylene *n* (VLDPE) Any polymer of ethylene with some higher-olefin content, in the density range from 0.90 to 0.915 g/cm^3, produced by the Union Carbide gas-phase process (Flexomer®). The materials have low moduli, with properties between those of low-density polyethylene and ethylene-propylene rubbers. They are useful for stretchable films.

Vestamides *n* Various nylon grades, manufactured by Hüls, Germany.

Vestan *n* Polycondensate from terephthalic acid and 1,4-dimethylol cyclohexane, manufactured by Hüs, Germany.

Vestolen A *n* Low-pressure poly(ethylene), manufactured by Hüls, Germany.

Vestolen P *n* Poly(propylene), manufactured by Hüls, Germany.

Vestolit *n* Poly(vinyl chloride), manufactured by Hüls, Germany.

Vestopal *n* Unsaturated polyester, dissolved in styrene, manufactured by Hüls, Germany.

Vestoran *n* Vinylchloride-vinyl acetate copolymer, manufactured by Hüls, Germany.

Vestyron *n* Poly(styrene), manufactured by Hüls, Germany.

VF Vulcan fiber.

VF$_2$/HFP *n* Abbreviation for Vinylidene Fluoride-hexafluoro-propylene Copolymer.

V$_i$ Symbol for volume fraction of component *i* in a blend or composite; or for a velocity component in the *i*-direction.

Viable \ˈvī-ə-bəl\ *adj* [F, fr. MF, fr. *vie* life, fr. L *vita*] (ca. 1832) Living; able to germinate or grow.

Vibrathane *n* Polyurethane elastomer, manufactured by Naugatuck, U.S.

Vibration Modes Vibrational energy due to change in length and angle in the molecule.

Vibration Welding A joining method in which two plastic parts are pressed together and one is vibrated through a small angular displacement in the plane of the joint. The frictional heat so generated melts the plastic at the interface. Vibration is stopped and pressure and alignment are maintained until the joint freezes.

Vibratory Feeder *n* A device for conveying dry materials from storage hoppers to processing machines, comprising a tray or tube vibrated by mechanical or electrical pulses. The frequency and amplitude of vibration control the rate of transport.

Vibroscope *n* An instrument for determining the mass per unit length of a fiber.

Vicara *n* Albumin fiber, manufactured by Virginia-Carolina Chemical, U.S.

Vicat Ester Resins *n* Thermosetting acrylated epoxy resins containing styrene reactive diluents. Cured by catalyzed polymerization of vinyl groups and crosslinking of hydroxy groups at room or elevated temperatures. Have good chemical, solvent, and heat resistance, toughness, and flexibility, but shrink during cure. Processed by filament winding, transfer molding, pultrusion, coating, and lamination. Used in structural composites, coatings, sheet molding compounds, and chemical apparatus.

Vicat Test *n* A test for determining softness of a polymer at temperature; an indentor under fixed load will penetrate a specified distance into the material.

Vicat Softening Point *n* The temperature at which a flat-nosed needle of 1-mm^2 circular cross section penetrates a thermoplastic specimen to a depth of 1 mm under a specified load using a uniform rate of temperature rise (www.astm.org). This test is used for thermoplastics such as vinyls, polystyrene, acrylic, and cellulosics that have no definite melting ranges.

Vicat Softening Temperature See ▶ Vicat Softening Point.

Vicinal *adj*. Description of substitution on the adjacent on atom (e.g., 1-2-chloro-) compared to ▶ Geminal substitution on the same on atom (e.g., 1,1-chloro-).

Vickers Hardness *n* A test similar to that of ▶ Brinell Hardness using an indenter in the form of a square-based diamond pyramid, with a vertex angle of 136° between the opposite faces. The result is expressed as the applied load divided by the projected area of the impression.

Viewing Angle *n* The angle between the viewing ray and a normal to the surface at the point of incidence. Also called *Observation Angle*.

Viewing Geometry *n* The geometry by which the incident illumination is projected onto the sample and by which the observed light is collected and transmitted to the detector. See ▶ Angle of Incidence, Angle of Viewing and ▶ Angle of Reflection.

Viewing Ray *n* The line connecting the point of incidence on the surface and the center of the receptor entrance stop.

Vinal \ˈvī-nal\ *n* [poly*vin*yl *al*cohol] (ca. 1939) Generic name for a manufactured fiber in which the fiber-forming substance in any long-chain synthetic polymer is composed of at least 50% by weight of vinyl alcohol units ($-CH_2CHOH-$), and in which the total of the vinyl alcohol units and any one or more of the various acetal units is at least 85% by weight of the fiber.

Vinal Fiber *n* A manufactured fiber in which the fiber-forming substance is any long chain synthetic polymer composed of at least 50% by weight of vinyl alcohol units and in which the total of the vinyl alcohol units and any one or more of the various acetal units is at least 85% by weight of the fiber (FTC definition). Vinal fibers show good chemical resistance but soften at comparatively low temperatures. Vinal fibers are used for apparel, industrial goods, and fishnets.

Vine Black *n* Intense blue-black pigment made by the partial burning of vine cuttings.

Vinidur *n* Poly(vinyl chloride) film, manufactured by BASF, Germany.

Vinnipas *n* Poly(vinyl acetate), manufactured by Wacker, Germany.

Vinnol *n* Poly(vinyl chloride), manufactured by Wacker, Germany.

Vinoflex Vinyl chloride/vinyl ether copolymer, manufactured by BASF, Germany.

Vintahite See ▶ Gilsonite.

Vinyl \ˈvī-n°l\ *n* [ISV, fr. L *vinum* wind] (1863) The unsaturated, univalent radical $CH_2:CH-$ derived from ethylene which is the basis for all vinyl plastics. The name vinyl is used when the open bond is filled by anything but H (ethene) or a hydrocarbon radical (olefin).

•HC≡

Vinyl Acetate *n* H₂C=COOCH₃. A colorless liquid obtained by the reaction of acetylene and acetic acid in the presence of a catalyst such as mercuric oxide.

It is the monomer for polyvinyl acetate, and a comonomer and intermediate for many members of the vinyl plastics family.

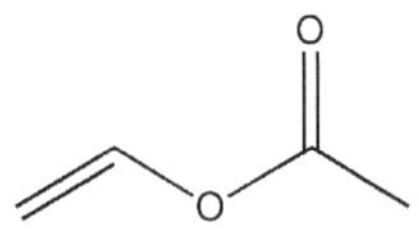

Vinyl Acetate Plastics *n* Plastics based on resins made by the polymerization of vinyl acetate or copolymerization of vinyl acetate with other unsaturated compounds, the vinyl acetate being in greatest amount by mass. See ▶ Polyvinyl Acetate.

Vinyl Alcohol *n* (1873) (ethanol) H₂C=CHOH. A conceptual compound, the theoretical monomer of ▶ Polyvinyl Alcohol but unknown in the free state. All attempts to synthesize it have led instead to its tautomer, acetaldehyde (CH₃CHO).

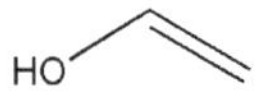

Vinyl Alcohol Plastic See ▶ Polyvinyl Alcohol.

Vinylation *n* The process of forming a vinyl derivative by reaction of alcohols, amines, or phenols with acetylene. Such derivatives are intermediates for polymers.

Vinyl Benzene See ▶ Styrene.

Vinyl Butyrate *n* CH₂=CHOOC₃H₇. A volatile liquid monomer for polymers used in water-based paints.

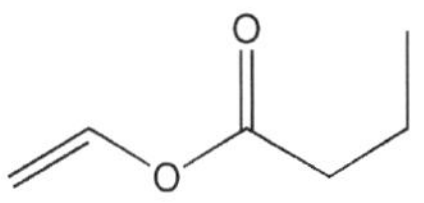

9-Vinylcarbazole *n* (*N*-vinylcarbazole) A tricyclic tertiary amine, H₂C=CHN(C₆H₄)₂, with the structure shown below.

This monomer, derived from acetylene and carbazole, is used in the production of ▶ Poly(*N*-Vinylcarbazole).

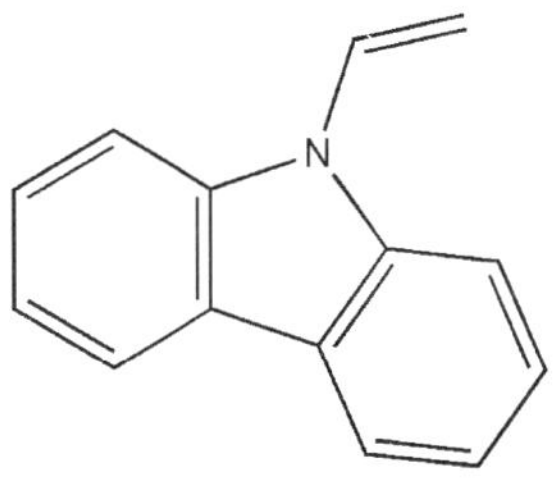

Vinyl Chloride *n* (1872) (chlorethylene, chloroethene, VC) H₂C=CCl. A colorless gas at normal temperatures and pressures that boils at −13.9°C, made by reacting ethylene with chlorine or hydrogen chloride to obtain ethylene dichloride, which is cracked to form vinyl chloride. (*Handbook of Polyvinyl Chloride* Formulating, Wickson, E. J., ed., John Wiley and Sons, Inc., New York, 1993)

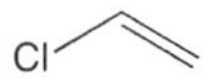

Vinyl Chloride-Ethylene Copolymer *n* Any copolymer of vinyl chloride with small percentages of ethylene. These resins possess superior heat stability and hot strength, and require lesser amounts of impact modifiers to achieve satisfactory impact strength than does straight PVC homopolymer. They are useful in producing films and bottles for packaging, since their better heat stability provides more latitude in selecting nontoxic stabilizers.

Vinyl Chloride Plastics *n* These are plastics based on reins prepared by the polymerization of vinyl chloride or copolymerization of vinyl chloride with other unsaturated compounds, vinyl chloride being in greatest amount by mass (*Handbook of Polyvinyl Chloride* Formulating, Wickson, E. J., ed., John Wiley and Sons, Inc., New York, 1993). See ▶ Polyvinyl Chloride.

Vinyl Coating *n* One in which the major portion of binder is of the vinyl resin family.

Vinyl Cyanide See ▶ Acrylonitrile.

Vinyl Ester Resin *n* Any of several epoxy-related resins in which the epoxides groups have been replaced by ester groups, typically acrylic. When cured, they have excellent resistance to strong chemicals such as chlorine and caustics.

Vinyl Ether See ▶ Vinylethyl Ether, ▶ Vinylisobutyl Ether, and ▶ Vinylmethyl Ether.

Vinylethylene Syn: ▶ Butadiene.

Vinylethyl Ether *n* (EVE, ethylvinyl ether) H₂C=CHOC₂H₅. A colorless monomer that can be polymerized either in the liquid or gaseous state. In plastics, it is used as a comonomer and intermediate.

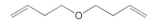

Vinyl Fluoride *n* (fluoroethylene, fluoroethene) H₂C=CHF. A colorless gas, the monomer for ▶ Polyvinyl Fluoride.

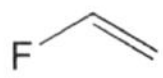

Vinyl Foam *n* Although cellular vinyls can be produced by many methods, including mechanical frothing and leaching-out of soluble additives, the most widely used procedure is chemical blowing. From 1 to 2% of a blowing agent such as azobisforamdie is incorporated in a vinyl compound or dispersion, remaining inert until it is decomposed by processing heat to release a gas.

Vinylformic Acid See ▶ Acrylic Acid.

Vinyl Group The unsaturated univalent radical $CH_2{:}CH{-}$.

Vinylidene \vī-ˈni-lə-ˌdēn\ *n* [ISV *vinyl* + *-ene*] (1898) Indicating a bi-substituted ethylene in which both hydrogen atoms on one carbon atom have been replaced, i.e., $H_2C{=}CXY$, or the vinylidene group, $H_2C{=}C{=}$. X and Y are usually the same element.

Vinylidene Chloride *n* $CH_2{:}CCl_2$. A monomer. Bp, 37°C. A colorless, volatile liquid that is produced by the dehydrochlorination of 1,1,2-trichloroethane. It is a monomer for ▶ Polyvinylidene Chloride and is a comonomer with vinyl chloride (see ▶ Saran) and other monomers such as acrylonitrile. Also known as *1,1-Dichloroethylene*.

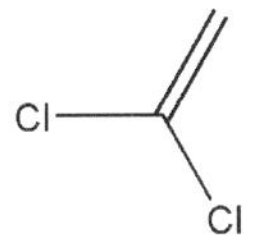

Vinylidene Chloride-Acrylonitrile Copolymer *n* Any VC copolymer containing 5 to 15% acrylonitrile, and mainly used as coatings for cellophane, paper, and films or other polymers. They are comparable with saran in their low permeability to oxygen and carbon dioxide, and have good chemical resistance, toughness, transparency, and heat-sealability. They have also found some applications as low-flammability fibers.

Vinylidene Chloride Plastics *n* Plastics based on polymer resins made by the polymerization of vinylidene chloride or copolymerization of vinylidene chloride with other unsaturated compounds, the vinylidene chloride being in the greatest amount by mass. See ▶ Polyvinylidene Chloride and ▶ Saran.

Vinylidene Fluoride *n* (1,1-difluoroethylene) $H_2C{=}CF_2$. A colorless, nearly odorless gas prepared by the dehydrohalogenation of 1-chloro1,1-difluoroethane, or by the dehalogenation of 1,2-dichloro-1,1-difluorethane. It polymerizes readily in the presence of free-radical initiators to produce the homopolymer polyvinylidene fluoride, and is also copolymerized with olefins and other fluorocarbon monomers to make fluorocarbon elastomers. See ▶ Polyvinylidene Fluoride.

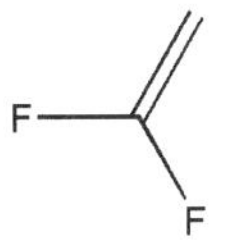

Vinylidene Fluoride-Hexafluoropropylene Copolymer *n* (VF_2/HFP) Any of a family of chemical- and heat-resistant, vulcanizable elastomers containing 60 to 85% VF_2 (DuPont's Viton® A). Terpolymers with small amounts of tetrafluoroethylene are also available.

Vinylidene Fluoride Hexafluoropropylene Tetrafluoroethylene Terpolymer *n* Thermosetting elastomeric polymer of vinylidene fluoride, hexafluoropropylene, and tetrafluoroethylene having good chemical and heat resistance and flexibility. Used in auto parts.

Vinylisobutyl Ether *n* (isobutylvinyl ether, IVE) $H_2C{=}CHOCH_2{-}CH(CH_3)_2$. A colorless, flammable liquid used to make polymers and copolymers used in coatings, adhesives, and lacquers; and modifiers for alkyd resins and polystyrene. See ▶ Polyisobutylvinyl Ether.

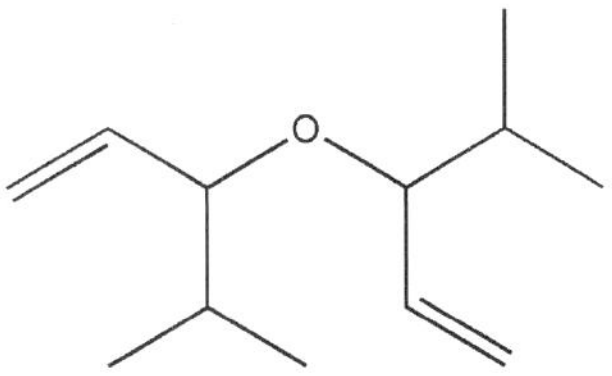

Vinylite, Vinyon *n* Vinyl chloride/vinyl acetate copolymer, manufactured by Carbide & Carbon Chemical, U.S.

Vinylmethyl Ether *n* (methyl vinyl ether). A low-boiling liquid (6°C) or gas, polymerizable to Poly(Vinylmethyl Ether). It is also used as a modifier for alkyd resins and polystyrene.

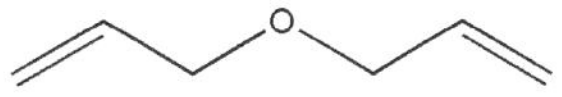

Vinylon *n* Poly(vinyl alcohol) (fiber), manufactured by Synthetic Fiber Manufacturers Group, Japan.

Vinyl Plastics *n* Plastics based on resins made from monomers containing th vinyl group, $CH_2{=}CH{-}$.

Vinyl Propionate *n* A volatile liquid, the monomer for emulsion-paint polymers.

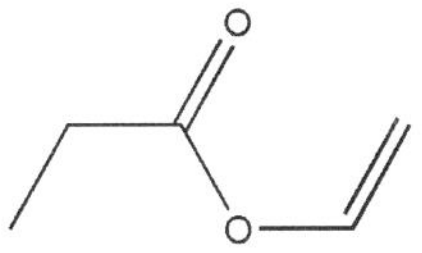

1-Vinyl-2-Pyrrolidone *n* (*N*-vinyl-2-pyrrolidone) A cyclic monomer derived from acetylene and formaldehyde, with the structure below. See ▶ Poly(1-Vinylpyrrolidone).

Vinyl Resins *n* (1934) According to common chemical nomenclature, all resins and polymers made from monomers containing the vinyl group, $H_2C=CHX$. In the chemical literature, polystyrene, polyolefins, polymethyl methacrylate and many other styrenic, ethenic, and acrylic copolymers are classified as vinyls (Mishra, M. K. M., and Yagci, Y., *Handbook of Vinyl Polymerization*, Marcel Dekker, New York, 1998). In the plastics literature, the above materials are given their own classifications and the term vinyl is restricted to compounds in which X, above, is not H, a hydrocarbon radical, nor an acrylic-type ester. In daily use, the term vinyl plastics refers primarily to ▶ Polyvinyl Chloride and its copolymers, and secondarily to the following: ▶ Polyvinyl Acetal, ▶ Polyvinyl Acetate, ▶ Polyvinyl Butyral, ▶ Polyvinyl Dichloride, ▶ Polyvinyl Formal, ▶ Polyvinylidene Chloride, ▶ Polyisobutylvinyl Ether, and ▶ Poly(1-Vinylpyrrolidone).

Vinyl Stearate *n* $H_2C=CHOOCC_{17}H_{35}$. A white, waxy solid, used as an internal plasticizer by means of copolymerization at low levels.

Vinylstyrene Syn: ▶ Divinylbenzene.

Vinyl Thermoplastic Elastomers *n* Vinyl resin alloys having good fire and againg resitance, flexibility, dielectric properties, and toughness. Processed by extrusion. Used in cable jackets and wire insulation.

Vinyltoluene *n* $H_2C=CHC_6H_4CH_3$. A colorless liquid, the commercial forms comprising a 60:40 mixture of the *m*- and *p*- isomers, used as a solvent and as a polymerizable monomer in place of styrene in the production of polyester resins.

Vinyltrichlorosilane *n* $H_2C=CHSiCl_3$. A coupling agent used in glass-reinforced polyesters.

Vinyltriethoxysilane *n* $H_2C=CHSi(OC_2H_5)_3$. A coupling agent used in glass-reinforced polyesters, polyethylene, and polypropylene.

Vinyl-tris(β-Methoxyethoxy)silane *n* $H_2C=CHSi(OCH_3OC_2H_5)_3$. A silane coupling agent used in glass-reinforced polyester and -epoxy structures.

Vinyl Wallcoverings *n* At present there are six types: 1 — Vinyl laminated to paper, 2 — Paper laminated to lightweight woven cloth and vinyl-coated, 3 — Vinyl laminated to lightweight woven cloth, natural or synthetic, 4 — Vinyl laminated to lightweight Nonwoven cloth, natural or synthetic, 5 — Vinyl laminated to Nonwoven paper/fabric web, and 6 — A man-made base impregnated with vinyl. (Harris, C. M., *Dictionary of Architecture and Construction*, McGraw-Hill Co., 2005)

Vinyon *n* Generic name for a manufactured fiber in which the fiber forming substance is any long-chain synthetic polymer composed of at least 85% by weight of vinyl chloride units, $-CH_2CHCl-$ (Federal Trade Commission).

Vinyon Fiber *n* A manufactured fiber in which the fiber-forming substance is any long chain synthetic polymer composed of at least 85% by weight of vinyl chloride units (FTC definition).

Virgin Material *n* Any plastic compound or resin that has not been subjected to use or processing other than that required for its original manufacture.

Veridian \və-ˈri-dē-ən\ *n* [L *viridis*] (1882) A chrome green that is a hydrated oxide of chromium.

Viscoelastic \ˌvis-kō-ə-ˈlas-tik\ *adj* [*visc*ous + -*o*- + *elastic*] (1935) Having both viscous and elastic properties.

Viscoelasticity *n* (1) A mechanical property involving a combination of elastic and viscous behavior; conforming to neither just simple elastic nor simple viscous behavior; also, strongly dependent on temperature. See ▶ Shift Factor. (2) The property of a polymer that characterizes it as neither an ideal solid nor a viscous liquid, but seeming to have the character of both. In addition, to having some of the characteristics of elastic solids, they possess some of the characteristics of viscous liquids. Polymeric materials show creep under a certain load and stress relaxation if stretched. In other words, the tendency of a plastic to respond to stress (or strain) as if it were a combination of an elastic solid and a viscous liquid (). This property, possessed by all plastics to some degree, dictates that while plastics have solid-like characteristics such as elasticity, strength, and form stability, they also have liquid-like characteristics such as flow over time that depend on temperature, pressure, and stress. As a result, the response to stress depends on both the rate of application of the stress and the time for which it is maintained. (Shenoy, A., V., *Thermoplastics Melt Rheology and Processing*, Marcel Dekker, 1996; Shah, V., *Handbook of Plastics Testing Technology*, John Wiley and Sons, New York, 1998). See ▶ Maxwell Model, ▶ Voigt Model, and ▶ Time-Temperature Equivalence.

Viscometer (Viscosimeter) \vis-ˈkä-mə-tər\ *n* [*viscos*ity + -*meter*] (ca. 1883) An instrument for measuring the viscosity and other flow properties of fluids, emulsions and dispersions having low to moderate viscosities. Instruments used with highly viscous materials, such as polymer melts, are usually called *rheometers*. A widely used type of viscometer is the rotational type, in which a rotor turns within a cup containing the liquid sample and the torque required to turn the rotor or to hold the cup stationary is measured, along with speed of rotation. Of the many other types, some measure the rate at which a bubble rises through the liquid, or a ball falls through it; others measure the time required for a known quantity of the liquid to drain by gravity through an orifice at the bottom of a cup. (*Paint and Coating Testing Manual (Gardner-Sward Handbook) MNL 17, 14thEd.*, ASTM, Conshohocken, PA, 1995; Patton, T. C., *Paint Flow and Pigment Dispersion: A Rheological Approach to Coating and Ink Technology*, John Wiley and Sons, New York, 1979; Van Wazer, Lyons, Kim, and Colwell, *Viscosity and Flow Measurement*, Lyons Kim, Colwell, Interscience Publishers, Inc., New York, 1963) ▶ Air-Bubble Viscometer, ▶ Brookfield Viscometer, ▶ Ford Viscosity Cups, ▶ Stormer Viscometer, ▶ Viscosity and ▶ Zahn Viscosity Cup.

Viscometers *n* Instruments for measuring viscosity including mechanical probe and torque types as the Brookfield Viscometer, capillary tube types as the Cannon-Fenske or Ostwald-Fenske, and flow through orifice types as the Ford Cup.

Viscose \ˈvis-ˌkōs\ *n* [*obs. viscose*, adj, viscous] (1896) (1) A solution of xanthated cellulose in dilute sodium hydroxide from which rayon fibers and cellophane films are formed. The xanthated cellulose is produced by reacting alkali cellulose, i.e., wood fibers or cotton linters treated with sodium hydroxide, with oxygen and carbon disulfide. Rayon produced by this method is known as *viscose rayon*. (2) Generic name for fibers from regenerated cellulose (prepared by the xanthate method). (*Fairchild's Dictionary of Textiles*, Tortora, P. G., ed., Fairchild Books, New York, 1997) See also ▶ Rayon.

Viscose Process *n* (1) One of the methods of producing rayon. (Also see ▶ Rayon Fiber). (2) The chemical process used in the manufacture of cellophane. Also see ▶ Viscose Solution.

Viscose Rayon One type of rayon. It is produced in far greater quantity than cuprammonium rayon, the other commercial type. Also see ▶ Rayon Fiber.

Viscose Solution The solution obtained by dissolving cellulose xanthate in caustic soda, from which viscose filaments and cellophane are produced.

Viscosimeter (Deprecated) See ▶ Viscometer.

Viscosity \vis-ˈkä-sə-tē\ *n* [ME *viscosite*, fr. MF *viscosité*, fr. ML *viscositat*-, *viscositas*, fr. LL *viscosus* viscous] (15c) The ratio of the shear stress existing between laminae of moving fluid and the rate of shear between these laminae. NOTE − A fluid is said to exhibit Newtonian behavior when the rate of shear is proportional to the shear stress. A fluid is said to exhibit non-Newtonian behavior when an increase or decrease in the rate of shear is not accompanied by a proportional increase or decrease in the shear stress. The resistance to flow; η = dyne.sec./cm^2, shear stress (dyne./cm^2) divided by shear rate (sec.$^{-1}$); where two opposite planes traveling at 0 and v velocities at d distance, at F force is expressed by,

$$F/A = \eta\, v/d$$

where,
F = force,
A = area,
v = velocity, and
d = distance
(Goodwin, J. W., Goodwin, J., and Hughes, R. W., *Rheology for Chemists*, Royal Society of Chemistry, August 2000; Patton, T. C., *Paint Flow and Pigment Dispersion: A Rheological Approach to Coating and Ink Technology*, John Wiley and Sons, New York, 1979)

Viscosity Coefficient n, η Resistance to flow, a fundamental property of fluids, first quantitatively defined by I. Newton in his *Principia*. A modern version of his equation of viscosity is:

$$\tau_{xz} = -\eta \frac{dv_z}{dx}$$

in which dv_z/dx is the rate of change of z-directed velocity at the coordinate location x in the fluid, x being directed perpendicular to z, η is the viscosity, a function of temperature and, more weakly, static pressure and τ_{xz} is the z-directed *shear-stress (force)* component acting on an imagined element of fluid surface normal to the x-direction. dv_z/dx, the *velocity gradient*, often written without the subscript, is also called the *shear rate*, usually symbolized as γ. The negative sign indicates that the stress in the fluid opposes the velocity. In the conceptually simplest case, that of steady flow between parallel plates separated by a distance h, one of which is moving with velocity v relative to the other, the stress is just the force F required to drag the moving plate divided by the plate area A, and Newton's equation reduces to: $F/A = -\eta\,(V/h)$, Goodwin, J. W., Goodwin, J., and Hughes, R. W., *Rheology for Chemists*, Royal Society of Chemistry, August 2000). For the circular-tube geometry of orifice-type rheometers, the shear rate and stress vary from zero at the tube axis to a maximum at its inside surface, while the velocity does just the reverse. Since for ordinary fluids, Newton's law holds throughout the fluid, it also holds at the tube wall (unless there is slip, rarely proved with polymeric solutions and melts). Rheologists have found it convenient to report their measurements in terms of the shear stress at the tube wall, $\Delta P \cdot R/2L$, and the Newtonian (or *apparent*) shear rate at the wall, $4Q/\pi R^3 (= 4V/R)$, which contain all the quantities they actually measure, i.e., the pressure drop from entrance to exit, the tube radius, and the steady flow rate, Q, which is equal to πR^2 times the average velocity V, (Parfitt, G. D., *Dispersion of Powders in Liquids*, Elsevier Publishing Co., New York,

1969. By Newton's law, the viscosity is given by the quotient of the shear stress and the shear rate, i.e., $\pi R^4 \cdot \Delta P/(8Q \cdot L)$. This expression is just an inversion of the Hagen-Poiseuille equation (Parfitt, G. D., *Dispersion of Powders in Liquids*, Elsevier Publishing Co., New York, 1969). Newton's equation applies accurately to ordinary fluids such as water, pure organic liquids, familiar oils and honey, even to dilute solutions of polymers, but not to concentrated polymer solutions or most molten plastics. Measurements show that these latter materials deviate from Newton's law in that their viscosities, as given by the quotient of shear stress/shear rate, diminish with rising shear rate (and stress). That is, they are nonnewtonian and pseudoplastic (Patton, T. C., *Paint Flow and Pigment Dispersion: A Rheological Approach to Coating and Ink Technology*, John Wiley and Sons, New York, 1979). Because some chemical engineers like to think of shear stress in its alternate identity, *momentum flux*, because many instruments were developed to measure viscosities related to specific industrial uses, and many viscosity units have evolved because viscosity ranges widely for different fluids and conditions such as temperatue. The intenational unit of viscosity is the SI unit of viscosity, the pascal-second (Pa·s). If, in the parallel-plate setup described above, the plate areas where 1 m^2, their separation 1 m, their relative velocity 1 m/s, and the drag force 1 N, the viscosity would be exactly 1 Pa·s. The dynamics of fluid flow over bodies is reviewed in Munson, B. R., Young, D. F., and Okiishi, T. H., *Fundamentals of Fluid Mechanics*, John Wiley and Sons, New York, 2005. For many years the *poise* has been used, equal to 0.1 Pa·s, and the centipoise, as their working units. Plastics engineers have also used the psi·s (1b$_f$·s/in^2, = 6895 Pa·s). The Pa·s may also be viewed, through the momentum-flux perspective, as 1 kg/(m·s). Many early instruments tried to gauge viscosity by the time required for a vessel full of liquid to drain through a short tube in its bottom. These, with suitable corrections, provide estimates of the ▶ Kinematic Viscosity. If mass-based viscosity units are used, kinematic viscosity will have the dimensions (length)/time or m^2/s in SI. Clearly, one must be careful, in using reported viscosities, to identify unambiguously the units used. In the older literature, and even today, the pound and kilogram may be either force or mass units, though in SI the kilogram, one of the seven base units, is strictly assigned to mass. To convert a kinematic viscosity to an equivalent absolute viscosity, one must know the liquid's density at the stated temperature. The old scientific unit, the Stokes, equal to 1 cm^2/s, is closely related to the poise. All liquid viscosities decrease with rising temperature, some much more steeply than others. The range of polymer-connected viscosities is very wide, from 0.001 Pa·s for very dilute solutions at room temperature to 100–5000 Pa·s for molten plastics, and many times more for cooler amorphous polymers (*Whittington's Dictionary of Plastics*, Carley, James F., (Ed.), Technomic Publishing Co., Inc., 1993). See ▶ Astm (www.astm.org) for the current and appropriate method for determining viscosity. See also the following viscosity-related terms.

Bingham Plastic	Intrinsic Viscosity
Brookfield Viscometer	Kinematic Viscosity
Capillary Rheometer	Laminar Flow
Capillary Viscometer	Melt-flow Index
Consistency	Newtonian Flow
Cup-flow Test	Pseudoplastic fluid
Dilatancy	Reduced Viscosity
Dilute-solution Viscosity	Relative Viscosity
Extrusion Plastometer	Rheology
Hagen-Poiseuille Equation	Rheometer
Inherent Viscosity	Rheopexy
Initial viscosity	Saybolt viscosity
Specific Viscosity	Viscometer
Stormer Viscometer	Viscous Flow
Thixotropy	Yield Value
Ubbelohde Viscometer	Weissenbert
Ultra-viscoson	Rheogoniometer
Viscoelasticity	Zahn Viscosity Cup

Viscosity, Absolute Dynamic *n* The force per unit area that resists the flow of two parallel fluid layers past one another when their differential velocity is 1 cm/(s)/cm separation. The viscous force is described by Newton's equation:

$$f = \eta A(dv/dx)$$

where A is the area (cm^2). (dv/dx) is the velocity gradient (s^{-1}), and η is the coefficient of absolute viscosity (poise).

Viscosity, Apparent *n* The quantity obtained by dividing the shearing force by the rate of shear. This is a term applied only to nonnewtonian materials. It is not a constant for a given material, because its value depends on the rate of shear selected in making the measurement. Apparent viscosity is obtained by "one-point" methods. It has no general scientific value, and it is doubtful whether it has any real technical value.

Viscosity-Average Molecular Weight $n\ (M_v, \overline{M}_v)$ An averaged molecular weight for high polymers that relates most closely to measurements of dilute-solution viscosities of polymers. The defining equation is

$$M_v = \left[\frac{\sum_{i=1}^{\infty} N_i M_i^{1+a}}{\sum_{i=1}^{\infty} N_i M_i} \right]^{1/a}$$

where N_i is the number of individual molecules having the molecular mass M_i. The exponent a, between 0.6 and 0.8 for many polymer/solvent systems, is best evaluated from measured viscosities of dilute solutions of narrow-molecular-mass fractions of polymers, determining the intrinsic viscosity $[\eta]$ for each fraction, then fitting the following equation to the data: $[\eta] = K^1 M^a$. For $a = 1$, occasionally seen, M_v = the weight-average molecular weight, M_w. (Kamide, K., and dobashi, T., *Physical Chemistry of Polymer Solutions*, Elsevier, New York, 2000; Slade, P. E., *Polymer Molecular Weights*, Vol. 4, Marcel Dekker, New York, 2001; Elias, H. G., *Macromolecules Vol. 1–2*, Plenum Press, New York, 1977)

Viscosity Coefficient *n* (1866) The shearing stress necessary to induce a unit velocity flow gradient in a material. In actual measurement, the viscosity coefficient of a material is obtained from the ratio of shearing stress to shearing rate. This assumes the ratio to be constant and independent of the shearing stress, a condition which is satisfied only by Newtonian fluids. (Kamide, K., and dobashi, T., *Physical Chemistry of Polymer Solutions*, Elsevier, New York, 2000)

Viscosity Cup *n* An efflux viscometer. See ▶ Viscometer.

Viscosity/Density Ratio *n*, $\frac{\eta}{\rho}$ where η is the viscosity of the polymer solution; ρ is the density of the polymer solution.

Viscosity Depressant *n* A substance that, when added in a relatively minor amount to a liquid, lowers its viscosity. Such materials, e.g., ethoxylated fatty acids, are often incorporated in vinyl plastisols to lower their viscosities without increasing plasticizer levels.

Viscosity Index *n* (1929) An arbitrary number assigned as a measure of the constancy of the viscosity of a lubricating oil with change of temperature with higher numbers indicating viscosities that change little with temperature.

Viscosity, Intrinsic See ▶ Intrinsic Viscosity.

Viscosity, Kinematic *n* Viscosity of a substance divided by the density of the substance at the temperature of measurement. Kinematic viscosity is commonly obtained from capillary and outflow viscometer data. The unit is the stokes. A liquid having a relative density of one has a kinematic viscosity of one stokes if its viscosity is one poise. (Kamide, K., and dobashi, T., *Physical Chemistry of Polymer Solutions*, Elsevier, New York, 2000)

Viscosity Number *n* The IUPAC term for reduced viscosity. $\frac{\eta - \eta_o}{\eta_o c}$ where η is the viscosity of the polymer solution; η_o is the viscosity of the pure solvent; c is the concentration of the polymer solution in grams of solute per milliliter of solution.

Viscosity, Plastic *n* Resistance to flow in excess of the yield value in a plastic material. Plastic viscosity is proportional to (Shearing Stress-Yield Value)/Rate of Shear. The coefficient of plastic viscosity is the force in excess of the yield value, tangentially applied, that will induce a unit velocity gradient. See ▶ Plastic Flow and ▶ Plasticity. (Shenoy, A., V., *Thermoplastics Melt Rheology and Processing*, Marcel Dekker, 1996)

Viscosity Ratio *n* The ratio of the viscosities of the polymer solution (of stated concentration) and of the pure solvent at the same temperature.

Viscosity, Relative (for Liquids) *n* See Viscosity, Specific (for Liquids).

Viscosity, Relative (for Suspensions and Solutions) *n* The ratio between the viscosity of a composition (η), divided by the viscosity of the liquid phase (η_o). It is designated η_{rel}.

Viscosity, Specific (for Liquids) *n* The ratio between the viscosity of a liquid and the viscosity of water at the same temperature. Specific viscosity is sometimes used interchangeably with relative viscosity for liquids.

Viscosity, Specific (for Suspensions and Solutions) *n* The viscosity of a composition (η) minus the viscosity of the liquid (η_o), divided by the viscosity of the liquid phase. The symbol is η_{sp}. (Kamide, K., and dobashi, T., *Physical Chemistry of Polymer Solutions*, Elsevier, New York, 2000)

Viscous \ˈvis-kəs\ *adj* [ME *viscouse*, fr. LL *viscosus* full of birdlime, viscous, fr. L *viscum* mistletoe, birdlime; akin to OHGr *wīhsila* cherry Gk *ixos* mistletoe] (14c) Having relatively great viscosity. A qualitative term denoting the material to which it is applied is "thick" and flows sluggishly, rather than being "thin" and flowing freely. The transition region between "free-flowing and "viscous" corresponds roughly to viscosities from 1 to 30 Pa·s.

Viscous Dissipation *n* (viscous-heat generation) In melt processing, wherever there is flow, the resistance of molecules to flow, i.e., ▶ Viscosity causes heat to be generated within the melt. The rate of dissipation equals the product of shear stress times shear rate, or viscosity times the square of the shear rate. Because both the viscosity and shear rate are high in processes such as extrusion, injection and transfer molding, and intensive mixing, viscous dissipation is a principal mechanism of heating plastics in those processes.

Viscous Flow *n* A type of fluid in which all particles of the fluid flow in a straight line parallel to the axis of a container pipe or channel, with little or no mixing or turbidity. This definition arises from O. Reynolds' classic experimental demonstration of the transition from viscous to turbulent flow, which is described in most elementary texts on fluid flow. See also ▶ Laminar Flow.

Visible Light *n* The narrow band in the electromagnetic spectrum that the human eye perceives, from about 380 nm (violet) to 760 nm (red). (Johnson, S. F., *History of Light and Colour Measurement: A Science In the Shadows*, Taylor and Francis, UK, 2001)

Visible Spectrophotometry *n* An analytical instrumental technique based on selective absorption of visible radiation from organic and inorganic substances which helps in their identification.

Vision \ˈvi-zhən\ *n* [ME, fr. OF, fr. L *vision-*, *visio*, from *vidēre* to see] (14c) Process of seeing; ocular perception.

Vision, Defective Color See ▶ Color Vision, Defective.

Vision, Normal Color See ▶ Color Vision, Normal.

Vistanex Poly(isobutylene). Manufactured by Standard Oil, U.S.

Visual \ˈvi-zhə-wəl\ *adj* [ME, fr. LL *visualis*, fr. L *visus* sight, fr. *vidēre* to see] (1603) Of or pertaining to sight; ocular.

Viton *n* Vinylidene fluoride/hexafluoropropylene copolymer, manufactured by DuPont, U.S.

Vitreous Enamel *n* Silicate glass fired on metal. *Also called Porcelain Enamel*.

Viton® *n* DuPont's trade name for a family of copolymer fluoroelastomers with a wide range of properties among them, but, in particular, good resistance to high temperatures and chemicals. They are used for gaskets, O-rings, oil seals, diaphragms, pump and valve linings, hose, tubing, and coating fabrics.

VLDPE *n* Abbreviation for Very-Low-Density Polyethylene.

VM & P Naphtha *n* Varnish Maker's and Painter's Naphtha. Any number of narrow-boiling-range fractions of petroleum with boiling points of about 93 to 149°C according to specific use. The term "benzine" is still used for VM&P Naphtha but is not preferred in modern nomenclature. *Known also as Painter's Naphtha*.

Void \ˈvóid\ *n* (1616) (1) In a solid plastic article or laminate, an unfilled space within the article large enough to scatter light (in transparent materials) or other radiant energy that might be used to detect such spaces. (2) In cellular plastics, a cavity unintentionally formed and substantially larger than the characteristic individual cells (ISO). (3) An empty volume within any material or liquid medium. See also ▶ Blister.

Void Fraction *n* The fraction or percentage of the volume of an article or material sample, such as fiber, powder or foam, that is within the material, and contains only vacuum or gas.

Voids *n* (1) See ▶ Holidays. (2) See ▶ Microvoids. (3) Interstitial space in media mill. Knowledge of the void volume is necessary to calculate the optimum charge.

Voigt Element *n* This is a Voight model which is a component, together with other Voight or Maxwell components, of a more complex viscoelastic model system, such as the standard linear solid.

Voigt Model *n* A conceptual, mechanical model useful as an analogy to the deformation behavior of viscoelastic materials. It consists of, side-by-side (in parallel), an elastic coil spring and a viscous dashpot rigidly connected at each end. When the ends are pulled apart, they will separate gradually and ever more slowly until the spring is stretched to a length corresponding in to the pulling force divided by the spring stiffness (spring constant), when motion stops. (Kamide, K., and dobashi, T., *Physical Chemistry of Polymer Solutions*, Elsevier, New York, 2000) Compare this with the ▶ Maxwell Model.

Voile Fabric \ˈvói(ə)l-\ A sheer spun cloth that is lightweight and soft. It is usually made with cylindrical, combed yarn. Voile is used for blouses, children's wear, draperies, bedspreads, etc. (Kadolph, S. J. J., and Langford, A. L., *Textiles*, Pearson Education, New York, 2001)

Volatile \ˈvä-lə-təl, *esp British* –ˌtīl\ *adj* [F, fr. L *volatilis*, fr. *volare* to fly] (1605) (1) The easily evaporated or vaporized components of any coating composition in contrast to the nonvolatile components. (2) Easily evaporated or vaporized at temperatures below about 40°C; low-boiling.

Volatile Loss The loss in weight, usually unintended, of a substance due to evaporation of one or more constituents.

Volatiles Content The percent weight loss, through loss of volatiles, from a plastic or impregnated reinforcement held at a specific temperature for a specified time (sometimes, under vacuum).

Volatile Thinner See ▶ Thinner.

Volatility *n* Of liquids and some solids, the tendency to evaporate when exposed to the atmosphere. This qualitative idea is closely related to ▶ Vapor Pressure. The *relative volatility* of two liquids at any temperature is the ratio of their vapor pressures at that temperature.

Volt \ˈvōlt, ˈvólt\ *n* [F *volte*, fr. It *volta* turn, fr. *voltare* to turn, from (assumed) VL *volvitare*, freq. of L *volvere* to roll] (1688) (V) The SI unit of electromotive force (emf), equal to the difference in electric potential between two points of a conductor carrying a constant current of one ampere when the power dissipated between these points equals, one watt (i.e., in SI, 1 V ≡ 1 W/A). It is also the potential difference required to cause a steady current of one ampere to flow through a conductor whose resistance is one ohm.

Volume \ˈväl-yəm\ *n* [ME, fr. MF, fr. L *volumen* roll, scroll, fr. *volvere* to roll] (14c) The space occupied by an article or sample of material, including any voids, within the defining surfaces. The SI unit of volume is the cubic meter, m^3, known in the past by the name *stere*, now deprecated (but alive and well in crossword puzzles). SI also allows the use of convenient subvolumes, e.g., m^3, cm^3. The exponent also operates on the abbreviated prefix in each case [i.e., 1 cm^3 = 1 $(cm)^3$ = 10^{-6} m^3, not 0.01 m^3] The special name *litre* (*liter* in the U.S.) has been approved for the cubic decimeter (dm^3) but is to be used only for volumetric capacity and dry and liquid measure. No prefix other than milli- (m) or micro- (μ) should be used with liter. Some conversions of other volume units to SI are given in the Appendix.

Volume Coefficient of Thermal Expansion *n* (cubical expansion coefficient) The rate of change of volume of a material with rising temperature, divided by the volume, i.e.,

$$\frac{\left(\frac{\partial V}{\partial T}\right)_P}{V}$$

where *P* subscript reminding one that the coefficient, which is mildly pressure-sensitive, is tied to the particular pressure at which it has been measured (most often, atmospheric). The SI unit is K^{-1}. Over a short internal of temperature, the expression above is very nearly equal to the change in volume divided by the change in temperature, i.e., $\Delta V/\Delta T$. In isotropic materials, the cubical expansion coefficient is three times the linear expansion coefficient, the quantity *usually* meant by ▶ Coefficient of Thermal Expansion. See also ▶ Dilatometer. (Groenewoud, W. M., *Characterization of Polymers by Thermal Analysis*, Elsevier Science and Technology Books, New York, 2001)

Volume Expansion *n* The change in volume of a test specimen under specified test conditions (ISO). This is usually expressed as a percentage of the original volume.

Volume Resistivity *n* (specific insulation resistance) The ratio of the potential gradient parallel to the current in a material to the current density. In SI, volume resistivity is numerically equal to the direct current resistance between opposite faces of a one-meter cube of the material, with the unit ohm-meters ($\Omega \cdot m$). The smaller cgs unit, $\Omega \cdot cm$, is still widely used. (Giambattista, A., Richardson, R., Richardson, R. C., and Richardson, B., *College Physics*, McGraw Hill Science, New York, 2003; Ku, C. C., and Liepins, R., *Electrical Properties of Polymers*, Hanser Publishers, New York, 1987)

Volume Solids, Percent *n* See ▶ Solids by Volume.

Vortex Cell See Benard Cell.

VSEPR Theory Valence-shell electron-pair repulsion theory.

Vulcanizate *n* rubber that had been irreversibly transformed from predominantly plastic to predominantly elastic material by vulcanization (chemical curing or crosslinking) using heat, vulcanization agents, accelerants, etc.

Vulcanizate Crosslinks *n* Chemical bonds formed between polymeric chains in rubber as a result of vulcanization.

Vulcanization \ˌvəl-kə-nə-ˈzhən\ *n* (1846) A chemical reaction in which the physical properties of a rubber are changed in the direction of decreased plastic flow, less surface tackiness, and increased tensile strength by reacting it with sulfur or other suitable agents. See also ▶ Curing (2) and ▶ Self-Vulcanizing.

Vulcanize *v* To subject to vulcanization.

Vulcanized Fiber *n* (1) Resin-free cellulosic plastic. Made by immersing cotton waterleaf paper in a solution of zinc chloride, applying slight heat and pressure and subsequently leaching the zinc chloride from the newly created product. At the completion of this process, vulcanized fiber is no longer a paper, nor is it what is commonly known as chem-board; it is a completely new form of matter with unusual physical characteristics. (2) A nearly homogeneous material consisting of hydrated cellulose, made by subjecting cellulose to a parchmentizing process.

Vulcanized Fiber *n* [fr. *Vulcanized Fibre*, a trademark] (ca. 1884) Cellulosic material that has been partly gelatinized by action of a chemical (usually zinc chloride solution), then heavily compressed or rolled to the required thickness, leached free of the zinc chloride, and dried. It has been used for electrical insulation, luggage, and materials-handling equipment.

Vulcanized Oil *n* Vegetable oil which has been reacted with sulfur, or sulfur chloride.

Vulcanizing *n* (vulcanization) The chemical reaction, usually accompanying by crosslinking, that induces extensive changes in the physical properties of a rubber or elastomer, brought about by reacting the material with sulfur and an accelerator. The discovery of vulcanization by C. Goodyear in 1839 was the beginning of a practical rubber-products industry. The changes brought about by vulcanizing include decreased plastic flow, reduced surface tackiness, increased moduli and resilience, much grater tensile strength, and considerably reduced solubility. Some thermoplastics, such as polyethylene can be modified to be vulcanizable. The associated crosslinking causes the final product to resist flow and deformation at temperatures above the melting range of the original polymer.

Vulcollan *n* Polyurethane. Manufactured by Bayer, Germany.

W

w \ˈdə-bəl-(ˌ)yü\ *n* {*often capitalized, often attributive*} (15c) Symbol for width or, in thermodynamics, work done by a system.

W (1) The SI abbreviation for WATT. (2) The chemical symbol for the element tungsten (from *wolframite*, the mineral in which it was first recognized, (Hibbard MJ (2001) Mineralogy. McGraw-Hill, New York)).

Wainscot \ˈwān-skət\ *n* [ME, fr. MD *wagenschot*, prob. fr. *wagen* wagon + *schot* shot, crossbar] (14c) (1) Wall paneling, usually of wood. (2) A decorative or protective facing applied to the lower three or four feet of an interior partition or wall, making use of new material different from that of the upper wall, and often consisting of wood paneling or other facing material. (Merriam-Webster's Collegiate Dictionary (2004) 11th Edn., Merriam-Webster, Springfield) See ▶ Dado.

Wale \ˈwā(ə)l\ *n* [ME, fr. OE *walu*; akin to ON *volr* staff and perhaps to ON *valr* round, L *volvere* to roll] (before 12c) (1) In knit fabrics, a column of loops lying lengthwise in the fabric. Number of wales per inch is a measure of the fineness of the fabric. (2) In woven fabrics, one of a series of ribs, cords, etc., running either warpwise or fillingwise. (Kadolph SJJ, Langford AL (2001) Textiles. Pearson Education, New York)

Wallboard \ˈwȯl-ˌbōrd\ *n* (1906) Such boards as pressed cellulose fibers, plasterboard, cement-asbestos board, plywood, used in place of plaster interior surfaces. (Kadolph SJJ, Langford AL (2001) Textiles. Pearson Education, New York)

Wallpaper *n* (1827) Paper, or paperlike material, usually decorated in colors, which is pasted or otherwise affixed to walls or ceilings of rooms.

Wall Stress (1) In a filament-wound pressure vessel, the stress calculated from the pressure or load divided by the entire cross-sectional area of the wall (*not* just that of the reinforcement). (2) In a fluid flowing through a channel (such as a die, tube, or extruder-screw channel), the shear stress at any channel surface. In the simplest case of a fluid in steady flow through a circular tube of radius R and length L, the wall stress is given by $\Delta P \cdot R/2L$, where ΔP is the pressure drop from inlet to outlet. (Munson BR, Young DF, Okiishi TH (2005) Fundamentals of Fluid Mechanics. John Wiley and Sons, New York)

Walnut Oil Vegetable semidrying or drying oil obtained from *Juglans regia*. It is used chiefly by artists, to whom its nonyellowing properties are attractive and in place of soya bean oil, in locations where walnut oil is economically available. It may contain as much as 63% of linoleic acid and 15% of linolenic acid. Sp gr, 0.926/15°C; iodine value, 140–150; saponification value, 195. (Langenheim JH (2003) Plant Resins: Chemistry, Evolution Ecology and Ethnobotany, Timber Press, Portland, OR, 2003; Paint: Pigment, Drying Oils, Polymers, Resins, Naval Stores, Cellulosics Esters, and Ink Vehicles, Vol. 3, American Society for Testing and Material, 2001)

Walnut-Shell Flour See ▶ Nutshell Flour.

Warm Forging The process of forming thermoplastic sheets or billets into desired shapes by pressing them between dies in a press, when the material and/or the dies have not been preheated. The blanks may be billets formed by extrusion, or may be die-cut from sheets. (Strong AB (2000) Plastics Materials and Processing. Prentice Hall, Columbus) See also ▶ Cold Forming and ▶ Solid-Phase Forming.

Warm-Setting Adhesive See ▶ Adhesive, Warm-Setting.

Warm-Up A milling operation to plasticize uncured rubber compounds before calendering, extruding, or molding.

Warp \ˈwȯrp\ [ME, fr. OE *wearp*; akin to OHGr *warf* warp, OE *weorpan* to throw, ON *verpa*] (before 12c) (1) A significant variation from the original, true, or plane surface. (2) In the textile industry, those threads of a cloth which are parallel to the selvage, i.e., running lengthwise in the loom. (3) To change shape spontaneously. This is seen particularly in flat surfaces such as sheet and sides of boxy shapes. Such changes are often traceable to creep caused by stresses generated during molding or forming, by uneven absorption of water or a solvent, by uneven heating, or, in fiber-reinforced thermosets, by unequal curing in thin and thick sections. (Tortora PG (ed) (1997) Fairchild's Dictionary of Textiles. Fairchild Books, New York)

Warpage Distortion caused by nonuniform change of internal stresses. See ▶ Dished.

Warp-Drawing See ▶ Draw-Warping. Warp-drawn fibers may be taken up on packages other than beam.

Warp Holding Place See Sticker (1).

Warping See ▶ Beaming.

Warp Knitting *n* (1946) See Knitting (1).

Warp Pile The extra set of warp yarns that forms the surface in a double-woven pile fabric, including types such as velvet and velour. Upholstery fabrics such as mohair, plush, and friezé are produced by this method. Also See ▶ Pile and ▶ Velvet Fabric.

Warp Printing See ▶ Printing.

Warp Sheet A sheet comprising up to several thousand ends that are combined to make up the warp during preparation for weaving or warp knitting. Also see ▶ Warp.

Warp Sizing See ▶ Slashing.

Warp Streaks A fabric fault that shows as bands or streaks running warpwise. Warp streaks should not be confused with reed marks.

Wash \wósh\ In reinforced-plastics molding, an area where the reinforcement placed in the mold has moved during closing of the mold, resulting in a resin-rich (and reinforcement-poor) region. (Murphy J (1998) Reinforced Plastics Handbook. Elsevier Science and Technology Books, New York; Ash M, Ash I, (1982–83) Encyclopedia of Plastics Polymers, and Resins. Vols. I-III, Chemical Publishing Co., New York)

Washability Ease with which the dirt can be removed from a paint surface by washing; also refers to the ability of the coating to withstand washing without removal or substantial damage. (Koleske JV, (ed) (1995) Paint and Coating Testing Manual, American Society for Testing and Materials)

Washable Distemper See ▶ Washable Water Paint.

Washable Water Paint Water paint which, in addition to a glue or casein binder, contains an emulsified oil or similar fixing agent, thus rendering it washable. Sometimes incorrectly described as *Washable Distemper*. (Weismantal GF (1981) Paint Handbook. McGraw-Hill, New York)

Wash-and-Wear *adj* (1956) A generic term applied to garments that satisfactorily retain their original neat appearance after repeated wear and home laundering with little or no pressing or ironing. A wash-and-wear garment is essentially free from undesirable wrinkles both during wear and after laundering and retains any original pressed-in creases or pleats. The garments should meet normal consumer demands for durability, color, stability, and shrinkage. The performance of a wash-and wear fabric or garment depends on several factors, including the types and amounts (percentages) of fibers used, the fabric construction, the finishing treatment, the presence of a colored pattern (either woven or printed), and the methods used for washing and drying. These factors determine, in any specific instance, if a fabric or garment's performance will meet customer requirements. Variable conditions result in the varying behavior of a specific fabric or garment. Garments are labeled to specify the appropriate care for optimal performance. (Humphries M (2000) Fabric Glossary. Prentice-Hall, Upper-Saddle River) Also See ▶ Ease-of-Care and ▶ Durable Press.

Washboard \wósh-bōrd\ *n* (1742) See ▶ Baseboard.

Wash Coat A very thin, semitransparent coat of paint; applied as a preliminary coating on a surface; acts as a sealer or guide coat.

Washfastness The resistance of a dyed fabric to loss of color or change in properties during home or commercial laundering.

Washing See ▶ Tinting.

Washing Out The removal of the original greasy ink on the design areas of etched lithographic plates.

Washout Inks Inks used on textiles which are easily removed by washing.

Wash Primer Priming paint usually supplied as one- or two-component systems. The paint contains carefully balanced proportions of an inhibiting chromate pigment, phosphoric acid, and a synthetic resin binder mixed in an alcohol solvent. On clean, light alloy or ferrous surfaces, and on many nonferrous surfaces, such paints give excellent adhesion, partly due to chemical reaction with the substrate, and give a corrosion-inhibiting film which is a good basis for the application of subsequent coats of paint. Although these materials are referred to as primes, the films which they produce as so thin that it is more correct to consider them as etching solutions and to follow them with an ordinary primer if maximum protection is required. (Tracton AA (ed) (2005) Coatings Technology Handbook. Taylor and Francis, New York; Weismantal GF (1981) Paint Handbook. McGraw-Hill, New York) *Also known as Pretreatment Primers, Etch Primers and Self-Etch Primers.*

Washup The process of cleaning the rollers, form or plate, and even the fountain of a press.

Waste \wāst\ *n* [ME *waste, wast*; in sense 1, fr. ONF *wast*, fr. *wast*, adj., desolate, waste fr. L *vastus*; in other senses, fr. ME *wasten* to waste] (13c) By-products created in the manufacture of fibers, yarns, and fabrics.

Wasted Loop Effect An intramolecular reaction involving gelation that does not occur at the predicted value because not all of the branching points on the growing chain are used effectively; many multifunctional units in the chain are wasted, and one theory is that the branch point leading out from the multifunctional repeating unit loops back to form a ring with another branch point on the same polymer molecule. (Odian GC

(2004) Principles of Polymerization. John Wiley and Sons, New York; Elias HG (1977) Macromolecules Vol. 1–2, Plenum Press, New York)

Water Absorption The percentage increase in weight of a plastic article when immersed in water for a stipulated time and at a specified temperature (usually room temperature). Most plastics absorb water to some extent, varying from almost zero in the case of polytetrafluoroethylene and polyolefins, to complete dissolution for some types of polyvinyl alcohol and polyethylene oxide. Water absorption can cause swelling, leaching of additives, plasticizing and hydrolysis, which in turn can cause discoloration, embrittlement, stress cracking, lowering of mechanical and electrical properties, and reduced resistance to heat and weathering. (Grellman W, Seidler S (eds) (2001) Deformation and Fracture Behavior of Polymers. Springer-Verlag, New York; Mark JE (ed) (1996) Physical Properties of Polymers Handbook. Springer-Verlag, New York; Zaiko GE (ed) (1995) Degradation and Stabilization of Polymers, Nova Science Publishers, New York)

Water-Based Coatings Latex coatings and coatings containing water-soluble binders. Latex coatings (Wicks ZN, Jones FN, Pappas SP (1999) Organic Coatings Science and Technology. 2nd edn. Wiley-Interscience, New York) Syn: ▶ Water-Borne Coatings, ▶ Water-Reducible Coatings.

Water-Based Inks Inks containing a vehicle whose binder is water-soluble or water-dispersible.

Water-Based Paint See ▶ Water Paint.

Water Blasting Blast cleaning of metal using high-velocity water.

Water-Borne Coatings See ▶ Water-Based Coatings.

Water Break The appearance of a discontinuous film of water on a surface, signifying nonuniform wetting and usually associated with a surface contamination.

Water-Break-Free The ability of the rise water to cover the entire surface in an unbroken film.

Water Color (1) An artists' colors in which the pigment has been mixed with gum as a binder (Gair A (1996) Artist's Manual. Chronicle Books LLC, San Francisco). (2) The term used to describe that form of transparent water painting in which the white of the paper furnishes the lights and in which no white pigment is used during the painting of the picture (Paint/Coatings Dictionary, Compiled by Definitions Committee of the Federation of Societies for Coatings Technology, 1978)

Water Glass n (1612) Sodium or potassium silicate that is soluble in water.

Water Imbibition See ▶ Imbibition.

Water-Jet Loom See ▶ Jet Loom.

Waterless Dyeing See ▶ Dyeing, ▶ Solvent Dyeing.

Water Paint (1) Paint, the vehicle of which is a water emulsion, water dispersion, or ingredients that react chemically with water. (2) A paint capable of being thinned or diluted with water, such as casein paint, latex paint, distemper, calcimine and whitewash.

Waterproof \ˈwȯ-tər-ˌprüf\ *adj* (1736) A term applied to materials that are impermeable to water; waterproof fabrics have had all their pores closed and are also impermeable to air and very uncomfortable.

Waterproofing Coatings Coatings which are formulated to prevent penetration of the substrate by water. These coatings include but are not limited to, bituminous roof and resilient type coatings.

Water-Reducible Coatings See ▶ Water-Based Coatings.

Water-Reducible Resins Water-soluble types or lattices or emulsions. Resins which can be diluted (reduced) with water, water-cosolvent mixtures and sometimes with alkali (alkali-soluble resins). (Wicks ZN, Jones FN, Pappas SP (1999) Organic Coatings Science and Technology. 2nd edn. Wiley-Interscience, New York)

Water Repellent *adj* (1896) A term applied to fabrics that can shed water but are permeable to air and comfortable to wear. These fabrics are produced by treating the material with a resin, wax, or plastic finish that is not completely permanent.

Water Repellents Materials or treatment for surfaces to provide resistance to penetration of water.

Water Resistance *adj* (1921) The ability of a coating to resist damage or degradation due to water.

Water-Soluble Polymers Macromolecules exhibiting solubility in aqueous solutions. Water-soluble polymers can be divided into four categories, biopolymers, nonionic, ionic, and associative polymers.

Water-Soluble Resins Any of several resin types that are produced by polymerization reactions in which the chain growth results from breaking of ring structure or double bonds of the monomers. Examples are alkyl- and hydroxyalkyl cellulose derivatives, carboxymethyl cellulose, polyvinyl alcohol, polyvinyl pyrrolidone, polyacrylic acid, polyacrylamide, polyethylene oxide, and polyethylene-imide. (Tracton AA (ed) (2005) Coatings Technology Handbook. Taylor and Francis, New York)

Water Spotting n (1939) Any change in the appearance of surface resulting solely from the action of water. See ▶ Rain Spotting.

Water Stain See ▶ Stain.

Water Swell *n* Expansion of material volume as a result of water absorption.

Water Vapor Transmission (WVT, WVTR) The rate of water vapor flow, under steady specified conditions, through a unit area of a material, between its two parallel surfaces and normal to the surfaces. Metric unit of measurements is 1 g/24 h·m^2. See ▶ PERM. (Paint and Coating Testing Manual (Gardner-Sward Handbook) MNL 17, 14th Ed., ASTM, Conshohocken, PA, (1995); www.astm.org)

Water Wax See ▶ Liquid Water Emulsion Wax.

Water White A material that approaches the colorless nature of water; generally applied to colorless transparent liquids and solids. Abbreviation: WW.

Watt \ˈwät\ *n* [James Watt † 1819] (1882) The SI unit of power, equal to 1 J/s (= 1 m·N/s), expressing the rate at which work is done or energy expended. In the case of alternating electric current the watt is computed as the product of voltage across the circuit, times the current flow in amperes times the cosine of the phase angle between the current and the impressed voltage. In purely resistive DC circuits, watts = volts × amperes. Some conversions of other power units to SI are given in the Appendix.

Wattle Gum Water-soluble or water-dispersible gum obtained from the Australian acacia tree.

Wave Equation *n* (1926) A mathematical equation describing the motion of a wave.

Wave Function The mathematical relation which solves a wave equation. Each correctly obtained electron wave function corresponds to a disc.

Wavelength \-ˌleŋ(k)th\ *n* (1850) Distance between two successive points of a periodic wave in the direction of propagation in which the oscillation has the same phase; designated as λ in spectrophotometry. It is usually measured in nanometers or, formerly, in Angstrom units (1 nm = 10A). See ▶ Frequency. (Giambattista A, Richardson R, Richardson RC, Richardson B (2003) College Physics. McGraw Hill Science, New York; Saleh BEA, Teich MC (1991) Fundamentals of Photonics. John Wiley and Sons, New York)

Wave Mechanics *n* [*plural but singular or plural in construction*] (1926) The branch of physics which describes the behavior of small particles by assigning wavelike properties to them, also known as Quantum Mechanics.

Wave Motion A progressive disturbance propagated in a medium by the periodic vibration of the particles of the medium. Transverse wave motion is that in which the vibration of the particles is perpendicular to the direction of propagation. Longitudinal wave motion is that in which the vibration of the particles is parallel to the direction of propagation.

Wave Number (1873) The number of waves per unit distance of radiant energy of a given wavelength: the reciprocal of the wavelength.

Wavy Cloth See ▶ Baggy Cloth.

Wavy Grain Grain in which the fibers and other longitudinal elements collectively take the form of waves or undulations.

Wavy Selvage See ▶ Slack Selvage.

Wax \ˈwaks\ *n* [ME, fr. OE *weax*; akin to OHGr wax, Lithuanian *vaškas*] (before 12c) Any of various unctuous, viscous or solid heat-sensitive substances, consisting essentially of high molecular weight hydrocarbons or esters of fatty acids (C_{16} to C_{30}), characteristically insoluble in water but soluble in most organic solvents.

Wax Solid, low-melting substances that may be of plant, animal, mineral, or synthetic origin. Waxes are generally slippery (though beeswax is somewhat sticky), plastic when warm, and, because their molecular weights are rather low, fluid when melted.

WAXS Abbreviation for Wide-Angle-X-Ray Scattering.

Wax Set Ink A printing ink designed to set and dry instantly upon immersion of the print in a bath of molten wax.

Wb SI abbreviation for ▶ Weber.

Weak (Electrolyte) Only partially dissociated.

Weak Web A web of fiber that, when being transferred from the card doffer to the calender rolls to form sliver, does not have sufficient strength from fiber cohesion or clinging entanglement to hold itself together while forming a continuous bridge in processing.

Weal Yarn A yarn that is found to be either below standard breaking specifications or to be weak enough to cause an abnormally high degree of stops in textile processing.

Wear Cycles In abrasion resistance tests using the Taber Abraser, the number of cycles of abrasion required to wear a film of specified thickness through to the test plate under a specified set of test conditions. (Federal Standard 141a, Method 6192).

Wear Index In abrasion resistance tests using the Taber Abraser, it is the loss in weight in milligrams per 1,000 cycles of abrasion under a specific set of test conditions. (Federal Standard 141a, Method 6192).

Wear Test A test for fabric wear, abrasion, flexibility, washing, crushing, creasing, etc., in which the fabric is made into a garment, worn for a specific time, then assessed for performance.

Weather \ˈwe-thər\ *v* (15c) To age, deteriorate, discolor, etc., as a result of exposure to the weather.

Weatherboarding \-ˌbōr-diŋ\ *n* (1632) See Siding.

Weathering *v* (15c) (1) Behavior of paint films when exposed to natural weather or accelerated weathering equipment, characterized by changes in color, texture, strength, chemical composition, or other properties. Natural outdoor weathering tests are normally carried out at selected exposure sites, on painted panels, generally exposed either vertically or at 45° facing south in the northern hemisphere. See ▶ Accelerated Weathering. (2) The process of disintegration and decomposition as a consequence of exposure too the atmosphere, to chemical action, and the action of frost, water, and heat. (3) A broad term encompassing exposure of plastics to solar or ultraviolet light, temperature, oxygen, humidity, snow, wind, and air-borne dust and biological or chemical agents, such as smog. (Paint and Coating Testing Manual (Gardner-Sward Handbook), (1995) MNL 17, 14th Ed., ASTM, Conshohocken, PA)

Weather-Ometer An apparatus, used to estimate the life of coatings, in which specimen materials can be subjected to artificial and accelerated weathering tests which simulate natural weathering by the use of controlled cycles of ultraviolet radiation, light, water, and heat. Electric arcs, water spray, and heating elements are used to simulate the natural conditions of sun, rain, and temperature changes. (Paint and Coating Testing Manual (Gardner-Sward Handbook) MNL 17, (1995) 14th Ed., ASTM, Conshohocken, PA, 1995) See ▶ Accelerated Weathering.

Weather Resistance The ability of a material, paint film, or the like to withstand effects of wind, rain, sun, etc., and retain its appearance and integrity.

Weather Stripping *n* (1846) Metal, wood, plastic or other material installed around door and window openings to prevent air infiltration.

Weather Testing Experimental tests that aim to predict the lifetime of manufactured articles.

Weave \ˈwēv\ *n* [ME *weven*, fr. OE *wefan*; akin to OHGR *weban* to weave, Gk *hyphainein* to weave, *hyphos* web] (1581) A system or pattern of intersecting warp and filling yarns. There are three basic two dimensional weaves: plain, twill, and satin. All other weaves are derived from one or more of these types. Also see ▶ Plain Weave, ▶ Twill Weave, and ▶ Satin Weave.

Weaving The method or process of interlacing two yarns of similar materials so that they cross each other at right angles to produce woven fabric. The warp yarns, or ends, run lengthwise in the fabric, and the filling threads (weft), or picks, run from side to side. Weaving can be done on a power or handloom or by several hand methods. Also see ▶ Loom and ▶ Woven Fabric.

Web \ˈweb\ *n* [ME, fr. OE; akin to ON *vefr* web, OE *wefan* to weave] (before 12c) (1) A continuous film or fabric in process in a machine. In extrusion coating, the *molten web* is that which issues from the die and becomes the coating, and the *substrate* (*web*) is the material being coated. (2) A continuous length of sheet material handled in roll form as contrasted with the same material cut into short lengths.

Webbing (British) *v* Development of wrinkles, often in a well-defined pattern, in the surface of a paint or varnish during drying. This condition results from the irreversible swelling of a partially dried surface skin and may be aggravated by impure gas fumes during stoving in a gas oven in which case it is termed "gas checking." Webbing is generally regarded as a paint defect but is made use of in some paint finished to give a textured coating (e.g., wrinkling) which obscures minor faults and indentations in the surface to be coated. (Paint/Coatings Dictionary, (1978) Federation of Societies for Coatings Technology, Blue Bell, PA, 1978)

Web Coating Any of a number of processes by which coatings are applied to continuous substrates such as papers, cloths, and metal foils, including ▶ Calender Coating, ▶ Extrusion Coating, ▶ Flow Coating, ▶ Gravure Coating, ▶ Roller Coating, ▶ Spread Coating plus other listed at ▶ Coating Methods.

Weber \ˈwe-bər, ˈvā-bər\ *n* [Wilhelm E. *Weber* † 1891 German physicist] (2892) (Wb) The SI unit of magnetic flux, defined a the magnetic flux which, linking a circuit of one turn, produces in it an electromotive force of one volt and the flux is reduced to zero at a uniform rate in one second. Therefore, 1 weber = 1 volt·second. The former unit of flux, the Maxwell, part of the so-called "absolute" system of electrical units, equals 10^{-8} Wb.

Weft \ˈweft\ *n* [ME, fr. OE; akin to ON *veptr* weft, OE *wefan* to weave] (before 12c) (woof, fill, filler yarn) In the textile industry, the transverse threads or fibers in a woven fabric; those fibers running crosswise to the ▶ Warp. See ▶ Filling.

Weft Insertion (1) Any one of the various methods, shuttle, rapier, water jet, etc., for making a pick during weaving. (2) A marriage of warp knitting and weaving brought about by inserting a length of yarn across the width of the knitting elements and fastening the weft yarn between the needle loop and the underlap.

Weft-Knit Fabric *n* (1943) See ▶ Circular-Knit Fabric.

Weft Knitting *n* (1943) See Knitting (2).

Weigh Feeder Syn: ▶ Gravimetric Feeder.

Weighing Equipment Equipment for measuring the mass and/or weight of objects.

Weight \ˈwāt\ *n* [ME *wight, weght*, fr. OE *wiht*; akin to ON *wætt* weight, OE *wegan* to weight] (before 12c) The force with which a body on or near the earth's surface is attracted to the earth, equal to the body's mass × g/g_c, where g is the acceleration due to gravity and $1/g_c$ is the proportionality constant in Newton's second law of motion (the law of momentum change). In the SI system, g_c = 1.00000 kg·m/(N·s^2) while g varies slightly (at sea level) from 9.78039 m/s^2 at 0° latitude to 9.83217 m/s^2 at 90° latitude because of the earth's spin and consequent flattening at the poles, with its "standard value", at about 45° N latitude, being 9.80655 m^2/s. Thus the abhorred kg_f is equated to 9.80655 N. See also ▶ Force. (Giambattista A, Richardson R, Richardson RC, Richardson B (2003) college Physics. McGraw Hill Science, New York)

Weight-Average Molecular Weight $(M_w, \overline{M}_w)$ For a sample with distributed molecular weights (all commercial polymers), the defining equation is

$$M_w = \frac{\sum_{i=1}^{\infty}(N_iM_i)(M_i)}{\sum_{i=1}^{\infty}N_iM_i} = \frac{\sum_{i=1}^{\infty}N_iM_i^2}{\sum_{i=1}^{\infty}N_iM_i}$$

where N_i is the number of individual molecules having molecular mass M_i and N_iM_i equals the mass of the N_i molecules in the sample with molecular mass M_i. The numerator and denominator quantities are also known as the *second* and *first original moments* of the distribution. M_w may be determined from measurements of Light Scattering or ▶ Size-Exclusion Chromatography. And it is M_w upon which melt viscosity in thermoplastics is strongly dependent. (Allcock HR, Mark J, Lampe F (2003) Contemporary Polymer Chemistry. Prentice Hall, New York; Slade, P. E., *Polymer Molecular Weights*, Vol. 4, Marcel Dekker, New York, 2001; Coleman, M. M., and Strauss, S., *Fundamentals of Polymer Science: An Introductory Text*, CRC Press, Boca Raton, FL, 1998) See also ▶ Molecular Weight, ▶ Molecular-Weight Distribution, ▶ Number-Average Molecular Weight, ▶ and Viscosity-Average Molecular Weight.

Weight-Average Molecular Weight A polymer molar mass average which is the mean value of the weight distribution of molecular sizes, defined a $M_w = \sum N_iM_i^2/\sum N_iM_i = \sum w_iM_i/\sum w_i$. Where N_i, M_i and w_i are the number of molecules, the molar mass and the total with of the molecular species.

Weighted Ordinate Method Method of calculating tristimulus values from measurements made at regular wavelength intervals. See ▶ Tristimulus Integration and ▶ Selected Ordinate Method.

Weighted Silk Silk that has been treated with metallic salts during dyeing and finishing to increase the fabric's weight and improve its drape. Over-weighting can cause deterioration of the fabric.

Weighting Equipment Equipment for measuring the mass and/or weight of objects.

Weight Loss The loss of mass of an article or test sample as a result of exposure to particular conditions as, for example, a few hours a 90°C in a vacuum oven, or two years outdoors on a weathering panel, often expressed as a percentage of the original mass. See also Volatiles Content, which may be the same as weight loss when the latter is due exclusively to loss of volatiles.

Weight-Per-Gallon Cup Essentially, a metal (less fragile than glass) pycnometer to simplify computations; the popular size holds 83.2 g of water at 77°F (25°C) or 100 g for the Imperial gallon, so that multiplying the weight of the contents by 0.1 represents the density expressed in pounds per gallon.

Weir \ˈwar\ *n* [ME *were*, fr. OE *wer*; akin to ON *ver* fishing place, OHGr *werien, werren* to defend] (before 12c) A simple device for controlling flow in open channels, consisting of a plate serving as a dam and having at its top center a V-notch or adjustable, rectangular, vertically sliding gate. Weirs can also be calibrated for flow measurement.

Weissenberg Effect A phenomenon sometimes encountered in rotational-viscometry studies of polymer melts and solutions at high speeds, characterized by the tendency of the polymer solution to climb the wall of the cup or the shaft of the rotor immersed in it.

Weissenberg Number (New) In the flow of viscoelastic liquids, the dimensionless Weissenberg number represents the ratio of the viscoelastic force to the viscous force and has sometimes been equated to $N_1/2\tau$, where N_1 = the first normal stress in a viscoelastic fluid flowing in simple shear and τ = the shear stress.

Weissenberg Rheogoniometer A vertical cone-and-plate rheometer (K. Weissenberg, 1948, and improvers since) for viscoelastic liquids in which the cone can twist through a measured angle while the plate is rotated at speeds providing shear rates to 100 s^{-1}. The cone and plate are enclosed in an oven that provides control of the sample temperature. The plate shaft rests on a ball bearing at its bottom which in turn rests on a force transducer. Thus it is possible to measure both the restoring torque on the cone and the normal force on the plate. The first provides an estimate of the liquid's viscosity while the normal force gives an estimate of the normal stress.

Weld \ ▌weld\ *v* [alter. of obs. E *well* to weld, fr. ME *wellen* to boil, well, weld] (1599) A metallic bond between like or unlike metals.

Weldbonding A process developed in the former USSR and introduced in the U.S. by the Air Force Materials Laboratory. Developed for the aerospace industry, the process combines spot welding with adhesive bonding of aluminum structures. It has provided an economical and efficient means of laying up and oven curing large, epoxy-bonded assemblies.

Welding The joining of two or more pieces of thermoplastic by fusion at adjoining areas, either with or without addition of plastic from another source (such as welding rod). The term includes *heat sealing*, with which it is synonymous in some countries, but in the U.S. the term heat sealing is limited to film and sheeting. Welding is almost always done with two (or more) pieces of the same plastic, but it can be done with compatible plastics that melt in the same temperature range. The various welding methods are described at the entries listed below.

Butt Fusion	Impulse Sealing
Dielectric Heat-Sealing	Induction-Welding
Extruded-bead sealing	Jig Welding
Friction Welding	Spin Welding
Heat Sealing	Stitching
High-Frequency Welding	Thermoband Welding
Hot-Gas Welding Ultrasonic Welding	
Hot-Plate Welding	Vibration Welding

(*Whittington's Dictionary of Plastics*, Carley, James F., ed., Technomic Publishing Co., Inc., 1993; *Engineering Plastics and Composites*, Pittance, J. C., ed., SAM International, Materials Park, OH, 1990)

Weld Line (weld mark, flow line) A flaw on a molded plastic article marking the meeting of two flow fronts within the mold. Because the two fronts may have cooled and skinned over before meeting, or had too little time in the molten state for interdiffusion of molecular segments across the interface, the weld may be imperfect and weak. (2) In extrusion of pipe, tubing, and some profiles from end-fed dies in which the cores are supported by spiders, a line parallel to the product axis where the flow front was split by a spider leg and subsequently reunited downstream. Weakness at such weld lines can depress the hoop strength of the pipe or tubing. (*Whittington's Dictionary of Plastics*, Carley, James F., ed., Technomic Publishing Co., Inc., 1993; *Engineering Plastics and Composites*, Pittance, J. C., ed., SAM International, Materials Park, OH, 1990)

Weld Slag Amorphous deposits formed during welding.

Weld Splatter Beads of metal left adjoining weld.

Wenzel's Blue See ▶ Cobalt Blue.

Westphal Balance Special kind of balance used for the determination of the specific gravity of liquids or solids, by a direct weighting method.

Wet Abrasion Resistance See ▶ Scrub Resistance.

Wet Adhesion Test See ▶ Tape Test.

Wet and Dry Sandpaper Sandpaper that can be used with water or other lubricants, making possible the sanding of some plastics and metals that is not possible with dry sanding. The lubricant cools the workpiece, washes away swarf, reduces friction, produces finer finishes, resists rusting (with additives), improves abrasive life and combats loading of the coated abrasive. (Wicks, Z. N., Jones, F. N., and Pappas,, *Organic Coatings Science and Technology*, 2nd Ed., S. P., Wiley-Interscience, New York, 1999; Weismantal, G. F., *Paint Handbook*, McGraw-Hill Corporation, Inc., New York, 1981)

Wet Edge Edge of a wet painted area which remains workable. When painting large surfaces, it is generally necessary to join up to the edge of a paint film which has been left for an appreciable time; when this can be done by blending this edge with free-working paint without any lap showing, the film is said to present a wet edge.

Wet-Edge Time Length of time a coating remains wet enough to allow for brushing-in at the laps.

Wet Film Gauge Device for measuring wet film thickness of coatings.

Wet Film Thickness Thickness of the liquid coating film immediately after application.

Wet Flexural Strength The flexural strength measured after boiling a test specimen in water, usually less than the strength of the original, dry specimen. See also ▶ Flexural Strength.

Wet Forming The production of a nonwoven fabric web from an aqueous suspension of fibers by filtering the short fibers onto a screen belt or perforated.

Wet Layup In the reinforced plastics molding, the process of forming an article by first applying a liquid resin

Wet-on-Wet Coating (sometimes a special ▶ Gel Coat) to the mold surface, then applying a reinforcing backing layer with more resin.

Wet-on-Wet Coating Technique of painting whereby further coats are applied before the previous coats have dried, and the composite film then dries as a whole. The process requires specially formulated paints.

Wet-Out The degree to which an impregnating resin has filled the voids among the filaments being impregnated. This may be expressed quantitatively as $100 \times (1 - v_r - VF)/(1 - v_r)$ in which v_r = the volume fraction of reinforcing fiber in the laminate and VRF = the final ▶ Void Fraction.

Wet-Out Time The time required for a resin to completely fill the interstices of a reinforcement material and wet the surfaces of the fibers, usually determined by an optical or light-transmission method.

Wet Printing That process whereby another impression cleanly transfers over a previously printed wet film. Successful trapping depends upon the relative tack and thickness of the ink films applied. See ▶ Trapping.

Wet Rot Decay of timber caused by fungi which flourish in alternate wet and dry conditions.

Wet Sanding Process of sanding using waterproof papers with liquids, such as water, soluble oil (emulsion), straight cutting oils (mineral and fatty), mineral lard oil (mixture of mineral and lard oils, sulfurized and chlorinated cutting oils and wax).

Wet Scrub Resistance Ability of a paint film to withstand scrubbing in contact with water. See ▶ Scrub Resistance.

Wet Spinning The process of forming synthetic fibers by extruding or forcing polymers through spinnerets. In wet spinning the polymer is dissolved in a solvent prior to extrusion. Also, in wet spinning, the jet or spinneret is immersed in a liquid, which either diffuses throughout the solvent or reacts with the fiber composition.

Wet Storage Stain See ▶ White Rust.

Wet Strength The strength of an adhesive joint determined immediately after removal from a liquid (usually water) in which it has been immersed under specified conditions of time, temperature and pressure. See ▶ Strength, Wet. (*Handbook of Adhesives*, Skeist, I., ed., Van Nostrand Reinhold, New York, 1990)

Wettability *n* (1913) The ability of a solid surface to accept contact of and by a liquid, allowing it to spread freely and completely cover the surface. Wettability is closely linked to the equality of components of surface energy and ▶ Surface Tension of the solid and liquid involved. If the surface is wettable by the liquid, the contact angle of a droplet on the surface will be less than 10°. (*Surface and Interfacial Tension*, Hartland, S. ed., CRC Press, Boca Raton, FL 2004; *Surface Analysis Methods in Materials Science*, O'Conner, D. J. J., Smart, R. S., and Sexton, B. A., Springer-Verlag, New York, 2003)

Wetted Out The condition of an impregnated reinforcement where in substantially all voids between the sized strands and filaments are filled with resin; 100% WET-OUT.

Wetting (1) Power a vehicle possesses of spreading uniformly and rapidly over the surface of pigment particles. A vehicle with good wetting properties assists in the grinding or dispersion of pigments and the ability to wet the surface to which the finished coating is applied. (2) Surrounding the pigment particles with varnish during the ink-making process. Pigments which wet out easily will grind more easily, form better ink bodies, and result in a finer dispersion. (Parfitt, G. D., *Dispersion of Powders in Liquids*, Elsevier Publishing Co., New York, 1969)

Wetting Agents *n* (1927) A compound that causes a liquid to penetrate more easily into, or to spread over the surface of, another material, usually by reducing the liquid's surface tension. Common wetting agents are soaps, detergents, and surfactants. They are widely used in polymerization reactions and in preparing emulsions of plastics. (Solomon, D. H., and Hawthorne, D. G., *Chemistry of Pigments and Fillers*, Krieger Publishing Co., New York, 1991; Parfitt, G. D., *Dispersion of Powders in Liquids*, Elsevier Publishing Co., New York, 1969).

Wet Winding A filament-winding process wherein the strand is impregnated with resin just prior to contact with the mandrel.

Wheatstone's Bridge \ˈhwēt-ˌstōn, ˈwēt-, *chiefly British* –stən-\ *n* [Sir Charles *Wheatstone*] (1872) If the resistances, r_1, r_2, r_3, and r_4 form the arms of a Wheatstone's bridge in order as the circuit (omitting cell and galvanometer connections) is traced, when the bridge is balanced,

$$\frac{r_1}{r_2} = \frac{r_4}{r_3} \quad \text{or} \quad \frac{r_1}{r_4} = \frac{r_2}{r_3}$$

(Giambattista, A., Richardson, R., Richardson, R. C., and Richardson, B., *College Physics*, McGraw Hill Science, New York, 2003)

Whipcord \ˈhwip-ˌkórd\ *n* [fr. its use in making whips] (14c) A compact woven fabric having a very steep twill on the face of the goods. Whipcord is used in dress woolens, worsteds, or wool blends and in many types of uniforms.

Whiskers \▍hwis-kər\ *n* [singular of *whiskers* mustache, fr. ²*whisk*] (ca. 1600) A colloquial term used for nearly perfect, single-crystal fibers produced synthetically under controlled conditions from inorganic materials such as aluminum oxide, beryllium oxide, boron, boron carbide, graphite, magnesium oxide, metals, quartz, silicon carbide, and silicon nitride. They range in diameter from 0.5 to 30 μm, and in length from 1 μm to several mm. Whiskers are available as loose fibers, mats, and felts. Having tensile strengths and moduli from 5 to 10 times those of glass, they impart extremely high strength and stiffness to reinforced-plastics structures. (Murphy, J., *Reinforced Plastics Handbook*, Elsevier Science and Technology Books, New York, 1998)

White \▍hwīt\ *adj* [ME, fr. OE *hwīt*; akin to OHGr *hwīz* white and prob. to OChurch Slavonic *světŭ* light, Sanskrit *sveta* white, bright] (before 12c) Most usually applied to neutral or near-neutral colors of high reflectance.

White Blast Blast cleaning to white metal.

White Bole (bolus alba) A Syn: Kaolin. See ▶ Aluminum Silicate.

White Earth Nonspecific description applied to several different types of white fillers.

White Goods *n* (ca. 1871) A broad term describing any goods that have been finished in the white conditions.

White Lac Shellac which has been chemically bleached.

White Lead *n* (15c) See ▶ Carbonate White Lead.

White Metal *n* (1613) Blasting metal to specified appearance such as SSPC-SP-10, NACE No. 1, or SA-3 (Swedish Standard SIS 05 59 00).

Whiteness *n* (before 12c) Perception of high lightness, high diffusion (scattering) and absence of hue generally applied to opaque or translucent solids or liquids.

Whiteness Index Any of several numerical indices used to indicate the degree of whiteness; examples are the following:

(1) W = 4B-3G where B and G refer to tristimulus colorimeter readings.
(2) W = 100-[$(100-L)^2 + (a^2 + b^2)]^{1/2}$ where L, a, and b refer to coordinates in Hunter's L, a, b Color Difference Equation.
(3) Chemstrand Whiteness Scale W = 10 $(Y - 2p^2)^{1/2}$ where Y is the CIE tristimulus value and p is the CIE excitation purity.

(*Colour Physics for Industry*, 2nd *Ed.*, ed., McDonald, Roderick, Society of Dyers and Colourists, West Yorkshire, England, 1997)

Whitening *n* (1601) A finely divided form of ▶ Calcium Carbonate obtained by milling high-calcium limestone, marble, shell, or chemically precipitated calcium carbonate. See ▶ Limewashing.

Whitening in the Grain Fault which sometimes develop in varnished or polished open-grained woods, filled or unfilled. It is manifested as a streaky white appearance.

White Portland Cement A Portland cement, produced from raw materials low in iron, which hydrates to a white paste; used to yield a concrete of considerable whiteness.

White Reference Standard See ▶ Standard.

White Reference Standard, Absolute See ▶ Perfect Diffuser and ▶ Standard.

White Reference Standard, Primary See ▶ Standard.

White Reference Standard, Secondary See ▶ Secondary Reference Standard.

White Rust *n* (ca. 1848) White corrosion products (zinc hydroxide and zinc oxide) on zinc-coated articles. They form when the parts are stored so close together that condensed moisture is entrapped between them and the air circulation is inadequate to assist drying. Known also as *Wet Storage Stain*. (*Corrosion Engineer's Handbook*, 3rd Ed, Baboian, R., NACE International – The Corrosion Society, Houston, TX, 2002)

White Spirit British name for mineral spirits.

Whitewash \-▍wósh\ *n* (1591) Cheap type of flat water paint based on lime or whiting loosely bound with glue, size, casein, or water-dispersible binders. Syn: ▶ Limewashing, ▶ Whitening.

White Zinc *n* (1847) Another name for zinc oxide.

Whiting *n* [ME, fr. gerund of *whiten* to white] (15c) Calcium carbonate powder of high purity. See ▶ Calcium Carbonate.

Wicking *vt* (1949) (1) Cord, loosely woven or braided tape, or tubing to be cut into wicks. (2) Dispersing or spreading of moisture or liquid through a given area, vertically or horizontally; capillary action in material.

Wide-Angle X-ray Scattering (WAXS) A technique for determining the amount of crystallinity and the sizes and perfection of crystals in polymers, in which diffraction patterns of X rays scattering at 20° to 50° from the incident beam are recorded on film and measured. X rays of wavelength from 0.1 to 0.3 nm are used to elucidate structural features with sizes from 0.1 to 2 nm. (Rhodes, G., *Crystallography Made Crystal Clear: A Guide for Users of Macromolecular Models*, Elsevier Science and Technology Books, New York, 1999)

Wide-Belt Sander A machine using the principle of a contact roll backup for a wide abrasive belt.

Width \ *width*\ *n* [¹*wide*] (1627) A horizontal measurement of a material. In woven fabric, it is the distance from selvage to selvage, and in flat-knit fabric, the distance from edge to edge.

Wien's Displacement Law When the temperature of a radiating black body increases, the wavelength corresponding to maximum energy decreases in such a way that the product of the absolute temperature and wavelength is constant.

$$\lambda_{\max} T = w$$

w is known as *Wien's displacement constant*. (Giambattista, A., Richardson, R., Richardson, R. C., and Richardson, B., *College Physics*, 3rd. ed., Giambattista, A, Richardson, R. and Richardson, B., McGraw-Hill Companies, New York, 2009; Saleh, B. E. A., and Teich, M. C., *Fundamentals of Photonics*, John Wiley and Sons, New York, 1991)

Wijs Method Analytical method for determining the degree of unsaturation of a material, which involves the addition of iodine to the existing double bonds. The iodine is derived from a solution of iodine monochloride in glacial acetic acid. See ▶ Iodine Test and ▶ Iodine Value; ▶ Hanus Iodine Number

Wilkinite See ▶ Bentonite.

Wilkinson's Blue See Iron Blue.

Williams-Landell-Ferry Equation (WLF equation) An empirical equation for the ▶ Time-Temperature Equivalence of creep and other properties, that has been successful with many plastics. It is

$$\log_{10} a_T = \frac{-17.44(T - T_g)}{51.6 + T - T_g}$$

where T and T_g = the temperature of interest and the glass-transition temperature of the polymer, and a_T is the ▶ Shift Factor, i.e., the ratio of the viscosities at the two temperatures. The equation holds over the range from T_g to about T_g + 100 K.

Williams Unit A wet-processing unit for open-width processing of fabric. The fabric passes up and down over rollers in the liquor. The unit is widely used for dyeing, washing, pretreating, and aftertreating.

Wilton Carpet Woven carpet in which the pile yarns are woven in as an integral part of the carpet, being held in place by the filling, usually made on a loom with a Jacquard head.

Winding \ *wīn-diŋ*\ *n* (before 12c) Winding is the process of transferring yarn or thread from one type of package to another to facilitate subsequent processing. The rehandling of yarn is an integral part of the fiber and textile industries. Not only must the package and the yarn itself be suitable for processing on the next machine in the production process, but also other factors such as packing cases, pressure due to winding tension, etc., must be considered. (Vigo, T., L., *Textile Processing, Dyeing, Finishing and Performance*, Elsevier Science, New York, 1994)

Window \ *win-(* *)dō*\ *n* [ME *windowe*, fr. ON *vindauga*, fr. *vindr* wind (akin to OE *wind*) + *auga* eye; akin to OE *ēage* eye] (13c) A globule of incompletely plasticated material in a thermoplastic film, sheet, or molding that is visible when viewed by transmitted light. It is equivalent to Fisheye except that the term window is usually employed to indicate a clear spot in an otherwise colored or opaque material. (*Rosato's Plastics Encyclopedia and Dictionary*, Rosato, D. V., ed., Hanser-Gardner Publications, New York, 1992)

Window Paning A fabric defect caused by nonuniform yarn. When thin sections of yarn become grouped together, the resultant increase in the transparency of the fabric is called window paning.

Window Sash See ▶ Sash.

Wind Ratio The number of wraps that an end or ends make in traversing from one side of a wound package to the other side and back to the first side.

Wine Gallon A standard U.S. gallon of 231 cubic inches (3.7854 litres). As commonly used, this term refers to alcohol at 60°F, which is the standard temperature prescribed by the Federal Government for gauging alcohol.

Winterized Oils Oils (e.g., fish oils) which have been stored in a cold place for a long time to allow the solid (i.e., mostly saturated) components to settle out.

Wipe-On Plate In offset lithography, a plate on which a light-sensitive coating is wiped on or applied with a coating machine.

Wire-Bar Application See ▶ Wire-Wound Rod.

Wire Brush (1) A hand cleaning tool comprised of bundles of wires. (2) The act of cleaning a surface with a wire brush including wire power brushes.

Wire Coating The application of a plastic, rubber, or enamel coating to a single- or multi-strand wire, or to a cable of many previously coated single wires. Most wire coating is done by extrusion from the melt, but some, such as magnet wire for electric motors, has been done by passing the wire through a solution of thermosetting resin, then evaporating the solvent and curing the resin in an oven. Lineal rates on extrusion-coating lines range from 0.5 m/s on a line over coating a large cable containing hundreds of wires to 30 m/s on a line coating hood-up wire. Over 500 Gg (0.55×10^6 tons) of leading thermoplastics were used to coat wire and cable in 1992.

Wire Gauge *n* (1833) (wire gage) (1) Any of several shorthand systems of consecutive numbers, each number relating inversely to a particular wire diameter. Steel producers in the U.S. use the Steel Wire Gauge, ranging from 7/0 (0000000), = 0.4900 in., to 0 = 0.3065 in., to 50 = 0.0044 in. In Britain, the British Standard Wire Gauge (Imperial Wire Gauge) has long been used, with diameters close to those of the Steel Wire Gauge. (This may be changing to metric.) Copper and aluminum wires, formerly given in Brown & Sharpe (B&S) wire gauge, are now specified in decimal-fractional inches. Contrarily, music- (piano-) wire sizes *increase* with their gauge numbers. The Standard for Metric Practice, ASTM E 380, has strangely omitted this important area of measurement. Presumably, in SI, there are no "gauges", and wire size are given in millimeters, as are screen sizes. (2) A metal plate perforated with graduated and labeled holes with which one may determine the size of a wire or drill bit by identifying the smallest hold through which the wire will pass.

Wire-Wound Rod A metal rod wound with a fine wire around its axis to that an ink or coating can be drawn down evenly and at a given thickness across a substrate. The thickness or gauge of the wire controls the depth of wet film applied by the rod. Syn: ▶ Equalizer Rod and ▶ Meyer Bar.

Withering \ ˈwith-riŋ\ *adj* (1579) The loss of gloss caused by varnishing open-pore woods without filling pores, use of improper undercoating, or applying topcoat before undercoat has dried.

Witherite \ ˈwi-thə-ˌrīt\ *n* [Gr *Witherit*, irreg. fr. William Withering † 1799] (1794) See ▶ Barium Carbonate.

WLF Equation Williams-Landell-Ferry Equation:

$$\log a_T = \frac{-17.4(T - T_g)}{51.6 + (T - T_g)}$$

where

a_T = shift factor
T = selected temperature
T_g = glass transition temperature

The shift factor is the shift in time scale corresponding to the difference between the selected and reference temperature, and the shift factor represents the temperature dependence of the rate of the segmental motion which underlies all viscoelastic behavior; the WLF equation demonstrates that all polymers, irrespective of their chemical structure, will exhibit similar viscoelastic behavior at equal temperature intervals ($T-T_g$) above their respective glass transition temperatures (T_g). (Elias, H. G., Macromolecules, Plenum Press, New York, 1977; *Physical Properties of Polymers Handbook*, Mark, J. E. ed., Springer-Verlag, New York, 1996) See ▶ Williams-Landell-Ferry Equation.

Wollastonite \ ˈwu-lə-stə-ˌnīt, ˈwä-\ *n* [William H. Wollaston] (1823) See ▶ Calcium Silicate.

Wood \ ˈwüd\ *adj* [ME, fr. OE *wōd* insane; akin to OHGr *wuot* madness] (before 12c) The hard, fibrous substance which composes the trunk and branches of a tree, lying between the pith and bark.

Wood Alcohol *n* (1861) CH_3OH. An impure alcohol historically obtained from the destructive distillation (pyrolysis in the absence of air) of wood, whose main constituent was ▶ Methanol. Today, methanol is synthesized from carbon monoxide and hydrogen. It is toxic and is used as a denaturant in ethanol to make it impotable.

Wood, Built-up Laminated An assembly made by joining layers of lumber with mechanical fastenings so that the grain of all laminations is essentially parallel.

Wood Chipboard See ▶ Particle Board.

Woodcut A wooden printing plate, the image of which has been left in relief by cutting away the background.

Wood Engraving *n* (1816) A relief printing surface consisting of a wooden block with a usually pictorial; design cut in the end grain.

Wood Failure The rupturing of wood fibers in strength tests on bonded specimens, usually expressed as the percentage of the total area involved which shows such failure.

Wood Fiber *n* (1875) Any of various fibers in or associated with xylem.

Wood Filler Heavily pigmented product used to fill the grain of wood before the application of undercoats or finishes.

Wood Finishing The planning, sanding, and subsequent staining, varnishing, waxing, or painting of a wood surface.

Wood Flour Very fine wood particles generated from wood reduced by a ball or similar mill until they resemble wheat flour in appearance, and of such size that the particles usually will pass through a 40-mesh screen.

Wood, Glued Laminated An assembly made by bonding layers of veneer or lumber with an adhesive so that the grain of all laminations is essentially parallel.

Wood Grain A fabric defect that consists of fillingwise streaks resembling the irregular appearance of wood grain in lumber. Wood grain is usually caused by strained filling in quilling, the tension being more pronounced near the butt of the quill.

Woodgraining A group of processes used to impart wood-like appearance to sheets or shaped articles. The substrates may be of plastic, wood, steel, or any other material. Among the processes used are conventional laminating techniques, multiple-coat painting, hot stamping, and introduction of several colors into the melt during molding.

Wood Oil See Tung Oil.

Wood Preservative Coatings Coatings which are formulated to protect wood from decay and insect attack.

Wood Pulp *n* (1866) The cellulosic raw material for viscose rayon and for acetate.

Wood Rosin Rosin obtained from pine stumps. See ▶ Rosin.

Wood Tar *n* (1857) A black, syruplike, viscous fluid that is a by-product of the destructive distillation of wood and is used in pitch, wood preserving oils, preservatives, an medicine.

Wood Turpentine *n* (ca. 1909) See ▶ Turpentine.

Wood Veneer A thin sheet of wood, generally within the thickness range from 0.01 to 0.25 in. (0.3 to 6.3 mm) to be used in a laminate. See ▶ Veneer.

Woof \wüf\ *n* [alter. of ME *oof*, fr. OE *ōwef*, fr. *ō-* (fr. *on*) + *wefan* to weave] (before 12c) Syn: ▶ Weft. See ▶ Filling.

Wool \wül\ *n* [ME *wolle*, fr. OE *wull*; akin to OHGr *wolla* wool, L *vellus* fleece, *lana* wood] (before 12c) The term is usually used for the fleece of sheep. More precisely defined as: The fiber from the fleece of the sheep or lamb or hair of the Angora or Cashmere goat (and may include the so called specialty fibers from the hair of the camel, alpaca, llama, and vicuna) which has never been reclaimed from any woven or felted wool product. Wool is used in a variety of blends in which it is combined with nearly all natural or manufactured fibers. (Schoeser, M., *World Textiles: A Concise History*, Thames and Hudson, 2003; *Elsevier's Textile Dictionary*, Vincenti, R., ed., Elsevier Science and Technology Books, New York, 1994)

Worbaloid Cellulose nitrate., manufactured by Worbla AG, Switzerland.

Work \wərk\ *n* [ME *werk*, *work*, fr. OE *werc*, *weorc*; akin to OHGr *werc*, Gk *ergon*, Avestan *varəzem* activity] (before 12c) The action of a force through a distance; the product of force times the distance. Also, the action of a torque through an angular displacement; the product of the torque times the displacement in radians. The SI unit of work, the same as that of energy, is the joule (J), equal to 1 n·m.

Work and Tumble To print one side of a sheet of paper, then turn the sheet over from gripper to back, using the same guide, and print the other side.

Work and Turn To print one side of a sheet of paper, then turn the sheet over from left to right and print the other side. The same gripper is used for printing both sides.

Work Hardening Alternate for strain hardening.

Working Distance The distance between the top of the coverslip and the nearest portion of the objective.

Working Life The period of time during which an adhesive, after mixing with catalyst, solvent, or other compounding ingredients, remains suitable for use. Syn: ▶ Pot Life and ▶ Service Life.

Working Loss The irrecoverable loss of weight or yardage of a textile material that occurs during a textile process.

Working Stress See ▶ Allowable Stress.

Working Time Period of time during which an adhesive or coating, after mixing with catalyst, solvent or other compounding ingredients, remains suitable for use. *Also called Working Life.*

Work Recovery The ratio of recoverable work to the total work required to strain a fiber a specified amount under a given program of strain rate.

Work-to-Break See ▶ Energy-to-Break.

Worsted \wər-stəd\ *n* [ME, fr. *Worsted* (now *Worstead*), England] (13c) A general term applied to fabrics and yarns from combed wool and wool blends. Worsted yarn is smooth-surfaced, and spun from evenly combed long staple. Worsted fabric is made from worsted yarns and is tightly woven with a smooth, hard surface. Gabardine and serge are examples of worsted fabrics.

Woven Fabric Generally used to refer to fabric composed of two sets of yarns, warp and filling, that is formed weaving, which is the interlacing of these sets of yarns. However, there are woven fabrics in which three sets of yarn are used to give a triaxial weave. In two-dimensional wovens, there may be two or more warps and fillings in a fabric, depending on the complexity of the construction. The manner in which the two sets of yarns are interlaced determines the weave. By using various combinations of the three basic weaves, plain, twill, and satin, it is possible to produce an almost unlimited variety of constructions. Other effects may be obtained by varying the type of yarns, filament or spun, and the fiber types, twist levels, etc. (Humphries, M., *Fabric Glossary*, Prentice-Hall, Upper-Saddle River, New Jersey, 2000)

Wrap-Around The phenomenon by which electrically charged paint droplets curve around to the rear side of the object being painted.

Wrinkle \riŋ-kəl\ *n* [ME, back-form. fr. wrinkled twisted, winding, prob. fr. OE *gewrinclod*, pp. of *gewrinclian* to wind, fr. *ge-*, perfective prefix + *-wrinclian* (akin to *wrencan* to wrench] (15c) (1) An imperfection in reinforced plastics that has the appearance of a wave molded into one or more plies of fabric or other reinforcing material (ASTM D 883). (2) In a plastic film or coated cloth, an inadvertent crease.

Wrinkle Finish (1) Type of finish characterized by the presence of wrinkles of fairly uniform dimensions. The effect is obtained by inducing a pronounced but controlled tendency toward rapid surface dry. This finish is generally baked rather than air dried. The size of the wrinkle can be adjusted by variation in the type and amount of drier, type and amount of pigment, thickness of applied film and baking temperature. (2) A varnish or enamel film which exhibits a novelty effect similar to skin wrinkles.

Wrinkle Mark See Steam Mark.

Wrinkle Recovery That property of a fabric that enables it to recover from folding deformations.

Wrinkle Resistance That property of a fabric that enables it to resist the formation of wrinkles when subjected to a folding deformation. Wrinkle resistance in a fabric is a desirable attribute, but it is not easily measured quantitatively. Wrinkle resistance varies from quite low in many fabrics to very high in resilient fabrics. In order to form a wrinkle, a fabric's wrinkle resistance must be overcome. The fabric may, however, produce strains and store potential energy that can become evident as wrinkle recovery under suitable conditions.

Wrinkling The distortion in a paint film appearing as ripples; may be produced intentionally as a decorative effect or may be a defect caused by during conditions or an excessively thick film. *Also called Crinkling Riveling.* See ▶ Gas Checking.

Wrong Color Pick See mixed end or filling.

Wrong Pick See Mispick.

Wurtzilite See ▶ Asphaltic Pyrobitumens.

WVTR Abbreviation for Water-Vapor-Transmission Rate.

WW Initial letters meaning "water-white" and applied to a pale grade in a color scale of rosin. The term is also applied to other resins, or solutions of the same, which are practically colorless.

X

X \eks\ *n* {*often capitalized, often attributive*} (before 12c) (1) Symbol for mole fraction, usually subscripted to indicate the component of interest. (2) Symbol for general variable, or independent variable, and the horizontal graphing axis, or abscissa. (3) One of the CIE chromaticity coordinates calculated as the fraction of the sum of the three tristimulus values attributable to the X Value:

$$x = \frac{X}{X+Y+Z}$$

(McDonald R (ed) (1997) Colour physics for Industry, 2nd edn. Society of Dyers and Colourists, West Yorkshire, UK). See ▶ Chromaticity Coordinates, CIE.

$\bar{x}$ *n* Special color matching functions of the CIE Standard Observer, used for calculating the X tristimulus value. (McDonald R (ed) (1997) Colour physics for Industry, 2nd edn. Society of Dyers and Colourists, West Yorkshire, UK). See ▶ Tristimulus Computation Data and ▶ Color Matching Functions.

X (1) One of the three CIE tristimulus values; the red primary. See ▶ Tristimulus Values, CIE. (2) Symbol, proceeded by a number, for power of magnification and, closely related, and sometimes used in place of x, the multiplication operator.

Xanthate *n* \zan-thāt\ *n* (1831) A sodium salt of a dithiocarbonic acid ester, in particular the one formed in the viscose-rayon process by the reaction between sodium hydroxide cellulose and carbon disulfide and having the structure shown below, called cellulose xanthate or *viscose*, The viscose is subsequently precipitated, filtered, extruded as filaments into dilute sulfuric acid, washed, and dried to make viscose that be spun into Rayon[R] fabric, and extruded thin sheets are called Cellophane[R]. (Kadolph SJJ, Langford AL (2001) Textiles. Pearson Education, New York; Ash M, Ash I (1982–1983) Encyclopedia of Plastics, Polymers, and Resins, Vols. I–III. Chemical Publishing, New York).

Xanthating *n* A process in rayon manufacture in which carbon disulfide is reacted with alkali cellulose to produce bright orange cellulose xanthate.

Xanthic Containing yellow, or pertaining to yellow color.

Xenon \zē-nän, ze-\ *n* [Gk, neuter of *xenos* strange] (1898) A heavy, colorless, and relatively inert gaseous element that occurs in air as about one part in 20 million by volume and is used especially in thyratrons and specialized flashtubes. (Whitten KW, Davis RE, Davis E, Peck LM, Stanley GG (2003) General Chemistry. Brookes/Cole, New York).

Xenon-Arc Aging *n* A test for evaluating the light stability of plastics, employing a xenon-gas-discharge lamp of special design that emits radiation duplicating the spectrum of natural sunlight more closely than most artificial sources. ASTM lists two such tests for plastics, D 4459 with dry specimens and G 26, in which the specimens may or may not be sprayed with water.

Xenon-Arc Lamp *n* A type of light source used in fading lamps. It is an electric discharge in an atmosphere on xenon gas at a little below atmospheric pressure, contained in a quartz tube.

XLPE *n* Abbreviation for Crosslinked Polyethylene. See ▶ Radiation Crosslinking.

XPS Abbreviation for Expandable or ▶ Expanded Polystyrene. See ▶ Polystyrene Foam.

X-Ray \eks-rā\ *n* {*often capitalized*} (1899) An electromagnetic radiation with wavelength in the range from 0.003 to 3 nm, produced by the bombardment of a target with cathode rays. Those with the shorter wavelengths are more energetic and are called *hard X-rays*, while those with the longer wavelengths are called *soft X-rays*.

X-Ray Diffraction *n* Crystals, whose interatomic spacings are commensurable with the wavelengths of some X-rays, can act as diffraction gratings for X-rays. When X-rays are directed obliquely at a crystal surface, and the resulting radiation is captured on photographic film, a symmetrical pattern of spots is observed that is related to the positioning of atoms in the crystal. This 80-year-old plus technique has been useful in studying crystalline structure in polymers (Suryanarayana C, Norton MG (2003) X-Ray Diffraction: A Practical Approach. Plenum, New York; Gooch JW (1997) Analysis and Deformulation of Polymeric Materials. Plenum, New York). An X-Ray Power Data File for materials is available from ASTM (Insert X-Ray Powder Data File table). See also ▶ Small-Angle X-Ray Scattering and ▶ Wide-Angle X-Ray Scattering.

Jan W. Gooch, *Encyclopedic Dictionary of Polymers*, DOI 10.1007/978-1-4419-6247-8,
© Springer Science+Business Media LLC 2011

X-ray powder data, lead carbonate Pb$_2$CO$_4$

hkl	d(Å)	I	Hkl	d(Å)	I	hkl	d(Å)	I
1.410	50		1.213	30				
1.371	30		1.192	50				
1.359	10		1.165	50				
1.339	20		1.150	20				
1.321	20		1.136	40				
1.308	10		1.112	40				
1.287	30		1.096	40				
1.276	10		1.088	40				
1.266	80		1.075	20				
1.255	10		1.067	30				
1.244	20		1.059	40				

X-ray diffraction spectrum of lead pigment specimen

X-Ray Microscopy *n* This instrument and technique is similar to an optical microscope except that ▶ x-rays are utilized to magnify and study an object instead of visual light. Images beneath a surface can studied because ▶ x-rays penetrate materials including metals. Defects located within the interior of an object (plastic pipe) can be observed without disturbing the structure or cutting a cross-section. The technique is useful for studying the structure of materials such as composites, fibers and plastics. (Cunningham, Davis, Graham (1986) X-Ray Microscopy. J Microsc., 144(3): 261–275).

X Units *n* X-ray wavelengths have been measured in two kinds of units. The older measurements are given in X units (XU) which are based on the effective lattice constant of rock salt being 2,814.00 XU. More recently X-ray wavelengths have been directly connected, through measurements with ruled gratings, to the wavelengths in the optical region and through them to the standard meter. It turned out that the XU which was originally intended as 10^{-11} cm was 0.202% larger than this value. It has become customary to give X-ray wavelengths in Angstrom units (Å) when the absolute scale is used (1 Å = 10^{-8} cm). The two are related by

$$1{,}000 \times U = (1.00202 + 0.00003)\text{Å}$$

and wavelengths given in XU must be multiplied by 1.00202 and then divided by 1,000 in order to concert them from Angstrom units. (Giambattista A, Richardson R, Richardson RC, Richardson B (2003) College Physics. McGraw-Hill Science/Engineering/Math, New York).

Xylene \ˈzī-ˌlēn\ *n* [ISV] (1851) C$_6$H$_4$(CH$_3$)$_2$. A commercial mixture of the three isomers, *o*-, *m*-, and *p*-xylene, used as a solvent for alkyd resins, polystyrene, natural resins, rubber, and polyisobutylene. *Also known as Xylol.* (Wypych G (ed) (2001) Handbook of Solvents. Chemtec Publishing, New York).

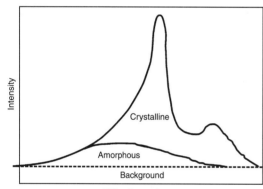

o-Xylene *n* (1,2-dimethylbenzene) 1,2-C$_6$H$_4$(CH$_3$)$_2$. A colorless liquid, insoluble in water, miscible with alcohol, ether, and many other organic liquids. Used as a feedstock in the production of phthalic anhydride. It can be extracted from the mixed isomers by distillation and can be isomerized to *p*-xylene. Mp, −25°C; sp gr, 0.897/20°C; bp, 144°C.

m-Xylene *n* (1,3-dimethylbenzene) C$_6$H$_4$(CH$_3$)$_2$. Colorless liquid, insoluble in water, miscible with alcohol, ether, and many other organic liquids. Bp, 139.3°C; sp gr, 0.−8684/15°C; mp. 47.8°C.

p-Xylene *n* (1,4-dimethylbenzene) 1,4-C$_6$H$_4$(CH$_3$)$_2$. A colorless liquid, insoluble in water, miscible with alcohol, ether, and many other organic liquids. Used in the synthesis of terephthalic acid and dimethylterephthalate, both of which are intermediates for polyester fibers and films. Mp, 13.2°C; bp, 137–138°C; sp gr, 0.8611/20°C.

p-Xylene-α,α′-Diol *n* C$_6$H$_4$(CH$_2$OH)$_2$. A white crystalline solid used as a crosslinking agent in polyurethanes, and in the production of polyesters and polycarbonates.

Xylenol Resin A phenolic-type resin produced by condensing xylenol (3,5-dimethylphenol) with an aldehyde. Polyphenylene Oxide is made from 2,6-xylenol.

Xylography \zī-ˈlä-grə-fē\ *n* [F *xylographie*, fr. *xyl-* + *-graphie*] (1816) The art of printing with wooden blocks.

(Leach RH, Pierce RJ, Hickman EP, Mackenzie MJ, Smith HG (eds) (1993) Printing Ink Manual, 5th edn. Blue print, New York).

Xylol \ˈzī-ˌlōl\ *n* [ISV] (1851) The term "xylol" is still used commercially for xylene, but is not preferred tin modern nomenclature. See ▶ Xylene.

Xylox Resin® *n* Trade name for a family of heat-resistant thermosetting resins made by the condensation of aralkyl ethers and phenols, resulting in hydroxyphenylene-*p*-xylene prepolymers that can be cured to hard, intractable resins by reaction with hexamethylenetetramine or epoxy compounds. These thermosetting resins have the good qualities of phenolics and epoxies, with superior mechanical and electrical properties at elevated temperatures.

p-Xylylene \ˈzīl-el-ˌlēn\ *n* (PX) $H_2C=C_6H_4=CH_2$. A highly reactive monomer from which parylene polymer is formed, $(-CH_2C_6H_4CH_2-)_n$. The gaseous monomer readily forms a stable solid dimmer, convenient for shipping, from which the monomer is easily regenerated by heating. PX polymerizes spontaneously in vacuum on any cool surface to form tough, uniform, impervious films. Dimers with ring-substituted chlorine are also available. See also Di-*p*-Xylylene and PARYLENE. *p-Xylylene.*

m-Xylylenediamine *n* A solid diamine useful as curing agent for epoxy resins.

Xylylene Diisocyanate *n* A mixture of the *m*- and *p*-isomers, used in the production of polyurethane coatings.

Y

y \ˈwī\ *abbreviation* (1) Abbreviation for Year. (2) Symbol for general dependent variable and the ordinate (vertical) axis in two-dimensional graphing. (3) One of the CIE chromaticity coordinates calculated as the fraction of the sum of the three tristimulus values, attributable to the Y value:

$$y = \frac{Y}{X + Y + Z}$$

See ▶ Chromaticity Coordinates, CIE.

ẙ *n* Special color matching function of the CIE Standard Observed used for calculating Y tristimulus value. See ▶ Tristimulus Computation Data, ▶ Tristimulus Values, and ▶ Color Matching Functions. (*Colour Physics for Industry*, 2nd Ed., McDonald, Roderick, Society of Dyers and Colourists, West Yorkshire, England, 1997)

Y *n abbreviation* (1) Chemical symbol for the element yttrium, which has been used as a catalyst for ethylene polymerization. (2) One of the three CIE tristimulus values, equal to the luminous reflectance or transmittance; the green primary. See ▶ Tristimulus Values.

Yacca Gum See ▶ Accroides.

Yardage \ˈyär-dij\ *n* (1867) The amount or length of a fabric expressed in yards.

Yard Denier *n* The denier of a filament yarn. It is the product of the denier per filament and the number of filaments in the yarn.

Yard Goods *n plural* (1905) Fabric sold on a retail basis by the running yard.

Yarn \ˈyärn\ *n* [ME, fr. OE *gearn*, fr. OHGr *garn* yarn, Gk *chordē* string, L *hernia* rupture, Sanskrit *hira* band] (before 12c) A generic term for a continuous strand of textile fibers, filaments, or material in a form suitable for knitting, weaving, or otherwise intertwining to form a textile fabric. Yarn occurs in the following forms: (1) a number of fibers twisted together (spun yarn); (2) a number of filaments laid together without twist (a zero-twist yarn); (3) a number of filaments laid together with a degree of twist; (4) a single filament with or without twist (a monofilament); or (5) a narrow strip of material, such as paper, plastic film, or metal foil, with or without twist, intended for use in a textile construction. (Kadolph SJJ, Langford AL, *Textiles*, Pearson Education, New York, 2001)

Yarn Construction *n* A term used to indicate the number of singles yarns and the number of strands combined for form each successive unit of a plied yarn or cord.

Yarn Dyeing See ▶ Dyeing.

Yarn Dyeing Differences *n* Variations in take-up of dyes by yarns, resulting in streaks in finished fabrics.

Yarn Intermediate *n* A generic term for products obtained during the conversion of fibers to yarn, including card webs, laps, slivers, rovings, and tops.

Yarn Number *n* A relative measure of the fineness of yarns. Two classes of systems are in use: (1) Direct yarn number (equal to linear density) is the mass per unit length of yarn. This system is used for silk and manufactured filament yarns. (2) Indirect yarn number (equal to the reciprocal of linear density) is the length per unit mass of yarn. This system is used for cotton, linen, and wool-type spun yarns. (Kadolph SJJ, Langford AL, *Textiles*, Pearson Education, New York, 2001) Also see ▶ Cotton Count.

Yarn Number, Equivalent Single *n* The number of a plied yarn or cord determined by the standard methods used for singles yarns. (Kadolph SJJ, Langford AL, *Textiles*, Pearson Education, New York, 2001)

Yarn Quality *n* Various grades of yarn designated by the producer with respect to performance characteristics, e.g., first quality, second quality, etc.

Yarn-to-Cord Conversion Efficiency In tire cord, this is a measurement relating tensile strength of untwisted yarn to tensile strength of cord. Increasing cord twist or increasing yarn diameter lowers conversion efficiency. (*Elsevier's Textile Dictionary*, Vincenti R (ed), Elsevier Science and Technology Books, New York, 1994)

Yarn Variation See Ring.

Year Ring *n* An annual ring of growth in timber.

Yellow Accroides *n* Bright, yellow-colored resin obtained from Australia. It resembles red accroides, and is used to a limited extent in some types of spirit varnishes. See ▶ Accroides.

Yellow Chrome See Chrome Yellow.

Yellowing *v* (1598) Development of a yellow color on aging; most noticeable in the dried films of white paints or clear varnishes. Primary cause of yellowing is the formation of color bodies due to oxidation of the oil or other unsaturated components of the binder.

Yellow Iron Oxide *n* $Fe_2O_3 \cdot xH_2O$. Pigment Yellow (77492). A natural or synthetic iron oxide. It has excellent bleed resistance, chemical resistance (both acid and alkali), lightfastness and dispersibility, as well as excellent toxicity ratings. However, these yellow oxides have only fair heat resistance since they are subject to color change bout 177°C due to loss of water of hydration. Syn: ▶ Ferrite Yellow and ▶ Limonite.

Yellow Limonite *n* See ▶ Iron Oxides, Natural.

Yellowness The attribute by which an object color is judged to depart from a preferred white toward yellow. (Billmeyer FW, Saltzman M, *Principles of Color Technology*, John Wiley and Sons, Inc., New York 1966, www.astm.org)

Yellowness Coefficient *n* Measure of the color of a molded acetate disc or dope solution. Cy = 1-T4400/T6400 where Cy is the yellowness coefficient; T4400 is the transmission at 4400A (blue); and T6400 is the transmission at 6400A (orange). (Billmeyer FW, Saltzman M, *Principles of Color Technology*, John Wiley and Sons, Inc., New York 1966)

Yellowness Index *n* (YI) A measure of the yellowing of a plastic, such as might occur after lengthy exposure to light. It is determined according to ASTM D 1925, and is therein defined as the deviation in chroma from whiteness or water-whiteness in the dominant wavelength range from 570 to 580 nm. The index is computed from the three tristimulus values measured with a spectrophotometer, relative to a magnesium oxide standard. An example is the following:

YI = 100 (R − B)/G where R, G, and B are tristimulus colorimeter readings, all normalized to 100. In CIE tristimulus terms, the equation becomes

YI = (128X − 106Z)/Y (ASTM D 1925). Abbreviation is YI. (*Colour Physics for Industry*, 2nd Ed., McDonald,, Roderick, Society of Dyers and Colourists, West Yorkshire, England, 1997)

Yellow Ocher *n* (15c) (1) A mixture of limonite usually with clay and silica used as a pigment. (2) A moderate orange yellow.

Yellow Ocher, Synthetic *n* A synthetic yellow iron oxide reduced with aluminum silicate or other extender pigments. See ▶ Ocher.

Yellow Pine *n* (1709) A hard resinous wood of the longleaf pine tree, having dark bands of summerwood alternating with lighter-colored springwood; used as flooring and in general construction.

Yield \\yē(ə)ld\ *n* (15c) (1) Number of linear or square yards of fabric per pound of fiber or yarn. (2) The number of finished square yards per pound of greige fabric.

Yield Point *n* In tensile testing, the first point on the stress-strain curve at which an increase in strain occurs without an increase in stress. This is the point at which permanent (plastic) deformation of the specimen begins. Many plastics do not exhibit an identifiable yield point. (Shah V, *Handbook of Plastics Testing Technology*, John Wiley and Sons, New York, 1998)

Yield Strength *n* The stress at which a material exhibits a specified limiting deviation from the proportionality of stress to strain. Unless otherwise specified, this stress will be the stress at the yield point (ASTM D 638 and D 638M). See also ▶ Offset Yield Strength. (Shah V, *Handbook of Plastics Testing Technology*, John Wiley and Sons, New York, 1998, www.astm.org)

Yield Value *n* A rheological term describing the flow properties of a printing ink, and defined as the minimum force required that produces flow, (i.e., or the minimum shearing stress which must be applied to a plastic material to initiate flow or shearing within the fluid. It may be measured in dynes/cm^2 or newtons/m^2 and is the shearing force producing infinitely slow rate of flow between adjacent planes. Also known as *yield stress*. (Patton TC, *Paint Flow and Pigment Dispersion: A Rheological Approach to Coating and Ink Technology*, John Wiley and Sons, New York, 1979; Coussot P, *Rheometry of Pastes, Suspensions and Grannular Materials: Applications in Industry and Environment*, John Wiley and Sons, New York, 2005)

Young's Modulus *n* A property of perfectly elastic materials, it is the ratio of change in stress to change in strain within the elastic limits of the material. The ratio is calculated from the stress expressed in force per unit cross sectional area, and the strain expressed as a fraction of the original length. Modulus so calculated is equivalent to the force required to strain the sample 100% of its original length, at the rate prevailing below the elastic limit. See ▶ Coefficient of Elasticity and ▶ Modulus of Elasticity. (Shah V, *Handbook of Plastics Testing Technology*, John Wiley and Sons, New York, 1998)

Z

z *n* \ˈzē, *Caadian, British, & Austral* ˈzed, *chiefly dial* ˈi-zərd\ *n* {*often capitalized, often attributive*} (1) In rectangular and cylindrical coordinate systems, the symbol for the vertical (axial) coordinate. (2) Symbol for Standard Normal Deviate. (3) One of the CIE chromaticity coordinates calculated as the fraction of the sum of the three tristimulus values attributable to the Z primary:

$$z = \frac{Z}{X+Y+Z}$$

(*Colour Physics for Industry*, 2nd Ed., McDonald, Roderick, Society of Dyers and Colourists, West Yorkshire, England, 1997) See Chromaticity Coordinates, CIE.

z̄ Special color matching function of the CIE Standard Observer used for calculating the Z tristimulus value. See Tristimulus Computation Data, Tristimulus Values, and Color Matching Functions. (*Colour Physics for Industry, 2nd Ed.*, ed., McDonald, Roderick, Society of Dyers and Colourists, West Yorkshire, England, 1997).

Z *n* (1) Symbol for atomic number, or electrical impedance. (2) One of the three CIE tristimulus values; the blue primary.

Zaffre *n* See Cobalt Blue.

Zahn Cup *n* A device for measuring viscosity of paints, varnishes, flexographic or gravure inks.

Zahn Viscosity Cup *n* A one-shot devices for obtaining quick measurements in a roughly linear way, of the ▶ Kinematic Viscosity of the test liquid, typically a free-flowing fluid. The cupful of test liquid is brought to the desired temperature in a bath of heating medium, then held over a collection vessel and allowed to empty through a short tube in its bottom. Measurements are reported as Zahn Number-X seconds, X indicating which of five different cups was used. (*Paint and Coating Testing Manual (Gardner-Sward Handbook) MNL 17, 14th Ed.*, ASTM, Conshohocken, PA, 1995).

Zanzibar Copal \ˈzan-zə-ˌbär ˈkō-pəl, -ˌpal; kō-ˈpal\ *n* A very hard fossil resin obtained from the eastern side of Africa, and in the past from the island of Zanzibar. Known also as Animi.

Z-Average Molecular Weight $n(M_z, \bar{M}_z)$ A higher-degree average than weight average or viscosity average, but closer to the former, and defined by the equation

$$M_z = \frac{\sum_{i=1}^{\infty} N_i M_i^3}{\sum_{i=1}^{\infty} N_i M_i^2}$$

M_z is more sensitive than the other averages to the largest molecules present in the sample. The sums in the numerator and denominator are also known as the *third* and *second original moments* of the molecular-weight distribution. (Slade, P. E., *Polymer Molecular Weights*, Vol. 4, Marcel Dekker, New York, 2001; *Physical Properties of Polymers Handbook*, Mark, J. E. ed., Springer-Verlag, New York, 1996; Elias, H. G., *Macromolecules Vol. 1–2*, Plenum Press, New York, 1977).

Z-Calender *n* A calender with four rolls arranged so that, as the web of material (fabric) passes through them, the cross section of its path has a shape resembling the letter Z.

Zein \ˈzē-ən\ *n* [NL *Zea*] (1822) A naturally occurring, high-molecular-weight protein, a polymer of amino acids linked by peptide bonds, derived from corn. It is considered to be a member of the protein family of plastics, the main member of which is casein plastic. Zein resins, rarely seen today, years ago were used for fibers (e.g., Vicara), films, and paper coating.

Zein Fiber *n* A manufactured fiber of regenerated protein derived from maize.

Zero-Point Energy *n* The energy of a solid at absolute zero due to its residual nuclear, electronic, atomic, and molecular motion. The lowest energy state of a substance. (Giambattista, A., Richardson, R., Richardson, R. C., and Richardson, B., *College Physics*, McGraw Hill Science/Engineering/Math, New York, 2003).

Zero-Twist *n* Twistless, devoid of twist.

Zeta Potential *n* The potential across the interface of all solids and liquids. More specifically, it is the potential across the diffuse layer of ions surrounding a charged colloidal particle, and is largely responsible for colloidal stability. (Becher, P., *Dictionary of Colloid and Surface Science*, Marcel Dekker, New York, 1989) Also known as *Electrokinetic Potential*.

Zeta Space *n* Color difference equation devised by Saunderson and Milner.

Ziegler Catalyst \ˈtsē-glər ˈka-tᵊl-əst\ Any of a large group of catalysts made by reacting a compound of a transition metal chosen from groups IV through VIII of the periodic table with an alkyl, hydride, or other compound of a metal from groups I through III.

A typical example is the reaction product of an aluminum alkyl with titanium tetrachloride or titanium trichloride. These catalysts were first discovered by the German chemist K. Ziegler (late 1940s) for the low-pressure polymerization of ethylene. Subsequent work by G. Natta (early 1950s) showed that these and similar catalysts are useful for preparing stereoregular polyolefins; thus, the family of catalysts is sometimes called *Ziegler-Natta catalysts*. (Odian, G. C., *Principles of Polymerization*, John Wiley and Sons, Inc., New York, 2004; Elias, H. G., *Macromolecules Vol. 1–2*, Plenum Press, New York, 1977).

Zimate *n* Trade name of the R. T. Vanderbilt Co., for a group of diallyl dithiocarbamates, useful as accelerators for curing rubber.

Zinc \ˈziŋk\ *n* [Gr *Zink*] (1651) A bluish white crystalline metallic element of low to intermediate hardness that is ductile when pure but in the commercial form is brittle at ordinary temperatures and becomes ductile on slight heating. It is used especially as a protective coating for iron and steel.

Zinc Baryta White Syn: ▶ Lithopone.

Zinc Borates *n* White, amorphous powders of indefinite composition, containing various amounts of zinc oxide and boric oxide. They are used ads flame retardants in PVC, polyvinylidene chloride, polyesters, and polyolefins, often in combination with antimony trioxide.

Zinc-Cadmium Sulfides *n* Combination of zinc sulfide and cadmium sulfide, along with special additives, manufactured by highly specialized procedures and used as fluorescent pigments.

Zinc Chromate *n* $4ZnO \cdot K_2O_4 \cdot CrO_3 \cdot 3H_2O$. Pigment Yellow 36 (77955). Bright yellow pigment which chemically is substantially zinc chromate, although its precise composition is rather complex. Its chief use is in anticorrosive paints and primers for steel. Density, 3.36–3.49 g/cm³ (28.1–29.1 lb/gal); O.A., 28–31; particle size, 0.2–5.0 μm. *Also known as Zinc Yellow.*

Zinc Chloride *n* (1851) A poisonous caustic deliquescent salt $ZnCl_2$ used as a wood preservative, drying agent, and catalyst.

Zinc Coated See ▶ Galvanizing.

Zinc Drier *n* Zinc salts of acids generally used for driers, such as naphthenic or 2-ethyl hexoic, are not driers. However, when used with driers such as cobalt it often prevents wrinkling and skinning. It acts as an auxiliary drier.

Zinc Dust *n* Finely divided zinc metal used as a pigment in protective paints for iron steel.

Zinc Ferrocyanide See ▶ Antwerp Blue.

Zinc Green See ▶ Cobalt Green.

Zinc Oxide *n* (1849) (Chinese white, flowers of zinc, zinc white) ZnO. Pigment White 4 (77947) An amorphous white or yellowish powder, used as a pigment in inks, rubber, paint and plastics for mildew resistance and film reinforcing properties. It is said to have the greatest power to absorb ultraviolet light of all commercially available pigments.

Zinc Palmitate *n* $Zn(OOCC_{15}H_{31})_2$. An amorphous white powder used as a lubricant in plastics.

Zinc Phosphate Coating *n* A thin, inorganic deposit formed on zinc treated with phosphoric acid.

Zinc Resinate *n* Zinc soap of rosin with a melting point higher than rosin.

Zinc-Rich Primer *n* Anti-corrosive primer for iron and steel incorporating zinc dust in a concentration sufficient to give electrical conductivity in the dried film, thus enabling the zinc metal to corrode preferentially to the substrate, i.e., to give cathodic protection. (*Coatings Technology Handbook*, Tracton, A. A., ed., Taylor and Francis, Inc. New York, 2005).

Zinc Ricinoleate *n* $Zn[OOC(CH_2)_7CH=CHCH_2CH(OH)C_6H_{13}]_2$. An amorphous white powder used as a stabilizer in vinyl plastics.

Zinc Silicate Primers *n* Inorganic zinc-rich primers that contain a silicate binder.

Zinc Stabilizer *n* Any of a group of zinc soaps of fatty acids, usually formulated in combination with barium and calcium soaps and organic phosphates in plasticized PVC compounds. (*Handbook of Polyvinyl Chloride Formulating*, Wickson, E. J., ed., John Wiley and Sons, Inc., New York, 1993).

Zinc Steareate *n* $Zn(OOCC_{17}H_{35})_2$. A white powder used as a lubricant and stabilizer in vinyl compounds. An insoluble zinc soap used to produce matteness, flatting or thickening in coatings.

Zinc Sulfate *n* (1851) A crystalline salt $ZnSO_4$ used in making a white paint pigment, in printing and dying, in sprays and fertilizers, and in medicine as an astringent, emetic, and weak antiseptic.

Zinc Sulfide *n* (1851) Pigment White 7 (77975) ZnS. Yellowish white pigment, soluble in acids, insoluble in water. Certain specially treated grades of this pigment are used as fluorescent pigments. Density, 4.0 g/cm³ (33.3 lb/gal); O.A., 13; particle size, 0.25 μm; refractive index, 2.37.

Zinc White *n* (1847) Another name for zinc oxide.

Zinc Yellow *n* A yellow pigment consisting essentially of zinc chromate. See ▶ Zinc Chromate.

Zircon \ˈzər-ˌkän, -kən\ *n* [Gr, mod. of F *jargon* jargon, zircon, fr. I *giargone*] (1794) $ZrO_2 \cdot SiO_2$. Occurs as natural double oxide. It is produced by physical separation of beach sand (Florida, Australia). Used in vitreous enamels and ceramics. Density, 4.56; particle size, 1.0–2.5 μm; refractive index, 2.0.

Zirconia \ˌzər-ˈkō-nē-ə\ *n* [NL, fr. ISV *zircon*] (1797) (zirconium oxide) ZrO_2. A white monoclinic powder used as a pigment when good electrical properties are required. (*Kirk-Othmer Encyclopedia of Chemical Technology: Pigments-Powders*, John Wiley and Sons, New York, 1996).

Zirconium Driers *n* Zirconium salts of various organic acids such as naphthenic or 2-ethyl hexoic. Used as a replacement for lead in fume-proof paints or other lead-free paints where equivalent through drying can be obtained. It is also used as an auxiliary drier where color and color retention are important.

Zirconium Oxide *n* ZrO_2. Pigment White 12 (77990). A natural mineral, baddeleyite. Present commercial grades are unsuitable as pigment grade in paint vehicles. Density, 5.68 g/cm³; particle size, 1.0–2.0 μm; refractive index, 2.1–2.2. Syn: ▶ Zirconia and ▶ Bad-Deleyite.

Zn *n* Chemical symbol for the element zinc.

Zr Chemical symbol for the element zirconium.

Z-Twist See Twist, Direction of.

Zunsober See ▶ Mercuric Sulfide.

Zwitterion \ˈtsvi-tər-ˌī-ˌän *also* ˌzwi-\ *n* [Gr, fr. *Zwitter* hybrid (fr. OHGr *zwitaran*, fr. *zwi-*) + *Ion* ion] (1906) A dipolar ion (positive and negative ends). (Odian, G. C., *Principles of Polymerization*, John Wiley and Sons, Inc., New York, 2004).

Zytel 31 *n* Nylon 6,10, manufactured by DuPont, U.S.

Zytel 101 *n* Nylon 6,6, manufactured by DuPont, U.S.